Advanced Functional Materials for Sustainable Environments

R. K. Kotnala · Anjali Sharma Kaushik ·
S. Shankar Subramanian · Amit K. Vishwakarma
Editors

Advanced Functional Materials for Sustainable Environments

Springer

Editors
R. K. Kotnala
Dr. KS Krishnan Marg Pusa
CSIR-National Physical Laboratory
New Delhi, Delhi, India

S. Shankar Subramanian
Department of Physics
Atma Ram Sanatan Dharma College
University of Delhi
New Delhi, Delhi, India

Anjali Sharma Kaushik
Department of Physics
Atma Ram Sanatan Dharma College
University of Delhi
New Delhi, Delhi, India

Amit K. Vishwakarma
Department of Physics
Atma Ram Sanatan Dharma College
University of Delhi
New Delhi, Delhi, India

ISBN 978-3-031-62622-7 ISBN 978-3-031-62620-3 (eBook)
https://doi.org/10.1007/978-3-031-62620-3

Jointly published with Capital Publishing Company
The print edition is not for sale in Afghanistan, Bangladesh, Bhutan, India, the Maldives, Nepal, Pakistan and Sri Lanka. Customers from Afghanistan, Bangladesh, Bhutan, India, the Maldives, Nepal, Pakistan and Sri Lanka please order the print book from: Capital Publishing Company

This Springer imprint is published by the registered company Springer Nature Switzerland AG
The registered company address is: Gewerbestrasse 11, 6330 Cham, Switzerland

Foreword

I am delighted to announce the successful organization of the 2nd International Conference on "Advanced Functional Materials and Devices (AFMD-2023)" by the Department of Physics, in collaboration with the Internal Quality Assurance Cell of ARSD. This achievement follows the exceptional outcomes of AFMD-2021. The college's strong foundation, coupled with adept leadership in adapting to new environments, has significantly contributed to its achievements. With an A++ rating from NAAC and a commendable 6th position in NIRF's All India ranking, the college's commitment to delivering quality and inclusive education is evident. The faculty's achievements over the past decade, harnessing available resources and staying aligned with global economic shifts, underscore the college's remarkable progress. ARSD is emerging as a premier research hub. Departments such as Physics, Chemistry, Biology, and Mathematics have secured funding from the Government of India under the DBT Star Scheme, a testament to the calibre of research within the institution.

The AFMD 2023 conference was a momentous gathering, featuring 29 Invited Lectures by esteemed researchers from around the globe. The response was overwhelming, with 196 abstracts submitted for oral and poster presentations, representing a diverse international participation. My sincere appreciation extends to the invited speakers and participants for generously sharing their invaluable knowledge and research insights. The conference's theme resonated with the forefront of technological advancement, as researchers globally work diligently to contribute to societal progress through technological innovation.

In an era characterized by rapid technological evolution, the pursuit of sustainable solutions stands at the forefront of scientific endeavours. The present book on, *Advanced Functional Materials for Sustainable Environments*, represents a significant milestone in our collective journey towards a greener and more harmonious world. This volume offers an encompassing view of cutting-edge research worldwide in the domain of functional materials, specifically directed towards energy harvesting, storage, and environmental monitoring. As the pace of technology accelerates, staying attuned to the latest advancements becomes a necessity, and this compendium provides an indispensable window into the forefront of material science.

The chapters within the book encapsulate the essence of electronic, magnetic, optical, adaptive, and dielectric materials, which serve as the building blocks for pioneering functionalities with heightened efficiency. Embracing topics like energy conversion, energy storage, and environmental sensing, the compilation exemplifies the comprehensive exploration of modern material science.

I am elated to announce that this publication marks the culmination of research papers presented at AFMD 2023. I extend my heartfelt congratulations to the authors whose papers successfully passed the rigorous review process. My sincere gratitude goes to the reviewers, editors, and the distinguished Dr. R. K. Kotnala, who led the Editorial board. The editorial team, comprising Dr. Anjali Sharma Kaushik, Dr. S. Shankar and Dr. Amit Vishwakarma, deserves recognition for their unwavering dedication.

To the convener, the organizing team, and all contributors, I extend my warmest wishes for their forthcoming pursuits. May their journey be marked by further remarkable achievements, adding to their legacy of success. I also convey my best wishes to the Department of Physics, hoping it continues to lead the vanguard in advancing knowledge within the society.

Prof. Gyantosh Kumar Jha
Principal
Atma Ram Sanatan Dharma College
University of Delhi
New Delhi, Delhi, India

Preface

The book *Advanced Functional Materials for Sustainable Environments* signifies a major achievement in the ongoing journey of humanity towards a greener and more ecologically balanced world. Within its pages, one will discover a comprehensive exploration of the most current global research in the realm of functional materials, with a specific focus on their applications in energy harvesting, storage, and environmental monitoring. In an era of rapid technological advancement, keeping up with the latest breakthroughs has become essential, and this compilation serves as an invaluable gateway into the forefront of materials science. The relentless progress of science and technology has propelled us into a realm where the frontiers of possibility continuously broaden. At the core of this transformative journey lies the pivotal significance of advanced functional materials. These materials possess extraordinary attributes that underpin groundbreaking innovations, fundamentally reshaping our lifestyles, occupations, and interactions with the world. They empower us to tap into energy from unconventional sources, adeptly store it, and meticulously observe our surroundings with unparalleled accuracy.

The chapters of the book encapsulate the core essence of electronic, magnetic, optical, adaptive, and dielectric materials, which serve as the fundamental building blocks of revolutionary functionalities characterized by elevated efficiency. Embracing topics like energy conversion, energy storage, and environmental sensing, the compendium exemplifies the comprehensive exploration of modern materials science. At the threshold of a new era, characterized by extraordinary challenges and equally remarkable possibilities, the center stage is taken by the importance of sustainable materials and technologies. The urgency of addressing issues like climate change, dwindling resources, and environmental deterioration demands a profound reevaluation of our technological underpinnings. Within this framework, the research presented in these pages emerges as of utmost significance. It not only enriches our understanding of materials science but also provides practical solutions to some of the most pressing global issues.

This book has been the result of a collaborative effort, with the expertise of researchers, scientists, and engineers from various corners of the globe being brought together. The spirit of inquiry and innovation that unites us in our quest for a more sustainable future is exemplified by this work. Each chapter provides a window into the intricate world of advanced functional materials, offering insights that will both inspire and inform, potentially serving as catalysts for the breakthroughs of tomorrow. Readers are invited to embark on a journey through these pages, where the wonders of materials science can be explored, and a vision of a world where technology not only fulfills our needs but also aligns harmoniously with the environment can be imagined. While exploring the contents of *Advanced Functional Materials for Sustainable Environments*, it's essential to remember that the pursuit of knowledge and innovation goes beyond a purely scientific endeavour; it signifies a shared dedication to creating a better, more sustainable world.

New Delhi, India

R. K. Kotnala
Anjali Sharma Kaushik
S. Shankar Subramanian
Amit K. Vishwakarma

About This Book

The quest for sustainable solutions has emerged as a paramount priority in the realm of scientific exploration. This book on *Advanced Functional Materials for Sustainable Environments* stands as a beacon of exemplary progress towards a greener and more harmonious global ecosystem. This book provides an extensive survey of cutting-edge research in the field of functional materials, with a particular emphasis on energy harvesting, storage, and environmental monitoring.

The chapters contained within delve into the intricate complexities of electronic, magnetic, optical, adaptive, and dielectric materials, laying the foundation for innovative functionalities. Encompassing domains such as energy conversion, storage, and environmental awareness, this compendium represents a meticulous exploration of contemporary materials science.

The collaborative efforts underlying this book underscore a shared commitment to foster a sustainable future. Each chapter offers insights that not only stimulate intellectual curiosity but also harbours the potential to catalyse future innovations. *Advanced Functional Materials for Sustainable Environments* not only enhances our understanding of materials science but also offers practical solutions to pressing global issues.

Contents

Part VI Materials for EMI Shielding

About the Editors

Dr. R. K. Kotnala is an Indian scientist, who is world-renowned for his spectacular work in Net Carbon Zero, Hydroelectric Cell, Water Splitting by Non-Photocatalytic process, Nanomaterials, Multiferroic, Spintronics, Magnetics and Biomedical Devices Metrology. With his constant hard work and dedication Dr. Kotnala, along with Dr. Jyoti Shah, has contributed exceptionally to a path-breaking invention of a green and clean energy device-Hydroelectric Cell. The hydroelectric cell is an astounding, miraculous, and incredible invention in the green energy domain and a panacea towards achieving Net Carbon zero. Dr. Kotnala is well known in the science fraternity for his outstanding contributions in the field of Solar cells, Multiferroics, Spintronics, Clean Energy, Environmental sciences, and Values of Scientific Ethics. He has written many books, chapters and delivered countless scientific lectures in diversified areas. He is also creating awareness for science among youth through numerous invited talks and YouTube videos on simple scientific concepts.

Dr. Anjali Sharma Kaushik received her B.Sc., M.Sc. and Ph.D. degrees in Physics in the year 2006, 2008 and 2013 respectively, from University of Delhi. At Present, she is serving as Assistant Professor at Atma Ram Sanatan Dharma College, University of Delhi. She is also working as Fellow of Institute of Eminence, University of Delhi. Her research interests are in gas sensor systems that include sensor characterization and development of metal oxide films for sensor coatings. She is working towards the fabrication of MEMS based Electronic-Nose for gas sensing applications. Her area of interest also includes the metal oxide based thermoelectric energy harvesters and fabrication of 2D materials like MoS_2 for sensing applications. She is recipient of Dr. G. C. Jain Memorial Award for best Ph.D. thesis by Material Research Society of India in year 2015 and Shri Ram Arora Award at 141st Annual meeting of The Metals, Minerals and Materials Society, Orlando USA in year 2012 for her excellence in Materials and Engineering. She has published more than 60 papers in journals of high repute with h-index of 24 and has one Indian patent (granted), one Indian patent (published) and four Indian patent (applied). She has been granted one Core Grant Research Proposal by DST-SERB. She is also working as Co-PI in three DRDO sponsored projects. She has been a member of organizing committee and session chair of various national and international conferences. She is also actively involved in many students and administrative activities of the college.

Dr. S. Shankar Subramanian has completed his Ph.D. from Department of Applied Physics, Delhi Technological University in 2020. He has published more than 35 publications in internationally reputed journals in the field of material science. He has also presented his work in many national and international conferences. His research areas include multiferroics, magnetoelectrics, composites, hydroelectric cell, green synthesis and energy storage studies. His doctoral thesis was focused on optimizing magnetoelectric bulk composites for energy storage applications. In addition, he is project investigator of one major research project (Rs. 16.5 lakhs funded by CSIR) and co-investigator of two completed research projects (one major project of Rs. 28 lakhs and one minor project of Rs. 5.5 lakhs funded by University of Delhi). Furthermore, he has supervised 15 Master dissertations of students from Physics and Chemistry disciplines of various universities. He was organizing member of national conference on "New Trends In Nanotechnology And Applications" (NTNA-2018 and NTNA-2020) and international conference "Advanced Functional Materials and Devices" (AFMD-2021) organized by Department of Physics & IQAC, ARSD College. He has also contributed as Session Chair in International Conference on "Advanced Functional Materials and Devices" (AFMD-2021) organized by Department of Physics & IQAC, ARSD College.

Dr. Amit K. Vishwakarma received his Ph.D. Degree form Delhi Technological University Delhi 2017. He has completed his UG and PG form VBS Purvanchal University Jaunpur in 2004 and 2006. Presently working as an Assistant Professor in Department of Physics at Atma Ram Sanatan Dharma College, University of Delhi. His research interests are in Luminescent Phosphors materials such as Phosphors, Nanophosphors that includes optical characterization and development of ceramics for lighting, solar cell and biomedical applications. His area of interest also includes preparation of the melt quenching glasses and glass ceramics for non-invasive biomedical application. He has received research excellence award twice from Delhi Technological University for his outstanding contribution in research. Also received best presentation award in conferences. He has published more than 14 papers in journals of high repute with h-index of 10. He has completed one National Project Implementation Unit (NPIU) sponsored research project as a Co-PI. He has more than 10 years teaching experience. He is guiding various UG and PG students in their projects and is also engaged in various administrative activities of his college.

Part I
Sustainable Materials

Chapter 1
Fundamentals of Sustainable Materials

Parul Devi

1 Introduction

This chapter initiates the fundamental science of sustainability by following the basic principles of sustainable materials that have been explained based on the introduction, conducted research, onward movement and remaining complexes in a selection of few categorized sustainable materials and afterward overall impact of materials for the sustainable environment.

In simple words, sustainability can be defined as what lasts in the world or creating a livable future for everyone on the earth. To lead a healthy life, sustainability is a matter of global issue that is concerned with continuous human activities recognition today, broadly referring to the system's capability in its institution i.e., Earth planet. Tracing modern civilization history along with the ecological state of health, industrialization bolstered commercialism (economic expansion) which has broadcasted the frightful demonstrations of unsustainability that we are observing apparently. Such factors are greenhouse gas emission (ozone depletion), soil infertility, biodiversity loss, ocean plastics, forever chemicals and pandemics, etc., which causes various regional and global impact on the native surroundings and health of humans (energy, strength, fitness, and well-being) (Kumar & Mani, 2022; Titirici et al., 2022).

Earlier, the outlook of sustainability was gleaned from efficiency which led to productive advantages to industries in the aspect of effectiveness of the process. As time evolves, this vantage point has been interchanged to execution-based products, that deal with the operational atmospheric concernment. At present, unprecedented resource consumption, driving manufacturing-dominated growth has arisen due to population increment, acquisitiveness, and upgrading lifestyles. The interdisciplinary domain of sustainability science can be illustrated by a dynamic interaction

P. Devi (✉)
Hochfeld-Magnetlabor Dresden (HLD-EMFL), Helmholtz-Zentrum Dresden-Rossendorf, 01328 Dresden, Germany
e-mail: phyparul@gmail.com

R. K. Kotnala et al. (eds.), *Advanced Functional Materials for Sustainable Environments*, https://doi.org/10.1007/978-3-031-62620-3_1

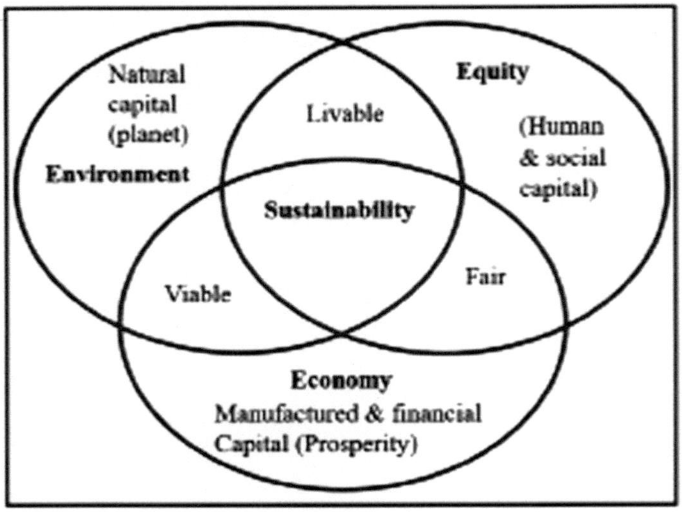

Fig. 1 The essentials of sustainability

between society (together economy) and the natural environment, which fundamentally requires a methodological approach. Sustainability and technology are dissembling yet intrinsically linked areas of research. The establishment of technology has the potential to significantly reduce the cost of manufacturing processes, making them more competitive with traditional manufacturing methods. The implication of the technology could also help to reduce the reliance on fossil fuels and promote the use of renewable energy sources as explained in Fig. 1.

Sustainable materials are the "heart" of sustainability while manufacturing those materials does not harm the environment. Sustainable material development is required for a community's progress which includes needs, necessities, and desires that influence the community's sustainability to varying degrees or levels (Mukherjee et al., 2021; Titirici et al., 2022).

2 Sustainable Materials

Sustainable or imperishable materials are those that can be used by our consumers and economically industrially as a consequence of higher production. However, the required properties for the production of those materials are less raw material, pollution prevention less energy, etc. To create more sustainable materials, we must use natural raw materials without detrimentally depleting the environment, to avoid the impact on future generations. The raw materials criticality depends on numerous viz. elemental scarcity, market concentration, supply disruption risk, substitutes, recycling, and temporal availability. These materials vary tremendously and are permitted to encompass bio-based polymers which are derived from polysaccharides or exceedingly reclaimable material e.g. wood or quartz that can be reprocessed indefinite number of times without demanding further mineral resources. Recently, “critical materials” have been the foremost component of renewable energy systems owing to their performance. For example, a permanent magnet made of rare earth metals (Nd, Gd, Dy, Pr) used in wind turbines, Li, Na and Co-ion in alkaline Li, Na-ion batteries, transition metals Pt and Ir in electrolyzes and fuel cells, and Si in solar cells. The literature comprises an uncounted number of papers on sustainable materials. This chapter gives an overview of nothing less but a few classes of sustainable materials that have been published until 2023, which is obviously a high claim, as this task comprises the analysis of more than thousands of papers. It only covers a small part of the entire literature. Following are few classes of sustainable materials.

(a) **Battery Materials (Li-ion Technology)**

In 2019, John Goodenough was awarded Nobel Prize in Chemistry for his work on batteries, which included the development of Li-ion batteries. The lightweight, powerful batteries sparked a revolution in the technology field, paving the way for modern portable electronics such as mobile phones and laptops as well as medical applications. Aiming at a green future, batteries share a large part of geothermal power in the electric power sector technology. Li-ion technology is still the predecessor amidst all the accessible battery technology concerning energy, power density, and cycle life. The performance of first-generation rechargeable Li-ion batteries has been improved considerably after advances in cell design and engineering, which were introduced by Sony in 1990. The execution metric required of the battery materials is becoming increasingly vigorous to meet large demand for inexpensive, everlasting, and denser devices from consumers. The limited ability of the cathode to store energy is the greatest barrier to enlarging the energy density of Li-ion batteries. A typical intercalation cathode is a lithium transition-metal oxide. At the anode, Si and Li metals can increase the storage capacity from the various routes. The charge storage space of a cathode is limited by crystal structure and chemistry. Indeed, theory promises large energy density in new battery chemistry like Li-S and metal-air, however, multiple challenges were demonstrated experimentally. Additionally, Li-S and Li-air batteries are associated with the engineering challenge of acquiring new technology for the industrial manufacturing of those batteries.

Although Li-ion batteries have been extensively used, the world still demands high enormously productive Li-ion batteries with huge capacity and immense charge-discharge for the utilization in current communication and convenient transportation. Material researchers also have investigated the capabilities of graphene and graphene-based materials in the solicitation of Li-ion batteries. The world still demands smaller and lighter-weight batteries as an implication to store more charge for hybrid electric vehicles that will complement the reduction of CO_2 emission during transportation and indeed, a cleaner environment. The future requires the advancement of battery materials to accomplish necessities given briefly as ample capacity, efficient power for consumer electronics, huge power for power tools, and hybrid electric vehicles. This can be sorted out based on different solid-state ionic materials that possess a contrasting combination of properties or several properties to optimize rechargeable Li-batteries for the diversity of succeeding markets (Xu et al., 2020).

(b) **Carbon Materials**

Carbon materials are the most ubiquitous because they are among the most versatile materials due to different kinds of structure, form, and properties, which is a consequence of the highly flexible coordination of carbon atoms and so, the valency. There are two well-known forms of carbon, viz. diamond and graphite. The hardness, extremely high refractive index, and high dispersion of light make it extremely useful for doing research such as diamond anvil cell to apply pressure that can go up to ~700 GPa, industrial applications as computer chip production, construction, drilling for minerals and jewelry as well. Unlike diamond, graphite is a conductor, which conducts electricity due to the delocalization of the pi-bond electrons above and below the planes of the carbon atoms. Many more forms of carbon also exist after reducing the dimension quantum mechanically such as fullerene (0D), carbon nanotubes (1D), and graphene (2D). However, fullerene and carbon nanotubes were discovered several decades ago. Recently discovered, a two-dimensional (2D) atomically thin crystal, graphene, composed of a single layer of carbon atoms packed in a hexagonal lattice with an interatomic distance of 0.142 nm. The discovery attributed the Nobel Prize in Physics for the year 2010 to Andre K. Geim and Konstantin S. Novoselov "after they made decisive contributions to the research of Graphene. Owing to the excellent properties of graphene and graphene-based materials it has become one of the most interested research topics in recent years. Graphene can also be used as an anode in battery materials. Different forms of carbon materials made them used in emerging technologies, as few of them are electrical energy storage devices, nuclear power plants, electromagnetic shields, electrocatalysis, photocatalysis, smart textiles, water purification and biomass, etc. (Gogotsi, 2013).

Hence, many applications of carbon materials have steadily increased the demand and will continuously increase in coming time. There are only a few carbon materials manufactured industrially from non-renewable precursors e.g., chemical vapor deposition, strong oxidants/reductants or electric-arc discharge techniques, etc. that involve relentless energy-intensive conditions. Thus, there is a need for the development of a more sustainable approach to the production of high-performance carbon-based sustainable materials, to complete the growing demand without compromising climate change, environmental pollution, alternative energy resources, or human health. Keeping all these factors in mind, carbon-based materials should be produced using more sustainable methods such as biomass/biomass wastes, and efficient, green manufacturing processes but not limited to only those methods. Integrating carbon material production into biorefinery schemes will help to gain full advantages of biomass for carbon production, reducing energy consumption and waste generation. After all great progress has been achieved at the research laboratory scale. Further entanglement and investment are required for the advanced commercialization of Industry (Gogotsi, 2013).

(c) Catalysts as Hydrogen Evolution Reaction (HER)

The rate of a chemical reaction can be increased by the addition of a catalyst. The process is known as catalysis. The involvement of catalysts in a reaction allows more sanctioned products to be assembled at a faster rate under more favorable conditions without consuming it. The contribution (~30–40%) to our global GDP of catalysts is huge in all aspects of modern life and economic impact. Approximately 85% of all manufactured products involve catalysis in their production chain and these products have a considerably huge impact on energy, healthcare, and pharmaceutical materials. Benjamin List and David Macmillan won the Nobel Prize in Chemistry in 2021 for the discovery of the development of asymmetric organocatalysis, a method that uses small organic molecules as catalysts instead of enzymes or metals. The innovations in molecular construction have prompted the development of catalysts that are less expensive and more environmentally friendly, opening up a new pathway in pharmaceutical research. Most of the chemists use metal-based catalysts; about half of the enzymes exclude the metals in their active sites as enzymes are enormously efficient and enantioselective. Small organic molecules can be highly selective and efficient catalysts, similar to metal complexes and biocatalysts. Consequently, asymmetric synthesis of organocatalysts is gaining more importance to complement bio and metal catalysts (Li et al., 2022).

Recently discovered Weyl semimetal compounds, the family of topologically non-trivial materials show high efficiency for HER. The band inversion of these materials in chemistry is referred to as the inert pair effect. A well-defined process for enhancing the catalysis efficiency of Weyl semimetals by applying a magnetic field has been explained in the patent (Mohamed et al., 2022). The most straightforward and effective strategy to boost the activity of catalysts is to increase the surface area, but it is limited to the number of catalytic active sites. The massless high mobility of electrons and their symmetrically protected metallic surface states make topological materials an ideal platform for catalysts. The catalyst efficiency of topological materials depends mainly on the surface electronic structure and bulk conductivity. The high chemical stability, high conductivity, and d-orbital-derived surface states (more localized than s and p-orbitals) increase their catalytic performance. Even under acidic conditions, topological materials were found to be stable for the HER process. Not only limited to the HER process, topological semimetals are also effective in activating multiple electron transfer reactions such as oxygen evolution reaction (OER) and nitrogen reduction reaction (NER). For the protection of the environment and a clean and more efficient energy system, heterogeneous catalyst reactions are vitally required for H_2 production and reduction of CO_2 and N_2. Interfacial sensitive electrochemical surface-enhanced Raman spectroscopy and electrochemical scanning tunneling microscopy could investigate the electrochemical reaction, enabling us to see the change of local density of states and bonding environments during the redox reaction at the atomic or molecular scale. Consequently, understanding the catalytic reaction at the atomic and electronic scale will open a new field for topological catalysts completely, for example, the separation of topological chiral fermions for asymmetry catalysis and enantiomeric recognition. However, the search for highly efficient and low-cost catalysts is still needed (Rajamathi et al., 2017).

(d) **Photovoltaic (PV) Materials**

Becquerel has observed an important way to convert solar radiation into electricity, known as the PV effect. In simple words, the PV effect is defined as the production of an electric voltage between two electrodes attached to a solid or liquid system upon shining the sunlight as shown in Fig. 2.

A photovoltage is developed across a p-n-junction in a semiconductor in all PV devices, known as solar cells. Commonly used materials in PV devices are indirect band gap semiconductors such as silicon, direct band gap semiconductors that as gallium arsenide (GaAs), metal chalcogenides, and organometallics. In recent years, organic solar cells have become a hot topic due to the simplified fabrication process resulting in cost, mechanically flexible, and large-area PV devices that allow relevant access to the sun's sustainable and clean energy in industrial research. Recent research has been devoted to increasing the power conversion efficiency of all classes of solar cells. Among these, organic solar cells are the only PV device that makes use of molecules to absorb photons and convert them into electric charges avoiding the involvement of intermolecular transport or electronic excitation. The use of bulk

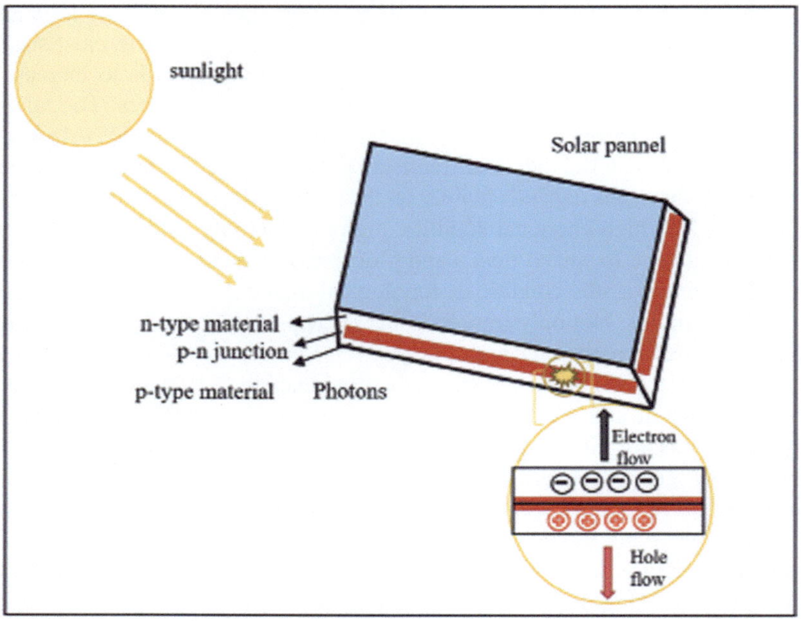

Fig. 2 The PV effect

hetero-junction structures, where the active layer is spin-coated from a mixed solution of donor and acceptor materials has given a breakthrough in the field of solar cells. Organic solar cells offer an advantage over conventional PV devices by separating the two functions, viz. are charge-carrier transport and light harvesting process (Khanafer & Vafai, 2013).

In 'next generation' solar cell device, several series of processes are accompanied which are respectively defined as photo absorption by the active layer, followed by the exciton migration to, and charge transfer at the interface, separation of the charges from their Coulomb attraction, and transport of separated charges to the electrodes must occur more efficiently. The current world demands electronic and optical properties (band gap and its position) of semiconductors, charge carrier mobility, and the recombination time of photodegraded charges, restricting substantially the alternative of suitable material so that it can perform as an efficient photovoltaic converter. Till now, perovskite solar cells based on organic–inorganic lead halide are the highest efficient solution-processed solar cells. Thus, offering a way to overcome efficiency limitations in other types of solar cells, by compensating the thermal losses of silicon solar cells in cooperation devices. In addition to that, the perovskite family can provide higher solar cell efficiencies by offering a rich multitude of crystal structures and substituents with the potential to leave unprotected new and exciting photophysical phenomena. The device's performance can be measured with the help of optical spectroscopy, which makes it a powerful tool in the field of PV. Optical probes are the most valuable tools to study the dynamics of charge transfer, the nature of the excited state, and recombination processes in donor–acceptor systems. The pump-probe process in optical spectroscopy gives an insight into the device's performance. Interrupted energy production, high investment costs, converting DC to AC, not-so-green, etc. are few factors that need to be considered to improve the PV technology (Lincot, 2017).

(e) **Hydrogen Liquefaction**

Nowadays, natural gases are one of the main sources of energy and their importance will increase in coming time. In the current scenario, there is an upward tendency in the production and consumption of this energy including hydrogen, oxygen, methane, etc. Natural gas as an energy source is rapidly booming with geopolitical importance. Gas has increased from a marginal fuel in regionally disconnected markets to an energy source that is transported across great distances for consumption in greatly diversified economic sectors. Natural gas is an alternative fuel choice for consumers seeking relatively low environmental impacts for an eco-friendly atmosphere. Subsequently, the world's natural gas consumption is projected to more than double over the next three decades, rising from 23 to 28% of the total primary energy demand by 2030 and surpassing coal as the world's number two energy source and potentially overtaking oil's share in many extensive industrialized economies (Key World Energy Statistics 2017, 2017).

Up to January 2016, the global nominal LNG (liquid natural gas) liquefaction capacity was 301.5 million tons per year. The consumption of liquid hydrogen as a longstanding replacement for hydrocarbon fuel for land and air transport is technically feasible nowadays. It is considered to be an ideal fuel in terms of a fully cyclic system serving as a "working substance" in a closed chemical and thermodynamic cycle. The three times higher amount of energy per unit weight compared to gasoline or any other hydrocarbon fuel makes liquid hydrogen particularly beneficial for aircraft and land-based transport. As an environmentally friendly fuel, this is a viable substitute for hydrocarbons in the twenty-first century. Fuel cells are among the most promising technologies depending on hydrogen sources. In this case, the useful energy output is quantized in electrical power produced by the system where the input energy is the energy stored in the fuel. According to the U.S. Department of Energy, hydrogen fuel cells have typically an energy efficiency (output divided by input energy) of 40-60%. This is somewhat beyond than other systems for energy generation. For example, the conventional internal combustion engine of a car can make use of only 25% of the stored energy in the gasoline. Under special conditions in large facilities, efficiencies of 70% and more are feasible nowadays. However, liquid hydrogen as an energy storage medium has not yet penetrated the market on a large scale. One of the reason is the high costs and energy losses upon liquefaction.

Hydrogen is the most important energy carrier; therefore, work on its liquefaction is required and has a direct prospect of introduction into the industry (Franco et al., 2012; Iea, n.d.).

(f) Metal–organic Frameworks (MOF) Materials

MOF are a hybrid class of materials, that epitomize the appeal of synthetic structures, and the capability of combining organic and inorganic chemistry; two disciplines of chemistry frequently regarded as diverse. The ultrahigh porousness and extensive internal surface areas altogether with the extraordinary degree of capriciousness for both the organic and inorganic components of their structures compose them potential candidates in applications ranging from membranes, thin-film devices, catalysts, gas repositories, and separation, sensing, energy storage, drug delivery, etc. MOF materials are built up of inorganic nodes, i.e., metal cations or metal-oxide clusters, interconnected by organic linkers especially polycarboxylic acids or heterocycles. The synthesis of MOF materials requires expensive reactants and ascetic conditions produce them expensive for large-scale applications. Their synthesis process can be modified by incorporating waste-derived reactants, which will manufacture them greener and inexpensive, thereby potentially enabling their applications. However, expensive reactants and harsh conditions constitute these materials extravagant for multipurpose applications. Their production can be constructed greener and cheaper by incorporating waste-derived reactants, definitely enabling their applications. In recent years, the debated topic in MOF research is carbon dioxide capture, which is directly related to clean energy and environmental protection. Most syntheses of MOF yield in powder form which makes them difficult to integrate into devices. Therefore, new synthesis tools need to be developed for their production as thin films, monolith fibers, etc. (Kreno et al., 2012).

(g) **Quantum Dots (QDs)**

QDs are crystals of a fluorescent semiconductor material with a diameter of ~2–10 nm. Their absorption and emission radiation spectral range can be modified by altering particle size and particle surface passivation under various ligands. The quantum confinement effect results in the production of QDs from bulk material as explained in Fig. 3. QDs present peculiar electronic properties among all bulk semiconductors and discrete molecules. The synthesis process controls the color of QDs, which is found to be size-dependent. Smaller dot emits higher energy bluer color light whereas larger dot emits lower energy red light. The implication is that QDs can emit light from ultraviolet to infrared regions. The bandgap of QDs can be tuned by changing their size based on the quantum confinement effect. These properties make them unique for potential applications in electronic/optoelectronic devices, quantum computing cancer diagnosis, etc. The entire spectral range is accessible through a targeted synthesis process, which makes their additional utilization in various technological applications, namely, luminophores in display technology, for up-conversion in photovoltaic devices, or as security attributes on banknotes. QDs can be synthesized via various chemical methods such as colloidal, plasma sintering, aqueous, hydrothermal, and solvothermal synthesis, etc. However, society still demands further development in QDs because of the potential drawbacks in biological applications, which arise due to their large physical size. The large physical size limits the application for diffusion across cellular membranes. Moreover, their toxic nature for the cell makes them inappropriate for biological applications. Not up to that, but QDs may blink or fade away and their quite extended lifetime may be a hindrance to certain applications that require QDs to biodegrade immediately after experimenting. Therefore, further development is required for the size reduction without compromising their unique properties.

After 2004, the discovery of carbon-based dots (CDs) brought attention to replace or limit the use of both types of QDs. Contrary to QDs, the fluorescence spectra of CDs are not only governed by the quantum confinement effect but by structural complexity as well. However, CDs facilitated from carbon-based structures also endow them their advantages such as low cost, effortless modification, large solubility, and less toxicity. CDs have an average size similar to QDs. Literature suggests two synthesis processes of QDs: the top-down method (cutting large carbon materials such as graphite into small fragments) and the bottom-up approach (growing small precursor molecules such as glucose into large-conjugated domains). The bottom-up synthesis process is deliberated to be a more feasible, ecological, and environmentally safe process. The bottom-up synthesis process facilitates numerous chemical structures ranging from crystalline to amorphous. These chemical structures have typically different functional groups on the surface. To complement QDs, it is vital to develop CDs with well-efficient optical properties for achieving a more conceivable life-cycle at all stages (Mohamed et al., 2022).

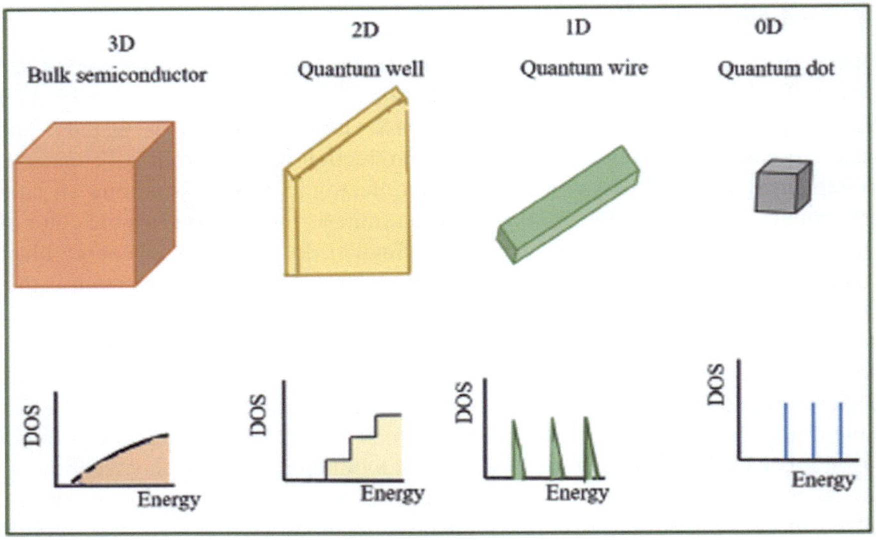

Fig. 3 Density of electronic states vs. energy (quantum confinement effect)

3 Impact of Materials for a Sustainable Environment

Nowadays, energy storage is more important than at any time in human history while keeping the environment safe for future generations and maintaining the economic perspective. Global warming is a great risk for humanity and the consequences are unpredictable. To stop climate change, we need to introduce the transformation towards feasible sustainable materials. To achieve this, new materials in physics and chemistry research is essential, through which we can find sustainable ways to manufacture or synthesize these materials. Figure 4 shows the cyclic process and necessary factors for achieving a sustainable environment.

MOF materials can be used as a replacement for plastic waste. However, their production cost hindered them as an alternative candidate for plastic waste. The economic growth production of MOF materials will eliminate plastic potentially for a greener or so-called healthy environment. Liquefaction of hydrogen is also necessary for several reasons as high specific energy density, mobility and storage reasons, renewable and environmentally friendly, and more efficient liquefaction at low temperatures. Graphene sheets could be an alternative anode for Li-ion batteries as their d spacing can be tuned by inserting CNT or C_{60}, which provide a progression advancement and reconciliation of lithium ions (Xu et al., 2020).

It is impossible to do justice to the many excellent contributions made in the active area of this sustainable material research. This artical reviewed only a few classes of sustainable materials. Lastly, to make this process truly environmentally friendly, care should be taken to develop processes for the reuse of alternative sustainable materials.

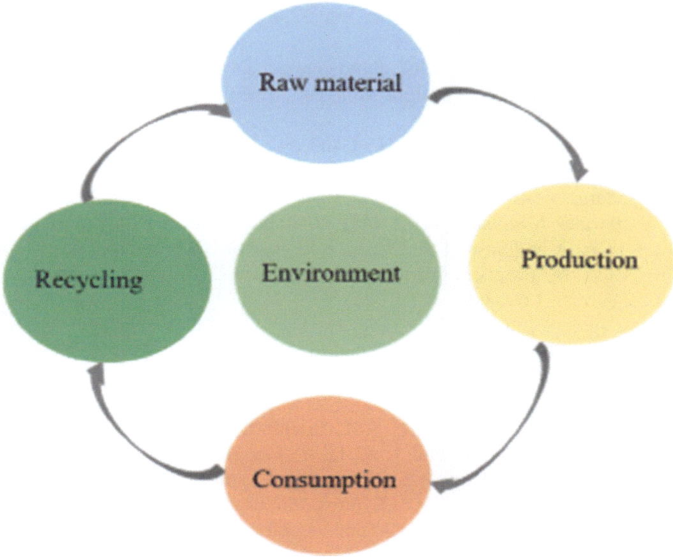

Fig. 4 Cyclic process for achieving sustainable environment

Acknowledgements Authors would like to thank Toni Helm and T. Hermannsdörfer for the discussion and also acknowledge the support of the HLD-HZDR, members of the European magnetic field laboratory.

References

Franco, V., Blázquez, J.S., Ingale, B. and Conde, A. (2012). The Magnetocaloric Effect and Magnetic Refrigeration Near Room Temperature: Materials and Models. *Annual Review of Materials Research*, **42(1)**: 305–342. https://doi.org/10.1146/annurev-matsci-062910-10035.

Gogotsi, Y. (2013). Carbon Nanomaterials (Y. Gogotsi & V. Presser, Ed's.). CRC Press. https://doi.org/10.1201/b15591.

Iea. (n.d.). Technology Roadmap Hydrogen and Fuel Cells. www.iea.org/t&c/.

Key World Energy Statistics 2017. (2017). OECD. https://doi.org/10.1787/key_energ_stat-2017-en.

Khanafer, K. and Vafai, K. (2013). Applications of Nanomaterials in Solar Energy and Desalination Sectors (pp. 303–329). https://doi.org/10.1016/B978-0-12-407819-2.00005-0.

Kreno, L.E., Leong, K., Farha, O.K., Allendorf, M., Van Duyne, R.P. and Hupp, J.T. (2012). Metal–Organic Framework Materials as Chemical Sensors. *Chemical Reviews*, **112(2)**: 1105–1125. https://doi.org/10.1021/cr200324t.

Kumar, M. and Mani, M. (2022). Sustainability Assessment in Manufacturing for Effectiveness: Challenges and Opportunities. *In:* Frontiers in Sustainability (Vol. 3). Frontiers Media S.A. https://doi.org/10.3389/frsus.2022.837016.

Li, G., Xu, Y., Song, Z., Yang, Q., Zhang Y., Liu, J., Gupta, U., Süß, V., Sun, Y., Sessi, P., Parkin, S., Bernevig, B.A. and Felser, C. (2022). Obstructed surface states as the descriptor for predicting catalytic active sites in inorganic crystalline materials Advanced Materials, **34(26)**. https://doi.org/10.1002/adma.202270196.

Lincot, D. (2017). The new paradigm of photovoltaics: From powering satellites to powering humanity. *In*: Comptes Rendus Physique **18(7–8)**: 381–390. Elsevier Masson s.r.l. https://doi.org/10.1016/j.crhy.2017.09.003.

Mohamed, W., El-Gawad, H.A., Handal, H., Galal, H., Mousa, H., El-Sayed, B., Mekkey, S., Ibrahem, I. and Labib, A. (2022). Remarkable Recycling Process of ZnO Quantum Dots for Photodegradation of Reactive Yellow Dye and Solar Photocatalytic Treatment Process of Industrial Wastewater. *Nanomaterials*, **12(15)**: 2642. https://doi.org/10.3390/nano12152642.

Mukherjee, A., Sarkar, D. and Sasmal, S. (2021). A Review of Green Synthesis of Metal Nanoparticles Using Algae. *In:* Frontiers in Microbiology (Vol. 12). Frontiers Media S.A. https://doi.org/10.3389/fmicb.2021.693899.

Rajamathi, C.R., Gupta, U., Kumar, N., Yang, H., Sun, Y., Süß, V., Shekhar, C., Schmidt, M., Blumtritt, H., Werner, P., Yan, B., Parkin, S., Felser, C. and Rao, C.N.R. (2017). Weyl Semimetals as Hydrogen Evolution Catalysts. *Advanced Materials*, **29(19)**. https://doi.org/10.1002/adma.201606202.

Titirici, M., Baird, S.G., Sparks, T.D., Yang, S.M., Brandt-Talbot, A., Hosseinaei, O., Harper, D.P., Parker, R.M., Vignolini, S., Berglund, L.A., Li, Y., Gao, H.L., Mao, L.B., Yu, S.H., Díez, N., Ferrero, G.A., Sevilla, M., Szilágyi, P.Á., Stubbs, C.J., … Anderson, P.A. (2022). The sustainable materials roadmap. *J Phys Materials*, **5(3)**. https://doi.org/10.1088/2515-7639/ac4ee5.

Xu, C., Dai, Q., Gaines, L., Hu, M., Tukker, A. and Steubing, B. (2020). Future material demand for automotive lithium-based batteries. *Communications Materials*, **1(1)**. https://doi.org/10.1038/s43246-020-00095-x.

Chapter 2
Sustainable Energy Systems: Challenging the Emerging Energy Technologies

Sanjay Kaul

1 Introduction

Sustainability is truly a meta discipline on nearly all the existing human knowledge in equal parts and importance and ranges from biological sciences to the physical sciences and to social sciences including politics (Bayliss et al., 2000). Nature itself does not measure in decades, but in millions and billions of years. The crucial question for humanity is whether the state of the world will be habitable for humans when the depletion of natural resource leads to economic crises (Bird et al., 2005). The long-term failure of liberal markets could incite severe social as well as national conflicts. Strategic alliances between proactive energy industry, and as well as global institutions would accelerate the commercialization of renewable energy generation and could prevent irreversible negative impacts to our social and environmental systems. We can extend sustainable to culture. In modern society. There are many different definitions of sustainability and also depends upon the area of interest. This chapter has tried to cover different aspects of sustainable energy in relation to modern context of technology (Brown L.R., 2006).

Sustainability is derived from two Latin words viz. *Sus* means Up and *Tenere* means hold. In modern concept…. It means to uphold human existence on planet earth for a very long time. United Nations definition for sustainability is: *Humanity has the ability to make development sustainable if we meet the needs of the present without compromising the ability of future generations to meet their own needs* (Cherp et al., 2014).

Sustainable development does not mean to be harmonious with nature only, but also is a process of change in

- exploitation of resources
- direction of investment

S. Kaul (✉)
Department of Engineering Technology, Fitchburg State University, Fitchburg, MA 01420, USA
e-mail: skaul01810@gmail.com

R. K. Kotnala et al. (eds.), *Advanced Functional Materials for Sustainable Environments*, https://doi.org/10.1007/978-3-031-62620-3_2

- orientation of technological development
- Institutional change that are made consistent with future as well as present needs.

It is difficult to predict that, can

- a civilization persists on its current path for any length and time into the future
- we continue to have modern economies/modern societies
- we expect course of this century same at end as it was in the beginning

Historically, humans used simple renewable energy technologies for providing energy services for heating, cooling, and cooking. Simple wind-ventilated systems were used for food drying and storage, and wood was burned for space heating and cooking. Solar radiation served for baking and drying purposes. Before the steam engine was invented, water and wind powered mills were used in small manufacturing (Biyik et al., 2017) & (Cornell, 2009).

These simple technologies using renewable resources were abandoned in the industrialization process. Premium energy sources such as oil, coal and natural gas yielded higher energetic output in compact energy units that could be easily transported (Holdren, 2007) (as shown in Fig. 1). Thus, industrial production became independent from local energy restrictions such as insufficient solar radiation or wind speed. The early renewable energy technologies could not have provided the vast amount of energy necessary for the rapidly emerging heavy industries (Estévez, 2020).

Nuclear power was the great hope of the twentieth century for cheap and unlimited power supply (Kaul, 2008). After a phase of euphoria and the construction of numerous nuclear power plants in several countries, the fear of nuclear accidents as well as economic aspects as rising construction and production costs stopped the

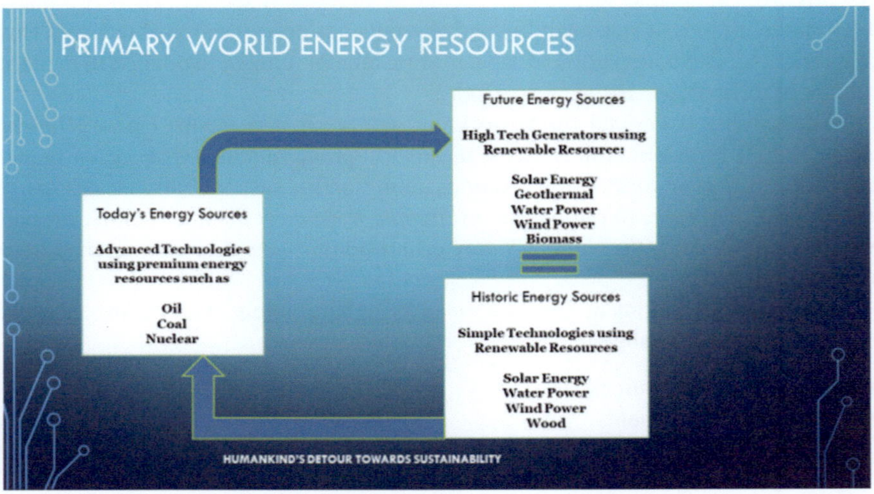

Fig. 1 Primary world energy resources

rise of nuclear power production (Kaul, 2000). New technologies finally produced cheaper energy in smaller units that could be added incrementally to the existing power system. While most industrialized countries experience a stabilization or even decline in the number of their population, numerous developing countries still struggle with an enormous population growth. These developing countries are aiming at reaching the same production and consumption levels as industrialized nations. This target is based on the influence of marketing and advertising campaigns of transnational corporations, television promoting the glamorous world of consumer societies, and political institutions such as the United Nations and the governments of industrialized nations advocating development though economic growth and free global trade. If this vision comes true, we will witness an unprecedented level of energy consumption in the next century (Nalley, 2021). If we try to base the future energy demand of this scenario on our current or only slightly modified energy system, the use of fossil fuels such as coal, oil and natural gas will eventually reach levels that may have irreversible impact on local economies in regard to resource exploitation and pollution as well as on global climate change. An energy forecast of the Energy Information Administration/U.S. Department of Energy in the International Energy Outlook 1997 claimed that global fossil fuel demand will almost double between 1990 and 2015 (Misbrener, 2021) (as shown in Fig. 2).

A growing world population that may well reach 10 or 20 billion in the next centuries will have an impact on the global environment (O'Shaughnessy, 2015). With developing countries and nations in transition imitating the consumption patterns of industrial countries, the use of energy will increase dramatically. Which forms of energy resources will eventually be used is yet unclear. Long-term forecasts claim the depletion of fossil fuels and climate changes due to high emissions of greenhouse gases.

While discussing solutions, we should also think about problems as we live in a world of interconnectedness. This brings some important issues:

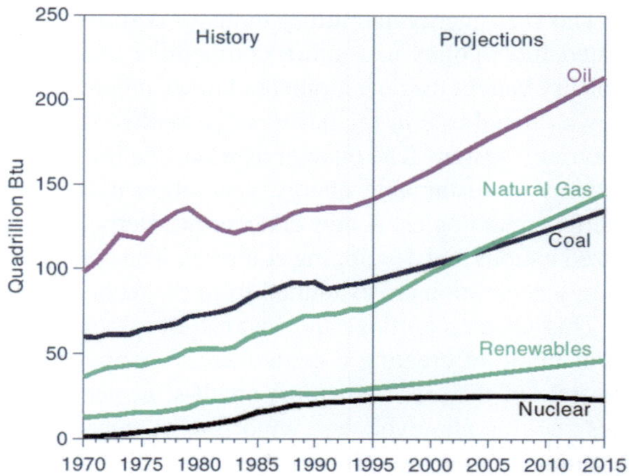

Fig. 2 World energy consumption by fuel type, 1970–2015

- can we continue to have modern life style
- can developing countries be able to catch up and experience increased life span and material well being.

Perhaps

- technology will make up for diminishing of fossil fuel, fix atmosphere
- may be, we find some new sources of energy

Technological advancement in renewable energy generation has made non-fossil fuels a viable option to significantly contribute to the world's growing energy demand. Prices of these technologies could be reduced by strategic partnerships and cooperation between utilities, consumers, state and federal representatives, and global advocate institutions such as UNO and the World Bank.

In order to support renewable energy production, the coordination of micro- and macroeconomic as well as global institutions and activities, information and project implementation is necessary. Non-profit organizations are intensively promoting clean energies, but their financial restrictions make them not very competitive with the marketing efforts and political influence of the fossil fuel industry.

Liberal markets tend to favor short-term oriented decisions. A highly dynamic competitive energy market will not reward long-term commitments nor social or ecological concerns without any additional incentives.

The concept of sustainable development offers a theoretical framework in which the conventional energy systems of industrialized countries and the emerging energy system of developing countries have to be shaped. A systematic and strategic approach is necessary to cope with the immanent environmental and social impacts of energy generation. Sustainable Energy Generation is an important prerequisite of global, long-term oriented economic development.

A new concept of sustainable energy system is based on renewable resources. Systemic, technical, economic, social, and behavioral aspects are incorporated into a new electricity system feasible for industrial and developing nations.

The U.S. energy and utility industry is shifting from a system of vertically integrated monopolies to a more competitive market. The new shape of the energy industry benefit the development of clean renewable energy options. Consumers will play a crucial role in the new energy market and determine the future shape of the electricity system. The chapter assesses the transition from fossil-fuel based energy systems to sustainable energy generation using advanced small-scale renewable energy technologies. A new concept of Micro-grids can be applied to both industrialized nations and developing countries. The concepts of Micro-grids and distributed energy generation are examined from the technical and economic perspective.

This chapter outlines the historic development of the U.S. energy system as an example of creating a centralized, monopolistic system and also introduces in the state-of-the-art of emerging small-scale renewable energy generation technologies such as photovoltaics, wind power, microhydro, and hydrogen fuel cells. A systemic approach is presented in the last part. Distributed energy generation is a cost-effective approach of reducing environmental and social externalities of electric power systems. Empirical studies indicate increasing interest of the utility industry in small-scale power systems.

2 Social and Behavioral Aspects of Energy Consumption

2.1 Sustainability Concerns

There is a compelling case to be made that need of people of today and leave behind the needs of people of tomorrow. This also brings political character in sustainability and that is how we weigh the need of people of today versus the people of future. This may be inherent stress in the whole concept that do we owe future generations anything.

The most common way to express population and consumption damaging environment-

$$\text{Impact (I)} = \text{Population (P)}^{*}\ \text{Affluence (A)}^{*}\ \text{Technology (T)}$$

Affluence is little complex to expand upon. Let us redefine it as

$$\text{Sustainability Impact (SI)} = \text{Population (P)}^{*}\ \text{Consumption/Person (C/p)}^{*}\ \text{Impact/Consumption (I/C)}$$

If we double the population, we double the sustainability impact or if we double the consumption, we double the Sustainability Impact on the other hand, we can reduce the impact if we develop new and appropriate technologies e.g. electric vehicles, energy saving devices.

Implementing energy efficiency programs or even reducing total energy consumption does not mean going back to the middle ages and reducing the current standard of living. Modern renewable energy technologies have advanced far during recent decades. Additional moderate modifications in social and behavioral energy consumption patterns could help reduce total energy demand and thus make it feasible to change the basis of our energy system from fossil fuel to clean and inexhaustible renewable technologies. Traditional economic analysis in the energy field has neglected the consumer's role in the end-use of energy. Scientific and commercial research primarily focused on adjusting energy supply to a given (and implicitly ever increasing) energy demand. New studies in social and behavioral research have indicated, that consumers are quite responsive to incentives and disincentives of using energy more efficiently.

Future economic research in the energy field will certainly find a vast potential of energy savings through minor changes in social and behavioral consumption patterns. Advocating consumer sovereignty and individual preferences in a world of distorted energy prices and other disincentives against clean and efficient energy use does not lead to economically optimal resource allocation. It is necessary to shift consumer preferences towards technologies and services that would lead to a more sustainable energy system.

3 Shaping a Sustainable Energy System

3.1 Alternative Grid Systems

The current U.S. electricity system features two alternatives to provide power for residential customers (as shown in Fig. 3). Large central grids connect a multitude of electric consumers nationwide, and power is provided by large central power stations. The size of these power stations requires sufficient reserve capacities to compensate power outages due to repair or failure of the large electric generating facilities. On the other hand, there are many off-grid residences located in remote areas. Small-scale energy systems like diesel engines, photovoltaic panels, wind or water-generators power these houses. These independent electricity systems are generally installed as hybrid systems including two or more generating technologies. There is another alternative to organize electric power systems, on a scale in between the central grid and stand-alone electric systems (as shown in Fig. 4).

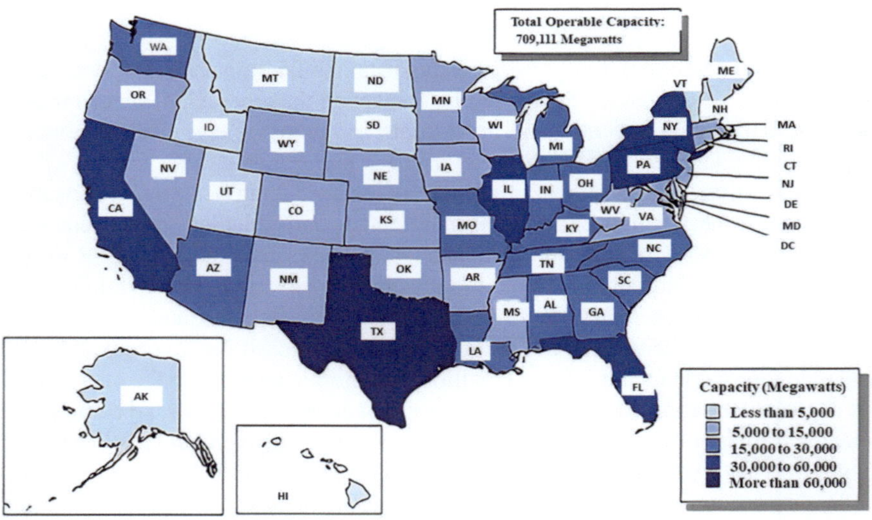

Fig. 3 Total energy capacity at U.S. electric utilities

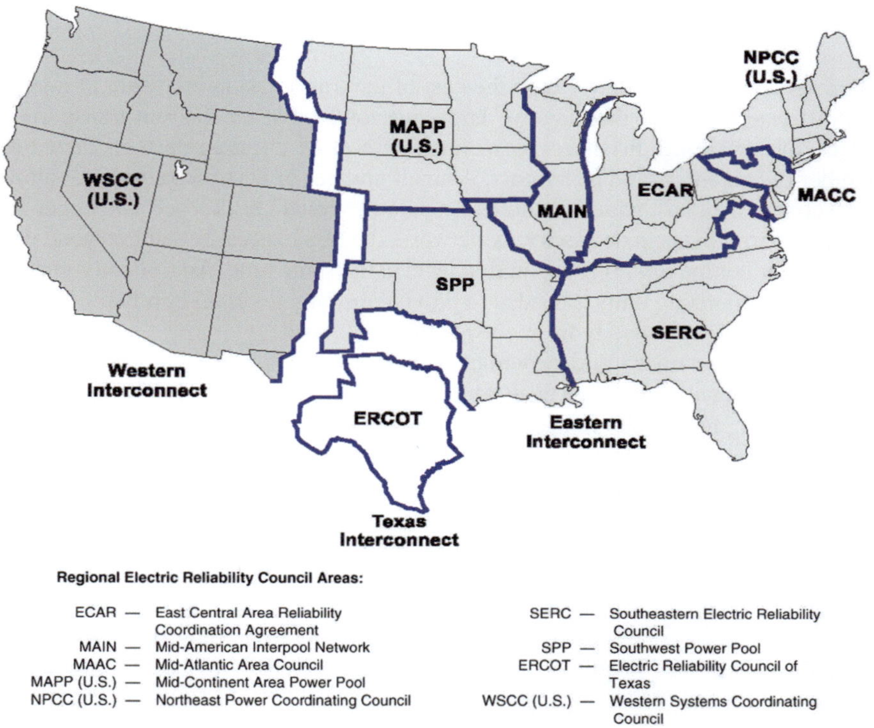

Fig. 4 Regions and interconnections of the North American Electric Reliability Council in the Contiguous United States

3.2 *The Concept of Micro-Grids*

The micro-grids are an electrically isolated set of generators, forming an independent grid that is not connected to the large central grid. The generators of a micro-grid would supply all of the demand of customers. Due to their modular character, small power generators such as photovoltaics and fuel cells would be used to power these micro-grids. Advances in hydrogen technologies could soon make it feasible to produce hydrogen by electrolysis with the surplus current generated by photovoltaics, store it safely and cost effectively and then reconvert it to electricity in a fuel cell. This power generating system would be both energetically efficient and ecologically friendly since it would not produce any emissions but water, nor would it use any fossil resources.

3.2.1 Micro-Grids: A Basic Model

Micro-grids are not burdened with the cost of the transmission hardware of a large central system, nor do they involve large line losses that result from transporting electricity over long distances. However, micro-grids do struggle with their low number of customers and generators. A small number of customers does not allow to levelize peak loads, which would lead to a lower system peak level. The larger the group of participants in an electricity network, the less likely it is that everyone will reach his or her maximum consumption level at the same time. Additionally, certain number of generating units is needed to provide the targeted level of reliability since there is no back-up capacity from outside the micro-grid.

The sizing of a micro-grid system phenomenon is due to the benefit of diversity; load balancing will occur with each additional customer entering the network. Also, the generating units that are necessary to provide a certain level of reliability (i.e. low reliability with 10 days of outage in 10 years vs. high reliability with only 1 day of outage in 10 years). Reliability gains are rather proportional to the number of generation units introduced into the electric power system.

Various economic analysis has shown that the micro-grid is cost-effective compared to a central grid solution. There might be a large number of cases in which the benefits of micro-grids could be notable, regarding high reliability; the absence of central-grid related costs; and environmental friendliness through the use of small-scale renewable power generation.

Micro-grids can offer an alternative to conventional central grid power systems in rural areas where a small number of people live in a neighborhood located far from a central power station (as shown in Fig. 5).

Developing countries with a low coverage of the central grid may be one of the top candidates for micro-grid systems. A non-market based energy system would not have to provide the profit share a utility would require for operating the system.

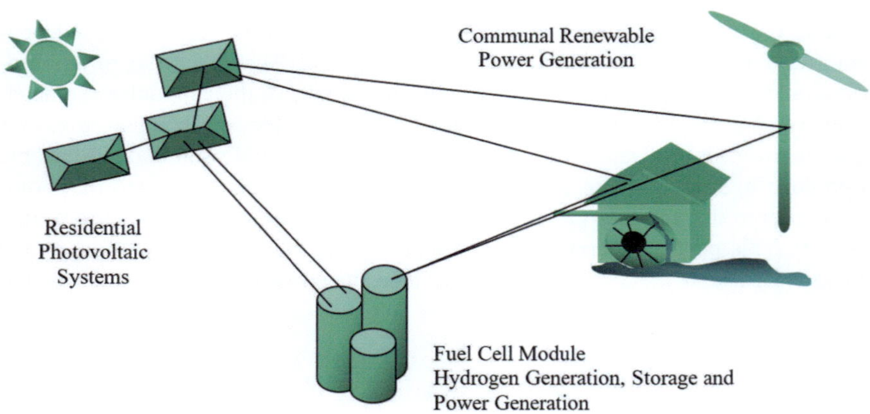

Fig. 5 Sizing of micro-grids system

However, depending on further technological advancement of small-scale power generation components, it could be economical to contract with a utility to help with financing, operation, and maintenance of the system.

4 Energy Consumption as a Part of Sustainability

4.1 Utility Demand Side Management Programs

Traditionally, utilities have focused on how to meet a given (mostly increasing) demand with additional generating capacities. New regulation and restrictions on installing further capacities reversed this way of supply-side oriented thinking to demand-side thinking. Energy demand and supply can be coordinated by either adjusting the total capacity or by adjusting the amount of energy that consumers actually consume. Demand side management (DSM) programs try to take the second approach. Utilities use DSM to improve energy efficiency on the customer side. The aim of DSM is to reduce the peak load of power demand by replacing high energy consuming appliances with more efficient ones.

Demand side management can be a low-cost alternative to investments in upgrading electricity generating facilities and the grid distribution lines. DSM programs should be conducted by utilities that have more accurate information on energy issues and available energy saving technologies than residential power consumers. Utility DSM programs have addressed the problem of relatively high initial capital costs for energy-efficient appliances. DSM options are economically viable in a long-term calculation. Customers, however, have a limited access to capital, higher discount rates, and tend toward a shorter decision making horizon than public or investor-owned utilities. Tenants who live have little incentive to invest in energy efficiency. Landlords may not be interested in investing in energy efficient appliances when tenants pay utility bills. Thus, DSM should be organized by the utility itself in cooperation with consumers who are willing to participate in DSM programs.

4.2 Load Management

How can we meet our energy needs with clean renewable energies? Although the earth receives much more solar energy every day than our current economies use, small-scale renewable power generation technologies convert but a fraction of this insolation to usable energy. There is no doubt that in the mid-term future, renewable energies will be able to provide the amount of energy modern industrialized nations consume today. Humans are not immune to laws of nature as humankind's actions are transforming the earth's landscapes, ocean and atmosphere. If we ask to write our most important sustainability concerns for next century, Invariably, we will write: Population, food security, ecosystem, energy, water, urban challenge. All these factors are interwoven.

If we approach the question from the consumer side, renewables are indeed capable of producing a remarkable share of the global energy demand. Two aspects in our current energy system have to be examined and eventually modified to make electricity systems based on renewables a feasible option. First, reducing the total amount of energy consumed by efficiency and demand-side management programs. The second is to use electric power only where necessary, and replacing energy with more appropriate forms of renewables where applicable.

The crucial question for renewable power systems is not how to meet today's energy demand with renewable technologies, but how to adapt our energy demand to a feasible amount of power that could be provided by renewable resources. Reducing the total energy demand, the "customer load", must be the first step in reorganizing the energy system. It is often the most cost-effective alternative to cut down the load side first before investing in renewable technologies (as shown in Fig. 6).

Photovoltaic systems are the most elegant way of producing electricity. This technology transforms solar radiation directly into electric power, without any moving

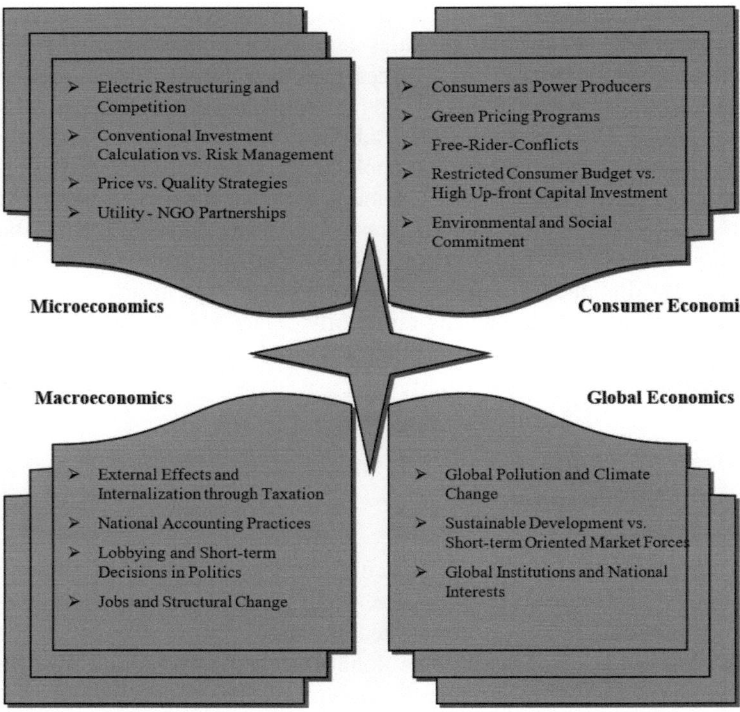

Fig. 6 Issues in renewable power generation and micro, macro, consumer and global economic levels

parts, and without using an intermediate source of energy that was once created by solar radiation. Electricity generated by photovoltaic panels transform solar radiation into electricity with the minimum technological and energetic effort possible, it also reflects the high value of this energy form.

An economic analysis has to determine where which form of energy is the most suitable. Consumers do not demand energy or electricity itself, but the services that energy provides, such as space and water heating, cooking.

5 Conclusion

Reaching a sustainable global economy is not possible without increasing the share of Renewables of today's Energy Base. Not only generation, but also use of consumption of energy has to be sustainable. Demand side management, efficiency and load reduction are crucial issues on the consumer side. Production and consumption of commodities has to be sustained. We have to keep the philosophy towards input –production pattern and stay away from end-of pipe production. Also, overcoming political, ethnical and religious conflicts as well as stabilizing the world population are additional prerequisities for sustainability.

References

Bayliss, K. and Hall, D. (2000). Independent power producers: A review of the issues. Public Service International Research Unit, University of Greenwich, London, U.K.

Bird, L. and Swezey, B. (2005). *Green power marketing in the United States: A status report* (No. NREL/TP-620–38994). National Renewable Energy Lab (NREL), Golden, CO, USA.

Biyik, E., Araz, M., Hepbasli, A., Shahrestani, M., Yao, R., Shao, L. and Atli, Y.B. (2017). A key review of building integrated photovoltaic (BIPV) systems. *Engineering Science and Technology*, an International Journal, Elsevier B.V., **20(3)**: 833–58.

Brown, L.R. (2006). Subsidizing climate change. Earth Policy Institute.

Center for Renewable Energy and Sustainable Technology (CREST), Annual Report (2007). Retrieved from https://ceert.org/. (Retrieved on November 20, 2021).

Cherp A. and Jewell J. (2014). The concept of energy security: Beyond the four As. *Energy Policy*. **75**: 415–21.

Cornell, P.E. (2009) Energy security as national security: Defining problems ahead of solutions. *Journal of Energy Security*, **19**.

Department of Energy (December 25, 2020). Federal agency use of renewable electric energy. Federal Energy Management Program. Retrieved from https://www.energy.gov/eere/femp/federal-agency-use-renewable-electric-energy (Retrieved on December 11, 2021).

Energy Information Administration (EIA) (2000). The changing structure of the electric power industry: An update. US Department of Energy, Washington, DC 20585.

Energy Information Administration (EIA) (2005). Electricity generation and environmental externalities. U.S. Department of Energy, Washington, DC.

Energy Information Administration (2020). International energy outlook 2020. U.S. Department of Energy.

Estévez, A.T. (2020). Sustainable living? Biodigital future! In sustaining resources for tomorrow. Springer, Cham, 137–162.

Holdren, J.P. (2007). Energy and Sustainability. *Science*, **315(5813):** 737–737.

Hyman, L.S. (1994). America's electric utilities: Past, present, and future. Arlington Virginia: Public Utilities Reports.

IEA (2019). Public energy RD&D budgets by country. Retrieved from https://www.iea.org/. (Retrieved on January 10, 2020).

International Renewable Energy Agency (November, 2019). Future of solar photovoltaic: Deployment, investment, technology, grid integration, and socio-economic aspects. IRENA, 1–88. Retrieved from https://www.irena.org/publications/2019/Nov/Future-of-Solar-Photovoltaic. (Retrieved on December 20, 2021).

IRENA – International Renewable Energy Agency (2019). Future of solar photovoltaic: Deployment, investment, technology, grid integration and socio-economic aspects.

Kaul, S. (2008). Environmental future – A time realization. Blog: www.aroundtoday.com.

Kaul, Sanjay (2004). Sustainable Mobility: Renewable Energies Powering Fuel Cell Vehicles (with Raphael Edinger), Praeger Publishers, Connecticut.

Kaul, Sanjay (2000). Renewable Resources for Electric Power: Prospects and Challenges (with Raphael Edinger), Quorum Books, Connecticut.

Misbrener, K. (September 17, 2021). SMUD plans to cut net-metering credits by 44% next year. Solar Power World. Retrieved from https://www.solarpowerworldonline.com/2021/09/smud-plans-to-cut-net-metering-credits-by-44-percent-2022/. (Retrieved on January 10, 2022).

Nalley, S. and LaRose, A. (October 6, 2021). International Energy Outlook. U.S. Energy Information Administration. Retrieved from https://www.eia.gov/outlooks/ieo/pdf/IEO2021_ReleasePresentation.pdf. (Retrieved on January 10, 2022).

O'Shaughnessy, E.J., Heeter, J.S., Liu, C. and Nobler, E. (2015). Status and trends in the US voluntary green power market (2014 data; No. NREL/TP-6A20–65252). National Renewable Energy Lab. (NREL), Golden, CO, USA. http://www.nrel.gov/docs/fy16osti/65252.pdf.

Our World in Data (OWID) (2019). Renewable energy consumption country wise. Retrieved from https://ourworldindata.org/. (Retrieved on January 10, 2020).

Perez, R., Burtis, L., Hoff, T., Swanson, S. and Herig, C. (2004). Quantifying residential PV economics in the US-payback vs. cash flow determination of fair energy value. *Solar Energy*, **77 (4)**: 363–368.

Prosper, T. (2008). CPUC raises CAP on cash incentives for customers installed clean energy. California Public Utilities Commission (R.08-03-008).

Schwartz, N.D. (May 11, 2008). A peak behind the price at the pump. The New York Times. Retrieved from https://www.nytimes.com/2008/05/11/business/11maker.html. (Retrieved on January 10, 2022).

Simshauser, P. (2020). Merchant utilities and boundaries of the firm: Vertical integration in energy-only markets. EPRG Working Paper 2008, Energy Policy Research Group, University of Cambridge.

Sonnichsen, N. (July 14, 2021). Global primary energy consumption by country 2020. Retrieved from https://www.statista.com/statistics/263455/primary-energy-consumption-of-selected-countries/ (Retrieved on January 10, 2022).

Timmons, D., Harris, J.M. and Roach, B. (2014). The economics of renewable energy. Global Development and Environment Institute. Tufts University, **52**: 1–52.

UNEP (2012). Global Trends in Sustainable Energy Investment 2012. Bloomberg, New Energy Finance, UNEP Collaborating Center for Climate & Sustainable Energy Finance. Retrieved from https://www.circleofblue.org/wp-content/uploads/2012/06/UNEP_global-trends-in-renewable-energy-investment-June-2012.pdf (Retrieved on November 10, 2022).

US Congress Senate Committee on the Judiciary (2008). Exploring the skyrocketing price of oil.

U.S. Energy Information Administration (October 29, 2021). Electric power annual. Retrieved from https://www.eia.gov/electricity/annual/ (Retrieved on January 10, 2022).

U.S. Energy Information Administration (November 5, 2021). Renewable energy explained: Incentives. Retrieved from https://www.eia.gov/energyexplained/renewable-sources/incentives.php (Retrieved on December 25, 2021).

US Environmental Protection Agency (November 2, 2021). Power plants and neighboring communities. EPA. Retrieved from https://www.epa.gov/airmarkets/power-plants-and-neighboring-communities. (Retrieved on January 15, 2022).

Yasukawa, K. and Anbumozhi, V. (2018). Assessment of necessary innovation for sustainable use of conventional and new-type geothermal resources and their benefit in East Asia. ERIA Research Project Report FY2017 (07). Economic Research Institute for ASEAN and East Asia.

Chapter 3
Two Dimensional (2D) Materials for Energy Storage Applications: A Review

Akash, Tanzeel Hameed, Gourav Singla, and Paramjyot Kumar Jha

1 Introduction

Two Dimensional (2D) materials and nanotechnology have received a lot of attention after the invention of graphene because of the numerous variations in their physical and chemical properties but they are getting incredible consideration primarily due to their nanosized thickness. As the stream of charge carriers, photons as well as heat are all constrained in that 2D plane; 2D materials tends to exhibit numerous changes in their opto-electronic properties. Metals, semi-metals, insulators, and plentiful semiconductors in the 2D structural form span the visible spectrum from ultraviolet to infrared due to their direct and indirect bandgaps and exhibit a wide range of electronic properties (Castro et al., 2009). Currently, 2D geometry is excellent for thin film manufacturing techniques as well in the semiconductor industry. Therefore, these materials can help to control electronics in nanoscale, optoelectronics, the assemblage of novel and in manufacturing more flexible and durable devices in the future (Wang et al., 2012; Wilson et al., 1969). Two-dimensional atomic crystal integration circuits, which include many components like a modulator, single memory, logic gate, amplifier, oscillator, mixer, and a switch, have also been successfully synthesized from 2D materials. The best material for making a wide range of electronic devices is a 2D material that is only a few atoms thick. Utilizing functional integration circuits made of 2D materials, we can overcome technological and fundamental obstacles in the electronic industry.

Due to their high anisotropy and crystal structure, 2D materials properties can be adjusted in a variety of ways, which includes disregarding dimensions, adjusting composition, fine-tuning external fields, and changing structural settings. The band structure of a material, for instance, will change when the thickness of the material

Akash · T. Hameed · G. Singla (✉) · P. K. Jha
Department of Physics, University Institute of Sciences, Chandigarh University, Ludhiana-Chandigarh State Hwy, Mohali, Punjab 140413, India
e-mail: gourav.e11805@cumail.in

R. K. Kotnala et al. (eds.), *Advanced Functional Materials for Sustainable Environments*, https://doi.org/10.1007/978-3-031-62620-3_3

is reduced from the bulk to the mono-layer (Morozov et al., 2008). We can demonstrate excellent performance in both contemporary technologies and their applications using a monolayered 2D material. This 2D material, which is only a few atoms thick, is capable of forming a wide variety of heterostructures without the use of lattice matching or their processing methods. The intrinsic properties of a material, such as in-plane anisotropy, interlayer interaction, and so on, can be revealed through the use of 2D material monomers (Lee et al., 2008).

Further, in order to prepare 2D materials, physical or chemical methods have been used to reduce the intermolecular forces that exist between the layers of materials. The properties of these materials will also change as the dimension is altered. The best example is graphene. This material gained popularity due to its numerous properties which includes high young's modulus, high specific surface area, high carrier mobility and spectral absorption (Zhu et al., 2010; Zhao et al., 2014).

Because of its remarkable properties, a greater number of researchers are working on this type of material. MXenes, alkene, nitride, transition metal dihalides, and organic materials are some of the two-dimensional organized elements that fall under the periodic table categories (Guardia et al., 2014). Numerous studies have demonstrated that each of these materials can be used for a variety of purposes. The properties and applications of a 2D material can be understood through an examination of its catalytic performance. Due to their structure and electron specificity, the 2D materials have performed better in photocatalytic activities like CO_2 reduction and hydrogen production (Kusior et al., 2018; Bafaqeer et al., 2019).

2 Exfoliation Approaches for Two Dimensional (2D) Materials

After the successfully discovery of graphene, 2D materials have been getting incredible consideration because of their better and unique properties which includes physical as well as chemical and these materials are becoming popular every year. During its synthesis process, the common shortcomings include reduction in ion-accessible surface area due to restacking which effected the research period and the applications of 2D material properties.

Currently, the mechanical exfoliation becomes more efficient and more mature method of exfoliation by which different kind of mechanical forces are being used to avoid the interaction between layered materials and some of the raw materials due to van der Waals force (Liu et al., 2016; Novoselov et al., 2004). After a couple of studies, a method is often explained about the concernment of preparation of 2D materials, which is often called the redox method because it uses less energy and produces more reaction products in the process. It has some flaws, like partially destroying the spatial structure of 2D materials during the redox reaction but this method can be used for a lot of different things.

2.1 Mechanical Exfoliation Method

Mechanical exfoliation is the most effective and one-of-a-kind method of exfoliation, surpassing all others in terms of product purity and yield. In this method, other than the assisting substances there is no requirement of any other substances in addition. During the exfoliation process, the Van der Waals forces between the various layers are balanced by employing the mechanical stripping principle. Some properties of auxiliaries are mostly used for the synthesis of 2D materials which improves the speed of exfoliation and to prevent the 2D material structure from being destroyed, which is mainly used in some graphene grinding process. (Zhao et al., 2010; used the ball milling technique to grind and centrifuge the pure material using dimethylformamide dispersion medium. In addition, research into a novel, effective exfoliation method for liquid-forming rich and higher-level graphene, h-BN, MoS_2, and WS_2 nanosheets was sparked by mechanical ultrasonic techniques.

2.2 Expansion and Exfoliation via Intercalation Method

The multiple thin layer 2D materials can also be prepared by some other ways which are not common, expansion and exfoliation via intercalation process. This way, the formation of layered structure material facilitates due to the decreasing of the van der Waals force between non-covalently bonded molecules or namely called polymers.

After the synthesis of graphene, researchers are also trying to prepare 2D materials by the method of an ionic interlayer (Zhao et al., 2015); have exfoliated nano sheets of transition metal dichalcogenides (MoS_2 and WS_2) via rapid chemical exfoliation method through the spreading of Na+ and OH– into the interlayer space of MoS_2 and WS_2 crystal, which often tends to form NaOH in the raw materials. The redox method's most distinctive advantages are its high productivity and this method consume less resources. However, environmental pollution is caused by the extensive use of oxidant-reducing agents in this process, which results in some flaws in this method for the manufacturing of 2D materials and a weakness in the preparation process. Due to their inherent chemical properties, synthesis of some 2D materials using the redox method is extremely challenging.

Another common method for exfoliating 2D materials is Hummer's method, in which graphite powder and potassium permanganate ($KMnO_4$) are combined in concentrated H_2SO_4 to produce graphene oxide flakes after clearing (Hou et al., 2020). The carbon–oxygen double bond and the hydroxyl group, which are produced by the reaction between the layered structures, are two examples of functional groups that are frequently involved in the exfoliation process.

Moreover, in the case of exfoliation of MXene it has been pointed out that the redox reaction can occur in the MXene particle, which contains nanoparticles of precious metals. MXene can be used as the reaction reducing agent without adding any other reducing agent, reducing the pollution caused by finding an exogenous reducing agent in this process because any metal cation can absorb the electrons provided by MXene (Cheng et al., 2020).

3 Most Emerging 2D Materials for Energy Storage Applications

3.1 Graphene

Throughout twentieth century, we have increased our understanding of the fundamental nature of materials. We are not only harvested the properties of existing materials but also, we have developed new methodologies in order to design new kinds of materials to acquire desired properties. This is because naturally occurring material seems to have attained saturation points of their abilities in compared to possible in science and technologies and it is a need of all to develop new kinds of materials and we are doing it very well. For example, we have developed nanomaterials in twentieth century that exhibit completely different properties compared to their bulk counter parts and we have witnessed applications of these kinds of materials in many fields like semiconductor, medicine, defense etc. Still many technologies such as virtual reality is replacing textbooks to transparent and flexible screens, still we cannot bend the screen for phones right or still our phone batteries cannot store the power that is sufficient for like weeks or months, space tourism, smart windows etc. There is a material on this earth that we can use it, which is cost affordable and can be used everywhere for any kind of applications that can bring all the science fiction kind of stuff to reality. In fact it is one single layer of atom that is thin enough and is the strongest known element on the earth than the diamond called "GRAPHENE". Graphene has been one of the first real 2D materials which is only one atom thick that has isolated in nature. Graphene is a single layer of graphite. The method used to isolate a single layer of this graphite is called as mechanical exfoliation which is also called as top-down approach. So mechanical exfoliation is also known as scotch-tape approach. But this method comes with so many disadvantages like so many defects in the material, these days we use several advanced technologies such as chemical vapor deposition, pulse laser deposition, molecule or beam extract. Hence, these are the different methods to synthesize 2-D materials. It has honeycomb like structure (Fig. 1) with the small atomic thickness of the order of 0.345 nm and all the carbon atoms in the graphene are sp^2 hybridized (Iijima, 1991).

Some of the properties of graphene includes its high stretch ability upto 20%, extreme high conductivity, optically transparent and highly dense, that it is impermeable to gases. Graphene is an inert material and it does not readily react with other atoms. However, it has the absorbing property, that is, it can absorb different atoms and molecules. This can lead to changes in the electronic properties and may also be used to make sensors or other applications. They have highest surface area and

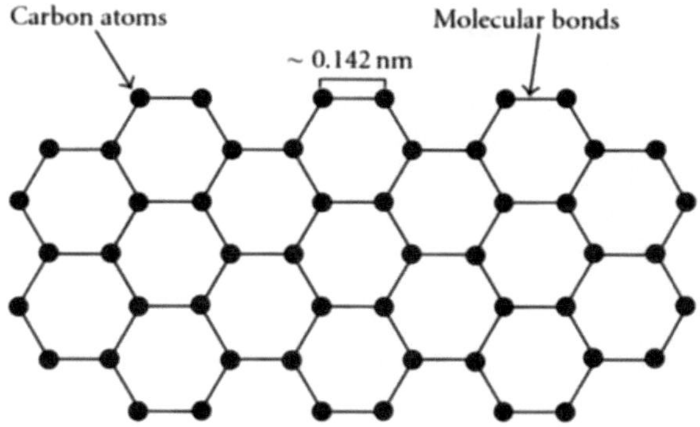

Fig. 1 Structure of graphene (Roberts et al., 2010)

can be 200 times stronger than steel. They are light-weight (0.77 mg) and exceptionally strong in tensile tests. It is possible to be piled, rolled into tubes, or formed into balls to revert to graphite. Its stiffness, strength and durability are one of the reasons that make graphene as individual material and a reinforcing agent also, that is, it can be combined along with other materials like polymers etc. It has high electrical current density (million times than copper) and its intrinsic mobility is also 100 times more than of silicon. These amazing properties and its multifunctionality make graphene suitable for a wide spectrum of applications ranging from electronics to optics, sensors, and many biodevices.

Graphene materials and its composites have shown better performance in the energy storage devices like Li-ion battery, Na-ion battery, supercapacitors, Li-S battery and Solar cells. It is due to the passage of electron, high surface area and flexibility of 2D planar structure of graphene material which make sure to give high utilization of carbon atoms.

3.1.1 Li/Na-Ion Batteries

The process of graphene hybridization through electroactive materials for both anode and cathode of Li-ion batteries notably improved their capacity, rate performance and cyclability (Cai et al., 2017). They can also be considered as a promising additive for both anode and cathode in Li-ion batteries due to its stable conducting network formation ability with low sedimentation threshold which results to increase the electrochemical performance of electrode.

3.1.2 Supercapacitors

Due to long life and high-power density, supercapacitors are counted under the best class of devices for the energy storage (Simon et al., 2008). Supercapacitors can be labeled as electric double layer capacitors (EDLCs) or pseudo capacitors based on their variations and distinctiveness in mechanism of energy storage (Zhang et al. 2009). Additionally, the surface area, pore structure, and size distribution of carbon materials all play a role in EDLC's specific capacitance, while pseudo capacitors' energy storage mechanism is dependent on Faradaic redox reactions between the electrode and electrolyte. The graphene material is typically combined with pseudocapacitive materials like polymers and transition metal oxides, sulfides and hydroxides.

3.1.3 Lithium-Sulphur Batteries and Solar Cells

Additionally, due to their higher hypothetical capacity of lithium (Li) and sulfur (S), Li-S batteries are in greater demand as the next generation of energy storage systems. Material made of graphene that is used to cover either the membrane or the cathode side of Li-S batteries to prevent polysulfide shuttling (Rana et al., 2019).

Crystal-clear electrodes made of graphene nanomaterial have shown great promise for the production of solar cells. Due to its superior conductivity, transparency, and mechanical strength, graphene is able to produce solar cells that are extremely flexible and thin. Currently, the researchers' primary focus is on functionalizing and doping graphene-containing solar cells to improve their efficiency.

3.2 MXene Materials

The 2D transition metal carbides and nitrides produced by MXenes are widely recognized. When we remove a specific atomic layer from MAX phase by use of various etchants that contain an acidic solution which contains fluoride in water, we get MXene material (Lengauer, 2000). The nitrides and carbides have a universal formulation $M_{n+1}AX_n$, where n ranges from 1 to 3, M stands for transition metals like chromium (Cr), scandium (Sc), vanadium (V), titanium (Ti), molybdenum (Mo), hafnium (Hf), zirconium (Zr), niobium (Nb), tantalum (Ta) and many more are the elements from the periodic table of group 13 and 14 and X belongs to either or both carbon and/or nitrogen (Zhang et al., 2006; Barsoum et al., 2011). In some cases, the etchant is used to break the metallic bonds of the MAX phase which leads to multifaceted $M_{n+1}X_nT_x$ materials. Where T_x is the functional groups on the surface that are formed during the etching, for instance O_2 and F. Further, adopting different techniques these multifaceted materials change into single layered by inserting some other organic metal ions which results in sonication in aqueous solution. The first MXene materials was proposed in 2011 by Gogotsi et al., 2017; which is $Ti_3C_2T_x$. MXene materials can be available in three distinct compositions, namely M_2X, M_3X_2, M_4X_3. For the synthesis of Mxene materials there are three key approaches, first, in which there is single transition element involve like Ti_4N_3, Ta_4C_3. In these materials there exist only one transition metal (Ti and Ta). In the second type two different transition elements such as $(Mo,V)_4C_3$ exist and are called double transition MXene

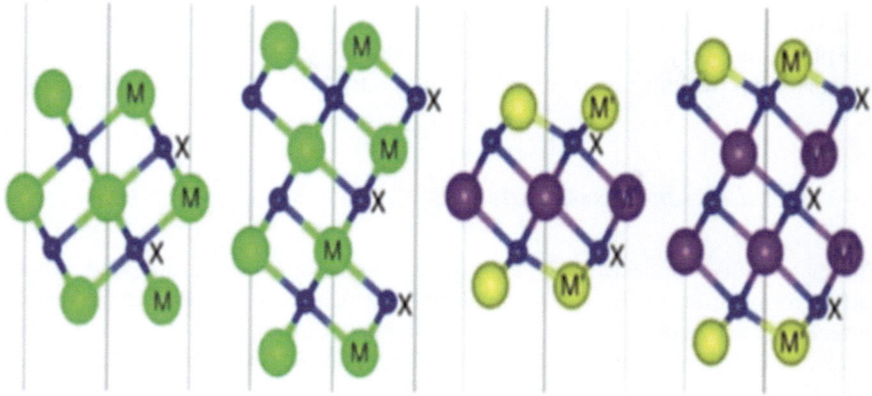

Fig. 2 Structure of MXene (Tahir et al., 2021)

materials. Under this instance, one transition metal fills the central M layers while a different transition metal absorbs the border layers. For example, in Cr_2TiC_2 the central layers are fill by Ti and outer M layers by Cr (Gogotsi et al., 2017). Due to the advantages of mechanical conductivity, its 2D structure, and its abundant biochemical processes, MXenes have attracted a lot of attention to be used as an electrode for rechargeable batteries and electrochemical capacitors. The most popular etchants, such as hydrofluoric acid or lithium fluoride combined with hydrochloric acid, are used to etch high quality MXene (as shown in Fig. 2).

3.3 *Black Phosphorous*

Black phosphorous (BP) is regarded as the most promising 2D material for energy storage due to its low density (2.69 g/cm^3), high theoretical capacity (2596 mAh/g for Li-ion batteries), low environmental impact, and high phosphorous content. It has a larger specific surface area due to its large lateral size and skeletal characteristics. Because of its high electronic conductivity (102 S/m), less conductive agent will be needed to make an electrode (Wu et al., 2018). Supercapacitors, magnesium ion batteries, Li-ion batteries, and Na-ion batteries all use this material as an excellent active material. Black phosphorous from the bulk BP has been taken for use in the energy storage field. Top-down synthesis via liquid exfoliation, mechanical cleavage, and bottom-up synthesis via chemical vapor deposition and wet chemistry are the most common methods for preparing black phosphorous (Brent et al., 2014). However, the main drawback of all of these approaches is their high cost and lengthy duration. Phosphor, BP's atomic layer, degrades when exposed to air, limiting its use in energy storage devices. Due to its capacitive properties, black phosphorous' use in capacitors is also restricted. The production of the electrode material is also influenced by the particular distribution of BP's pore sizes and surface area. Black phosphorus is a very attractive 2D material, but it also has a lot of problems that can be overwhelm by doing calculations and studies for better development in future.

3.4 Metal Oxide-Based Materials

The metal oxide nanosheet have gain incredible attention due to the well-defined surface terminations with oxide and hydroxide, which results to enhance the specific surface area and hence become attractive component for the fabrication of ultrathin porous electrode architectures for lithium-ion batteries and faradaic supercapacitors. The 2D nanosheets synthesized with the exfoliation of layered metal oxides and double hydroxides into single-crystalline platelets. The most common examples for the 2D metal oxide nanosheet-based electrodes are $Ti_{1-x}O_2$, $Ca_2Nb_3O_{10}$, MnO_2, V_2O_5 etc. (Elshof et al., 2016). Furthermore, depending on the chemically exfoliation process of the available layered compound different compositions can also be obtained. Depending on their 2D nature, these nanosheets look after to restack throughout electrode evolution, by lowering their surface area and by making the porosity to be slit-shaped. Such kind of restacking of nanosheets during charging/discharging process suppress the phase transformations and hence improve their life. The specific surface area and pore size distribution of the electrodes may be used to calculate the ion storage capacity of energy storage. In general, we can say that these large pore sizes accelerate high yield rate capacities and fast lithium-ion transport. Some additional entities like nanoparticles, nanowires and nanoplatelets have been incorporating and is known as pillaring in which these have shown logical approach to slow down their structural downfall and for the realization of structurally stable electrode architectures in which discharge capacity is approximately up to 860 mA h/g (Kang et al., 2012). Majority of 2D metal oxide nanosheets has been synthesized so far by the exfoliation on oxides which includes Ti, Ta or Nb in which their conductivity is very low and they are having a wide band gap.

3.5 Molybdenum Disulfide (MoS_2)

MoS_2 material has gained incredible attention in recent years due to its definite chemical, electronic, mechanical, magnetic and some optical properties. It is one of the transition metal dichalcogenides (TMDs). This material has been used in varieties of applications such as high-efficiency field effect transistors, sensing applications, and some energy and medical applications. This material has been in different crystalline structures like hexagonal, rhombohedral or tetrahedral. MoS_2 material naturally exist as 2H MoS_2 and semiconducting 2H and 3R phases and 1T metallic phase were the most popular structures but from these 2H is more stable and less conductive than 1T. This material has got attention to be used in energy application as this material

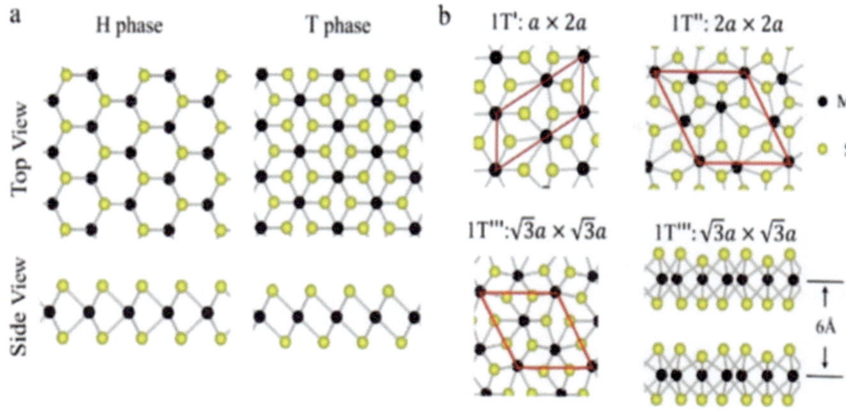

Fig. 3 Different MoS_2 crystal structures: (a) top and side view of monolayer MoS_2 for H and T phases and (b) different lattice structure of MoS_2 metallic phases (Shang et al., 2018)

was used as a substitute silicon in the electronics industry. 1T MoS_2 is the one which got incredible attention because of its high conductivity and has been mostly used in energy applications.

The 1T MoS_2, has been intercalated with some alkali metals like potassium or lithium and then this product is directed to be hydrated and then carry on to the oxidization process. Electrons transfer from those alkali metals (K or Li) to MoS_2, which will continue the octahedral coordination of Mo atoms. Electrons which came to be excess is then removed by the oxidation process but the coordination of octahedral remain same with some distortion. The ion named Ligand, which is attached to molybdenum through a coordinate bond is responsible for the metastable structure and metallic conductivity of 1T MoS_2 in which three d orbitals which are degenerated are settled by two electrons. An early theoretical study showed that monolayer MoS_2 can have a higher Na adsorption when compared to bulk MoS_2. It is perfect as an anode electrode in Na-ion batteries, with a theoretical capacity of 335 mA h g^{-1} (Fig. 3).

This article proposes the idea of studying the 2D materials in future studies for efficient and environmentally friendly approaches. Table 1 summarizes some of the 2D structured based materials applications in energy storage applications.

Table 1 Energy storage application of different 2-D structured materials

Lithium - ion	1T-MoS_2	Discharge capacity = 666 mA h g^{-1} at current density = 3500 mA g^{-1}	Liu et al., 2015
Sodium - ion	1T-MoS_2	Capacity of 175 mA h g^{-1} at a high current density of 2 A g^{-1}	Geng et al., 2017
Li - S batteries	Graphene/CNTs	Capacity of 1286 mAh/g at 0.2 C	Shi et al., 2019
Lithium - ion	MoS_2/Graphene	Capacity of 466 mAh/g at a current density of 4 A g^{-1}	Cao et al., 2013
Lithium - ion	MXene($VNbCT_X$)	Capacity of 520 mAh/g at current density of 0.1 A g^{-1}	Cheng et al., 2023
Lithium - ion	Black phosphorous/ Graphene	Capacity = 920, 501, 141 mAh/g at current density = 0.1, 0.5, 2.5 A g^{-1}	Zhang et al., 2016
Lithium - ion	V_2O_5	Specific capacity = 219 mAh/ g at current density = 1 A g^{-1}	Li et al., 2013

4 Conclusion

The 2D structured materials with advantages of better conductivity are emerging as excellent electrode materials for batteries and electrochemical capacitors. The innovations in the materials synthesis, their electrode structure design, and a better understanding about the charge storage mechanism have greatly improved the electrochemical performance of 2D electrodes material. Development of methods to improve 2D based electrodes density and columbic efficiency on the first cycle will greatly improve the application in energy devices. Despite several challenges, it can be provisioned that 2D materials will become competitive electrode materials from the practical point of view and will develop fast in energy storage applications.

References

Anasori, B., Lukatskaya, M. and Gogotsi. Y. (2017). 2D metal carbides and nitrides (MXenes) for energy storage. *Nature Reviews Materials*, **2:** 16098.

Bafaqeer, A., Khan, A.A., Amin, N.A.S. and Tahir, M. (2019). Indirect Z-scheme assembly of 2D ZnV_2O_6/RGO/g-C_3N_4 nanosheets with RGO/pCN as solid-state electron mediators toward visible light-enhanced CO_2 reduction. *Industrial & Engineering Chemistry Research*, **58**: 8612–8624.

Barsoum, M.W., Radovic and M. Annu (2011). Elastic and mechanical properties of MAX phases. *Annual Review of Materials Research*, **41**:195–227.

Brent, J.R., Savjani, N., Lewis, E.A., Haigh, S.J., Lewis, D.J. and O'Brien, P. (2014). Production of few-layer phosphorene by liquid exfoliation of black phosphorus. *Chemical Communications*, **50:** 13338–13341.

Cai, X., Lai, L., Shen, Z. and Lin, J. (2017). Graphene and graphene-based composites as Li-ion battery electrode materials and their application in full cells. *Journal of Materials Chemistry A*, **5**: 15423–15446.

Cao, X., Shi, Y., Shi, W., Rui, X., Yan, Q., Kong, J. and Zhang, H. (2013). Preparation of MoS_2–coated three-dimensional graphene networks for high-performance anode material in lithium-ion batteries. *Small*, **9**: 3433–3488.

Castro Neto, A.H., Guinea, F., Peres, N.M.R., Novoselov, K.S. and Geim, A.K. (2009). The Electronic Properties of Graphene. *Reviews of Modern Physics,* **81:** 109–162.

Cheng, R., Hu, T., Hu, M., Li, C., Liang, Y., Wang, Z., Zhang, H., Li, M., Wang, H. Lu, H., Fu, Y., Zhang, H., Yang, Q-H. and Wang, X. (2020). MXenes induce epitaxial growth of size-controlled noble nanometals: A case study for surface enhanced Raman scattering (SERS). *Journal of Materials Science & Technology*, **40**: 119–127.

Cheng, Y., Yang, L. and Yin, S. (2023). Synthesis and lithium-ion storage performance of novel two-dimensional vanadium niobium carbide ($VNbCT_X$) MXene. *Composites Communications*, **40:** 101588.

Elshof, ten, J.E., Yuan, H. and Rodriguez, P.G. (2016). Two-dimensional metal oxide and metal hydroxide nanosheets: synthesis, controlled assembly and applications in energy conversion and storage. *Advanced Energy Materials*, **6**: 1600355.

Geng, X., Jiao, Y., Han, Y., Mukhopadhyay, A., Yang, L. and Zhu, H. (2017). Freestanding Metallic 1T MoS_2 with Dual Ion Diffusion Paths as High-Rate Anode for Sodium-Ion Batteries. *Advanced Functional Materials*, **27**: 1702998.

Guardia, L., Paredes, J.I., Munuera, J.M., Villar-Rodil, S., Ayán-Varela, M., Martinez-Alonso, A. and Tascón and Juan M.D. (2014). Chemically exfoliated MoS_2 nanosheets as an efficient catalyst for reduction reactions in the aqueous phase. *ACS Applied Materials & Interfaces*, **6**: 21702–21710.

Hou, Y., Lv, S., Liu, L. and Liu, X. (2020). High-quality preparation of graphene oxide via the Hummers' method: understanding the roles of the intercalator, oxidant, and graphite particle size. *Ceramics International*, **46**: 2392–2402.

Iijima, S. (1991). Helical microtubules of graphitic carbon. *Nature*, **354**: 56–58.

Kang, J., Paek, S. and Choy, J. (2012). Porous SnO_2/layered titanate nanohybrid with enhanced electrochemical performance for reversible lithium storage. *Chemical Communications*, **48**: 458–460.

Kusior, A., Zych, L., Zakrzewska, K. and Radecka, M. (2018). Photocatalytic activity of TiO_2/SnO_2 nanostructures with controlled dimensionality/complexity. *Applied Surface Science*, **471**: 973–985.

Lee, C., Wei, X., Kysar, J.W. and Hone, J. (2008). Measurement of the Elastic Properties and Intrinsic Strength of Monolayer Graphene. *Science*, **321**: 385–388.

Lengauer, W. (2000). Transition Metal Carbides, Nitrides, and Carbonitrides. *In*: Handbook of Ceramic Hard Materials, 1 ed.; Riedel, R.; Wiley-VCH, Weinheim, Germany, 202–252.

Li, Y., Yao, J., Uchaker, E., Yang, J., Huang, Y., Zhang, M. and Cao, G. (2013), Leaf-Like V_2O_5 Nanosheets Fabricated by a Facile Green Approach as High Energy Cathode Material for Lithium-Ion Batteries. *Advanced Energy Materials*, **3:** 1171–1175.

Liu, Q., Li, X., He, Q., Khalil, A., Liu, D., Xiang, T., Wu, X. and Song, L. (2015). Gram-Scale Aqueous Synthesis of Stable Few-Layered 1T-MoS_2: Applications for Visible-Light-Driven Photocatalytic Hydrogen Evolution. *Small*, **11:** 5556–5564.

Liu, Y., Weiss, N.O., Duan, X., Cheng, H.C., Huang, Y. and Duan, X. (2016). Van der Waals Heterostructures and Devices. *Nature Reviews Materials*, **1**: 16042.

Morozov, S.V., Novoselov, K.S., Katsnelson, M.I., Schedin, F., Elias, D.C., Jaszczak, J.A. and Geim, A.K. (2008). Giant Intrinsic Carrier Mobilities in Graphene and its Bilayer. *Physical Review Letters*, **100**: 016602.

Novoselov, K.S., Geim, A.K., Morozov, S.V., Jiang, D., Zhang, Y., Dubonos, S.V., Grigorieva, I.V. and Firsov, A. A. (2004). Electric field effect in atomically thin carbon films. *Science*, **306**: 666–669.

Rana, M., Ahad, S.A., Li, M., Luo, B., Wang, L., Gentle, I. and Knibbe, R. (2019). Review on areal capacities and long-term cycling performances of lithium sulfur battery at high Sulphur loading. *Energy Storage Materials*, **18**: 289–310.

Roberts, M.W., Clemons, C.B., Wilber, J.P., Young, G.W., Buldum, A. and Quinn, D.D. (2010). Continuum Plate Theory and Atomistic Modeling to Find the Flexural Rigidity of a Graphene Sheet Interacting with a Substrate. *Journal of Nanotechnology*, **868492**: 1–8.

Shang, C., Fang, Y.Q., Zhang, Q., Wang, N.Z., Wang, Y.F., Liu, Z., Lei, B., Meng, F.B., Ma, L.K., Wu, T., Wang, Z.F., Zeng, C.G., Huang, F.Q., Sun, Z. and Chen, X. H. (2018). Superconductivity in the Metastable 1 T' and 1 T''' Phases of MoS_2 Crystals. *Physical Review B*, **98**:184513.

Shi, H.D., Zhao, X.J., Wu, Z.S., Dong, Y.F., Lu, P.F., Chen, J., Ren, W.C., Cheng, H.M. and Bao, X.H. (2019). Free-standing integrated cathode derived from 3D graphene/carbon nanotube aerogels serving as binder-free sulfur host and interlayer for ultrahigh volumetric-energy-density lithium sulfur batteries. *Nano Energy*, **60**: 743–751.

Simon, P. and Gogotsi, Y. (2008). Materials for electrochemical capacitors. *Nature Materials*, **7**: 845.

Tahir, M., Khan, A.A., Tasleem, S., Mansoor, R. and Fan, W.K. (2021). Titanium Carbide (Ti_3C_2) MXene as a Promising Co-catalyst for Photocatalytic CO_2 Conversion to Energy-Efficient Fuels: A Review. *Energy Fuels*, **35**: 10374–10404.

US Congress Senate Committee on the Judiciary (2008). Exploring the skyrocketing price of oil.

U.S. Energy Information Administration (October 29, 2021). Electric power annual. Retrieved from https://www.eia.gov/electricity/annual/ (Retrieved on January 10, 2022).

U.S. Energy Information Administration (November 5, 2021). Renewable energy explained: Incentives. Retrieved from https://www.eia.gov/energyexplained/renewable-sources/incentives.php (Retrieved on December 25, 2021).

US Environmental Protection Agency (November 2, 2021). Power plants and neighboring communities. EPA. Retrieved from https://www.epa.gov/airmarkets/power-plants-and-neighboring-communities (Retrieved on January 15, 2022).

Wang, Q.H., Kalantar-Zadeh, K., Kis, A., Coleman, J.N. and Strano, M.S. (2012). Electronics and Optoelectronics of Two-Dimensional Transition Metal Dichalcogenides. *Nature Nanotechnology*, **7**: 699–712.

Wilson, J.A. and Yoffe, A.D. (1969). The Transition Metal Dichalcogenides Discussion and Interpretation of the Observed Optical, Electrical and Structural Properties. *Advances in Physics*, **18**: 193–335.

Wu, S., Hui, K. and Hui, K. (2018). 2D Black Phosphorus: From preparation to applications for electrochemical energy storage. *Advanced Science*, **5**:1700491.

Yasukawa, K. and Anbumozhi, V. (2018). Assessment of necessary innovation for sustainable use of conventional and new-type geothermal resources and their benefit in East Asia. ERIA Research Project Report FY2017 (07). Economic Research Institute for ASEAN and East Asia.

Zhang, H.B., Zhou, Y.C., Bao, Y.W. and Wang, J.Y.J. (2006). Oxidation behavior of bulk Ti_3SiC_2 at intermediate temperatures in dry air. *Journal of Materials Research*, **21:** 402–408.

Zhang, L.L. and Zhao. X.S. (2009). Carbon-based materials as supercapacitors electrodes. *Chemical Society Reviews*, **38:** 2520–2531.

Zhang, X., Hou, L., Ciesielski, A. and Samorì, P. (2016). 2D Materials Beyond Graphene for High-Performance Energy Storage Applications. *Advanced Energy Materials*, **6**:1600671.

Zhao, G., Han, S., Wang, A., Wu, Y., Zhao, M., Wang, Z. and Hao, X. (2015). "Chemical weathering" exfoliation of atom-thick transition metal dichalcogenides and their ultrafast saturable absorption properties. *Advanced Functional Materials*, **25**: 5292–5299.

Zhao, J., Deng, Q., Bachmatiuk, A., Sandeep, G., Popov, A., Eckert, J. and Rümmeli, M.H. (2014). Free-Standing Single-Atom-Thick Iron Membranes Suspended in Graphene Pores. *Science*, **343**: 1228–1232.

Zhao, W.F., Wu, F.R., Wu, H. and Chen, G.H. (2010). Preparation of colloidal dispersions of graphene sheets in organic solvents by using ball milling. *Journal of Nanomaterials*, **528235**: 1–5.

Zhu, Y., Murali, S., Cai, W., Li, X., Suk, J.W., Potts, J.R. and Ruoff, R.S. (2010). Graphene and Graphene Oxide: Synthesis, Properties, and Applications. *Advanced Materials*, **22**: 3906–3924. Energy Finance.

Chapter 4
Nanofluids for Thermal Management in Defense Applications

Supreteti Das

1 Introduction

Thermal management in defense application is essential for maintaining the performance of all machines, military automobiles, manufacturing industries, aviation sector, weaponry and in diverse industries related with defense. The basic approach for enhancing heat removal is either by changing the geometry the addition of fins or by altering the thermophysical properties of the heat transfer fluids. Altering the geometry can result in heavy and unwieldy devices. The current study is focused on changing the thermal properties of the heat transfer medium by the addition of nanoparticles to the conventional heat transfer fluids (HTF) like water, ethylene glycol and oil. The fluid obtained by dispersing nanoparticles in a fluid is termed as 'Nanofluid'(NF). The practice of augmenting HTFs with high thermal conductivity solid particles has been an ancient practice (Ahuja, 1975). However, due to their size, these particles cause clogging of the channels and being heavy the walls of the pipe through which they flow, undergo general wear and tear. Sedimentation of these heavy particles due to gravity results in a non-homogenous dispersion which is detrimental for smooth heat transfer. These challenges are overcome by replacing the solid particles by nanoparticles. The negative impact on the channel walls is reduced and the small size particles reduces sedimentation and blocking of the pipes. With the advancement of technology, the requirement of compact devices has resulted in the miniaturization of electronic circuits in all the devices. While this meets the requirement for better communication and faster data transmission but it also leads to a huge heat generation. To address the burgeoning demand for quick heat removal, innovative thermal management has become essential. Nazari et al. (2014) have used carbon nano-tubes (cnt) and alumina NF with water and ethylene glycol (EG), for exploring the heat transfer in microprocessors and cooling of CPUs of computers.

S. Das (✉)
Gargi College, Delhi University, New Delhi, India
e-mail: supreeti25@gmail.com

R. K. Kotnala et al. (eds.), *Advanced Functional Materials for Sustainable Environments*, https://doi.org/10.1007/978-3-031-62620-3_4

Carbon nanotube nanofluid with volume fraction 0.25% and flow rate of 21 ml/s has been found to have the highest heat transfer. Another application of nanofluids for thermal management is, in electricity distribution grids. Nanofluids as coolants in transformers is highly effective in quick dissipation of heat. This results in increasing the life span of the devices by maintaining their operational temperature. High performance of these transformers is therefore assured. Similar applications of nanofluids for thermal management is in aviation industry for designing compact light weight heat exchangers. Recently a detailed review on thermo-physical properties and heat transfer characteristics of nanofluids have been carried out (Ali and Salam, 2020).

Graphene Nanofluids for Thermal Management

Graphene is a special allotrope of carbon with excellent mechanical, thermal and electrical properties. The suitability of graphene water nanofluid as a coolant has been discussed in detail for several applications (Akhavan-Zanjani et al., 2016; Tharayil et al., 2016; El Sayad et al., 2021). Experiments have been performed to evaluate the thermal conductivity of graphene water nanofluids (Das et al., 2019). Enhancement in thermal conductivity of NF for 0.1% graphene weight concentration at room temperature has been found to be 17% and 29% respectively for 25 °C and 45 °C. Graphene nanofluids for convective heat transfer has been experimentally studied by Baby et al. (2011). Zongpan et al. (2020) have performed experiments to study the flow and heat transfer characteristics of graphene-based lubricants flowing through a horizontal tube. Graphene water nanofluid as a heat removing medium in heat exchangers has been studied by Purbiya et al. (2019). Their conclusion is that a small concentration of graphene in HTF can increase the thermal performance with minimal increase in viscosity and pressure drop. Zolfalizadeh et al. (2023) have used graphene nanoplate in water as the heat transfer fluid in shell and tube heat exchanger. They have explored the effect of adding 0.01%, 03% and 0.06% weight percentage of graphene, on the thermal performance of the heat exchanger. Their conclusion is that the heat transfer efficiency increases by 8.8% for 0.06 wt.%. Bhattacharya et al. (2021) have investigated the fluid dynamics and heat transfer of graphene water nanofluid flowing through mini channels for the cooling of electronic circuits and concluded that the cooling performance of graphene nanofluid depends on the concentration of graphene and on its flow velocity. The defense sector uses a range of military vehicles from jeeps to huge trucks for negotiating difficult terrain. In these defense automotives, huge energy is wasted in overcoming the aerodynamic drag as structurally, the military vehicles have large radiators for dissipating the heat generated in the air. An improved thermal management in these heavy-duty engines, which requires lesser coolants due to better heat transfer properties, can result in higher fuel efficiency along with meeting the stringent carbon emission laws. Graphene and graphene oxide nanofluids have been studied for coolant applications in the automobile industry (Ijaz et al., 2020; Ali et al., 2015). Cooling of high-powered military devices like diode lasers and for weaponry industries, nanofluids can be an alternative to base HTFs. Nanofluids for solar stills have been investigated theoretically by Tiwari and Sahota (2017). For navy set-ups in coastal areas, if graphene nanofluids are used in the heat exchangers instead of water, the evaporation rate of saline water placed in solar stills can be increased resulting in enhanced yield of pure drinking water. Venkatesh et al. (2022) have studied the thermophysical properties of graphene water nanofluids and used it for cooling the PV module and concluded that addition of graphene nanoplatelets increases the thermal conductivity of the HTF and at 0.3% volume fraction of GNP, the decrease in temperature of the panel is 25 °C, at a flow rate of HTF at 0.085 kg/s. The energy efficiency enhancement is 23%. With all the investigations concluding the high suitability of graphene nanofluids for thermal management, the current work is focused on investigating thermophysical properties of graphene based nanofluids as a function of volume fraction employing separately, the base fluids, water, EG and oil (Equations 1–4). Further, this work will also investigate the heat transfer under natural convection, from a hot spot, to graphene water nanofluid flowing through a rectangular duct. For military applications, this study is significant as it is representative of extraction of heat from hot spots generated from microprocessors and other components in miniaturized electronic circuits which have limited area available for the dissipation of heat. Thus, the study on heat transfer using graphene nanofluids for thermal management in defense will be useful for the smallest defense devices to the largest fighter planes.

2 Simulation

2.1 Simulation Set-Up, Simplified Geometry and Boundary Conditions

Graphene water based nanofluid is circulated through a tube with a velocity of 5 mm/s at the inlet. As it flows through the channel it encounters the hot spot and extracts heat from this heat source and flows out of the cooling section. Temperature difference of the fluid between the inlet and outlet is evaluated. Since this system involves a coupling of heat transfer and fluid flow (Equations 5–7), it is described as a multiphysics problem involving natural convection. For simplification, the flow is assumed to be laminar. Comsol Multiphysics software which is based on the finite element method, has been used for obtaining the velocity and temperature distribution within the channel for different volume fractions of graphene. Similar application has also been used by Nasrin et al., 2018; Mahian et al., 2019, Mahian et al., 2019. Single phase flow assumption is justified as the nanofluid has a low concentration of graphene.

For reducing the computation time, the system is reduced to 2-d. This does not affect the process as the solutions are constant in the direction of the heating tube. End corrections are also ignored in the computation, for simplification (COMSOL Multiphysics Version 5.5, 2019). The final geometry is a rectangular domain with a curved region. This region represents the intersection of the tube and the hot spot. No-slip boundary condition is considered and the back flow is suppressed. The initial temperature at which the fluid enters is Tin, the ambient temperature. The temperature of the curved region (hot spot) is T_{heat}. All other walls are assumed to be thermally

Table 1 Properties of graphene

Density (kg/m³)	*Thermal Conductivity (W/mK)*	*Specific heat*
2200	3000	643

Table 2 Geometrical parameters

Width of channel (m)	*Height of channel (m)*	*Position of centre of hot rod (m)*	*Radius of hot rod (m)*
0.005	0.04	0.015	0.0025

insulated. The fluid gets heated as it flows past the hot spot. Natural convection sets in due to the buoyant force originating due to the temperature gradient. Table 1 refers to the properties of graphene and Table 2 refers to the geometrical parameters of the set-up.

2.2 Nomenclature and Governing Equations

$$\mathrm{p}_{nf} = (1 - f) * \mathrm{p}_f + f * \mathrm{p}_s \tag{1}$$

$$c_{nf} = ((1 - f) * \rho_{\mathrm{f}} * c_f + f * \rho_s * c_s)/\rho_{nf} \tag{2}$$

$$k_{nf} = k_f * (k_s + 2 * k_f - 2 * f * (k_f - k_s))/(k_s + 2 * k_f + f * (k_f - k_s)) \tag{3}$$

$$\mu_{nf} = \mu_f * (1 + 2.5 * f) \tag{4}$$

$$\frac{\partial \rho_{nf}}{\partial t} + \nabla.(\rho_{nf} u) = 0 \tag{5}$$

$$\rho_{nf} \frac{\partial u}{\partial t} + \rho_{nf} (u.\nabla) u = -\nabla \mathrm{p} + \nabla.[\mu_{nf} (\nabla u + (\nabla u)^T) - \frac{2}{3} \mu_{nf} (\nabla.u) I] + F \tag{6}$$

$$\rho_{nf}\, c_{pnf} \left[\frac{\partial T}{\partial t} + (u.\nabla) T \right] = k_{nf} \nabla^2 T + D \tag{7}$$

f volume fraction of the nanoparticle

μ_{nf} viscosity of the nanofluid, $(Pa - s)$

μ_f viscosity of the base fluid, $(Pa - s)$

c_{nf} specific heat of the nanofluid, $(J/kg\ K)$

c_f specific heat of the base fluid ($J/kg\ K$)

c_s specific heat of the solid particle ($J/kg\ K$)

c_p specific heat of the fluid

k_{nf} thermal conductivity of the nanofluid ($W/m - K$)

k_f thermal conductivity of the base fluid ($W/m - K$)

k_s thermal conductivity of the solid particle ($W/m - K$)

u velocity of the fluid (m/s)

p pressure (Pa)

F volume force (n/m^3)

I Identity matrix

D heat dissipation ($J/m^3\ s$)

p_{nf} density of nanofluid (kg/m^3)

p_f density of base fluid (kg/m^3)

p_s density of solid (kg/m^3)

3 Discussion of Results

Figure 1 represents the variation of density with volume fraction. For water, EG and oil the density increases with the volume fraction. The specific heat of NF is significant for heat transfer purposes since it denotes the cooling/heating capacity of fluid per unit change in temperature. A fluid with a higher specific heat therefore is a better HTF. Figure 2 shows variation of specific heat with volume concentration of graphene. Both experimentally as well as in the current work, specific heat of the base fluid, water, is higher than the graphene nanofluid. This is detrimental for thermal management.

Thermal conductivity variation is represented in Figure 3. The increase in thermal conductivity with volume fraction is almost linear. The ratio of graphene nanofluid thermal conductivity to that of base fluid is highest for oil. The rate of increase of this ratio with volume concentration is steepest for oil followed by EG. Figure 4 represents the velocity profile of the graphene nanofluid in the rectangular domain. Due to the curved shape assumed for the hot spot, velocity distribution is non uniform near the hot region. Near the edge of the wall above the hot spot the velocity is maximum and almost uniform. Near the curvature, the value becomes smaller. The inlet velocity is 5 mm/s. Heating results in the development of the buoyant force which pushes the fluid upwards. Natural convection sets in and the heat transferred now is by both the modes, conduction and convection. Motion of the fluid aids in the heat

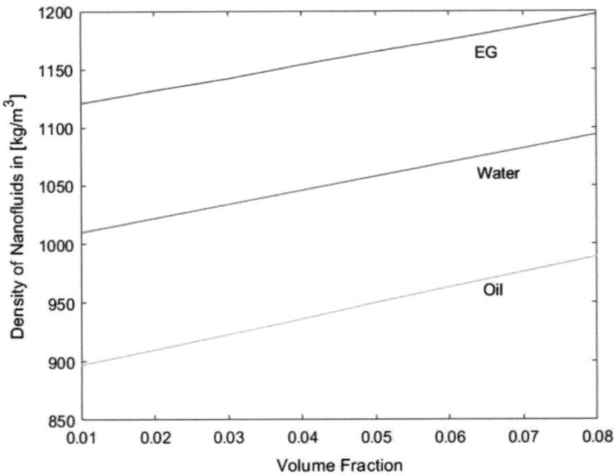

Fig. 1 Density variation of graphene nanofluid with volume fraction for different base fluids

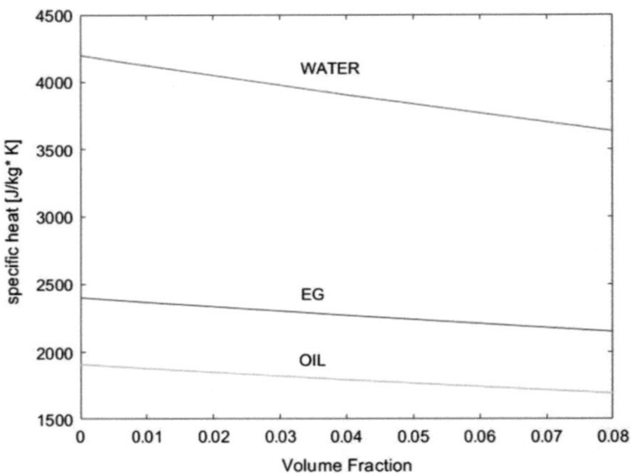

Fig. 2 Variation of specific heat of nanofluid with volume fraction for different base fluids

extraction from the hot spot resulting in efficient heat removal. Figure 5 shows the temperature distribution of the nanofluid within the domain. Temperature at the inlet is the ambient temperature. Due to the non- uniform velocity of the fluid near the hot spot, the temperature is significantly higher above the hot spot edge on the left compared to the other regions. The arrow represents the velocity of the fluid. Figure 6 represents the temperature difference between the inlet and outlet. With the increase in volume concentration of graphene, the difference of temperature increases which signifies higher heat extraction from the hot spot.

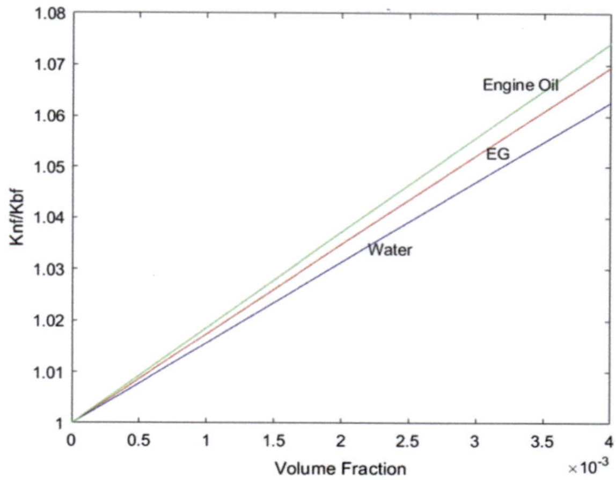

Fig. 3 Variation of thermal conductivity ratio with volume fraction for different base fluids

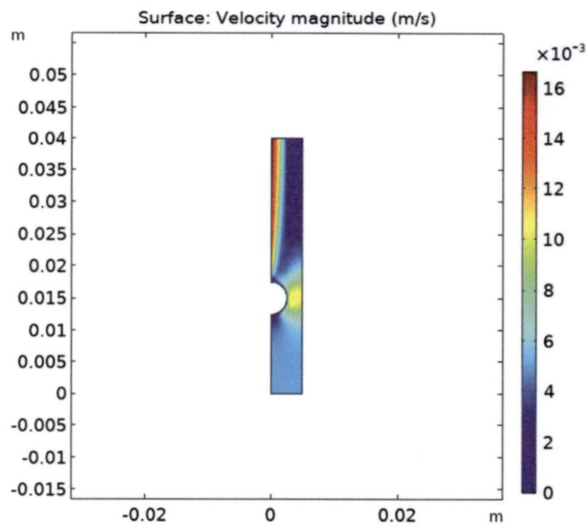

Fig. 4 Velocity distribution of graphene water nanofluid

4 Conclusions

For graphene nanofluids density, thermal conductivity and viscosity increases with the volume concentration while the specific heat decreases. Enhancement of thermal conductivity with very low volume fractions makes graphene nanofluids an excellent heat exchange medium for thermal management. From Equation 4, viscosity increases with volume fraction resulting in high pressure drop. However, since very low volume fractions are required for high heat transfer enhancement, hence the increase in viscosity is insignificant. The pumping cost increase due to pressure

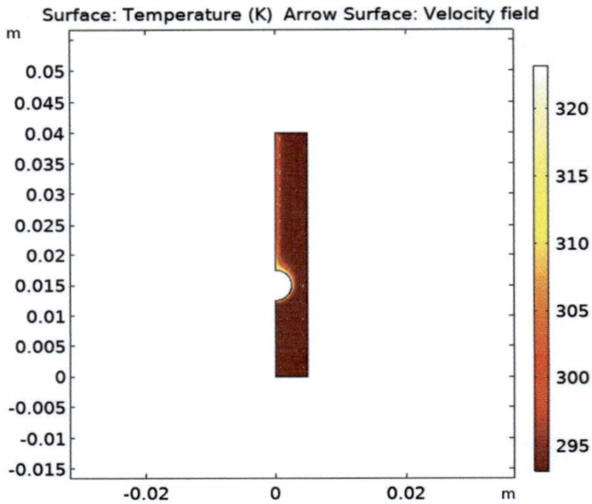

Fig. 5 Temperature profile for graphene water nanofluid

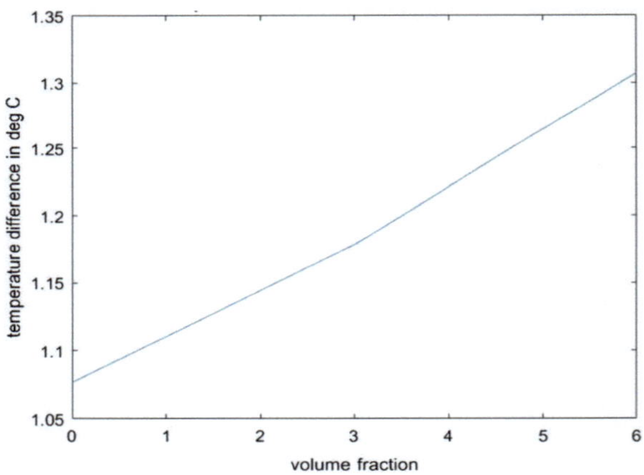

Fig. 6 Variation of temperature difference with volume fraction percentage

drop is minimal. The conclusion is that graphene as a nanofluid, can be an excellent substitute for conventional HTFs for thermal management in military.

5 Challenges

Optimization of heat transfer is a very important component in most industries. Stability of graphene nanofluid over a long period of time is required for industrial applications else there could be agglomeration of the particles which may finally result in sedimentation under the effect of gravity. The nanofluid will become inhomogeneous.

For simplification, only laminar flow has been considered. It is essential to consider the turbulent flow, since in this regime there will be an intense mixing of the fluid which may suggest a higher heat transfer rate and hence better thermal management for military devices, electronic equipment, automobiles, fighter planes and diverse industries.

A deeper understanding of the interplay of forces at the nanoscale is required since the theoretical models predict lower values than the experimental results, particularly for thermal conductivity. Since a lot of variables are involved for accurately investigating the thermophysical properties of graphene nanofluids, there is a need for intelligent computing. Prediction of these properties invoking AI is currently in its nascent stage and research in this direction needs to be carried forward for more effective thermal management in the defense sector.

References

Ahuja, A.S. (1975). Augmentation of heat transport in laminar flow of polystyrene Suspensions, Experiments and results. *J. Appl. Phys.,* **46(8):** 3408–3416.

Akhavan-Zanjani, H., Saffar-Avval, M., Mansourkiaei, M., Sharif, F. and Ahadi, M. (2016). Experimental investigation of laminar forced convective heat transfer of Graphene-water nanofluid inside a circular tube. *Int. J. Therm. Science,* **100:** 316–323.

Ali, A.R.I. and Salam, B. (2020). A review of nanofluid: preparation, stability, thermophysical properties, heat transfer characteristics and application. *S N Applied Sciences* **2(10):** 1636.

Ali, H., Azhar, M., Saleem, M., Saeed, Q. and Saeed, A. (2015). Heat Transfer Enhancement of Car Radiator Using Aqua based Magnesium oxide nanofluids. *Thermal Science,* **19(6):** 2039–2048.

Baby, T. and Sundara, R. (2011). Enhanced convective heat transfer using graphene dispersed nanofluids, *Nanoscale Res. Lett.,***6:** 289.

Bhattacharya, R. and Pranab Samanta, P. (2021). Simulation based Study of Graphene-water Nanofluid flow through Microchannel Heat sink. *IOP Conf. Ser.: Mater. Sci. Eng.* **1206:** 012008.

COMSOL Multiphysics User Guide Version 5.5 (2019).

Das, S., Giri, A., Samanta, S. and Kanagaraj, S. (2019). Role of graphene nanofluids on heat transfer enhancement in thermo-syphon, *Journal of Science: Advanced Materials and Devices,* **4(1):** 163–169.

Elsaid, K., Abdelkareemb, M.A., Hussein, M. et al. (2021). Thermo-physical properties of graphene-based nanofluids. *International Journal of Thermo-fluids,* **10:** 100073.

Ijaz, H., Raza, H., Gohar, G.A., Ullah, S. et al. (2020). Effect of graphene oxide doped nano coolant on temperature drop across the tube length and effectiveness of car radiator – A CFD study. *Thermal Science and Engineering Progress,* **20:** 2451–9049.

Mahian, O., Kolsi, L., Amani, M. et al. (2019a). Recent advances in modelling and simulation of nanofluid flows-Part I: Fundamentals and theory. *Physics Reports,***790:** 1–48.

Mahian, O., Kolsi, L., Amani, M. et al. (2019). Recent advances in modelling and Simulation of nanofluid flows-part II: Applications, *Phys. Rep.,* **791:** 1–59.

Nasrin, R., Rahim, N. A., Fayaz, H. and Hasanuzzaman, M. (2018). Water /MWCNT nanofluid Based cooling system of PVT: Experimental and Numerical research. *Renew. Energy,* **121:** 286–300.

Nazari, M., Karami, M. and Ashouri, M. (2014). Comparing the thermal performance of water, Ethylene Glycol, Alumina and CNT nanofluids in CPU cooling: Experimental study. *Experimental Thermal and Fluid Science,* **57:** 371–377.

Purbia, D., Khandelwal, A., Kumar, A. and Sharma, A.K. (2019). Graphene-water nanofluid in heat exchanger: Mathematical modelling, Simulation and economic evaluation. *International Communications in Heat and Mass Transfer*, **108:** 104327.

Tharayil, T., Asirvatham, L.G., Ravindran, V. and Wongwises, S. (2016). Thermal performance of miniature loop heat pipe with graphene-water nanofluid. *Int. J. Heat Mass Transfer*, **93:** 957–968.

Tiwari, G.N. and Sahota, L. (2017). Advanced solar distillation systems: Basic Principles thermal modelling and its application, 1st edition. *Springer*: ISBN 978-981-10-4672-8.

Venkatesh, T., Manikandan, S., Selvam, C. and Harish, S. (2022). Performance enhancement of hybrid solar PV/T system with graphene based nanofluids. *International Communications in Heat and Mass Transfer*, **130:** 105794.

Zolfalizadeh, M., Heris, S.Z., Pourpasha, H. et al. (2023). Experimental Investigation of the Effect of Graphene/Water Nanofluid on the Heat Transfer of a Shell-and-Tube heat exchanger. *International Journal of Energy Research,* **2023:** 3477673.

Zongpan, C., Tiang, M. and Zang, G. (2020). Experimental Study on the Flow and Heat Transfer of Graphene-Based Lubricants in a Horizontal Tube. *Processes*, **8(12):** 1675.

Chapter 5
Realization of KNN-PVDF Cantilever for Mechanical Energy Harvesting

Babita Sharma, Reema Gupta, Mallika Verma, Bilasini Naorem, and Monika Tomar

1 Introduction

Energy harvesting includes the conversion of waste energy to the useful one. Hence, dependency on the other power sources and the batteries may be replaced by the energy harvesters. Vibrational energy is the most common and abundant source of energy through which electrical energy can be generated through capacitive, electromagnetic and piezoelectric effect. Cantilever beams are mostly employed to harness vibrational energy using piezoelectric effect in which there is generation of the electric charges when mechanical stress is given to the material (Choi et al., 2006). A cantilever is simply a rigid structural beam which is anchored at one end and free from the other ends to vibrate freely in any direction. Due to the stress in the beam, the beam vibrates with its resonant frequency. Several piezoelectric materials are available which can be employed for the realization of energy harvester such as Aluminum Nitrate (AlN), Zinc Oxide (ZnO), Polyvinyledene Fluoride (PVDF), Potassium Sodium Niobate (KNN), Lead Zirconium Nitrate (PZT), etc. (Huang & Chen, 2016; Kumar & Kim, 2012; Zhao et al., 2015) (Kanno et al., 2012) (Kang et al., 2016). Lead based piezoelectric like PZT and their derivatives have been extensively explored for energy harvesting application because of their high piezoelectric properties. However, lead based piezoelectrics are highly toxic in nature and

B. Sharma
Department of Physics and Astrophysics, University of Delhi, Delhi 110007, India

B. Sharma · M. Verma · B. Naorem · M. Tomar (✉)
Department of Physics, Miranda House, University of Delhi, Delhi 110007, India
e-mail: monikatomar@gmail.com

M. Tomar
Institute of Eminence, University of Delhi, Delhi 110007, India

R. Gupta
Department of Physics, Hindu College, University of Delhi, Delhi 110007, India

R. K. Kotnala et al. (eds.), *Advanced Functional Materials for Sustainable Environments*, https://doi.org/10.1007/978-3-031-62620-3_5

in order to cope up with the environmental and health issues there is a need to replace lead based materials with the lead free. Thus, out of all these materials Potassium Sodium Niobate ($K_xNa_{(1-x)}NbO_3$ or KNN) which is having a perovskite structure and has good piezoelectric properties has captivated attention as a capable lead free piezoelectric and electro optic material (Minh et al., 2011).

Recent demand for flexible electronics requires integration of KNN with the lead free. Polymer based piezoelectric materials provide high flexibility and ease of fabrication. PVDF is one such material which is flexible shows some piezoelectricity. Hence, the materials having high piezoelectric coefficients can be incorporated with the PVDF matrix to enhance the energy harvesting efficiency along with flexibility and better mechanical strength (Alluri et al., 2017). Utilization of piezoelectric cantilever beam is the most prevailing approach for transforming vibrational or mechanical energy into electrical energy using direct piezoelectric effect (Choi et al., 2006). Energy harvesting efficiency of piezoelectric cantilever relies on optimizing the properties of utilized piezoelectric material, geometry and resonating frequency of cantilever (Friswell & Adhikari, 2010; Hosseini & Hamedi, 2015).

Factors such as ease of applications, suitable methods for thin film growth, greater output power, suitable methods for thick film growth at geometric scale, simple structure and large power density make piezoelectric energy harvesting, a method of utmost importance (Elvin & Erturk, 2013). Output performance of the piezoelectric energy harvesters devices can be enhanced by optimizing the shapes and geometries of the cantilever beams.

Xiong et al. (2020) depicted a cantilever beam with a rectangular hollow middle layer, which bring down the resonance frequency of the cantilever beam and increased the output voltage. Wang, L. et al. (2019) considered a hollow triangular piezoelectric cantilever beam favorable for the random low frequency vibration environment. The rectangular hole, triangular hole, and trapezoidal hole were correlated (Wang, B. et al., 2021) and the PEH with a trapezoidal hole was proved to have the better performance. Hashim et al. (2021) have found the trapezoidal crossection cantilever to be the most efficient one. Qiuyu Xu et al. (2022) designed a PEH with trapezoidal hollow structure and obtained natural frequency and output for the peak voltage to be 23.29 Hz and 40.4 V, respectively.

Wang, B. et al. (2021) developed a piezoelectric cantilever beam with a trapezoidal hole, and achieved a power density of 8.93 mW/cm^3 at an acceleration of 1.5 g. In addition to the geometry optimization, material optimization and substrate replacement can be the other two methods to optimize PEH. Mateu & Moll (2005) compared two different shapes of cantilever (a triangular and a rectangular) in which the bigger end of the cantilevers was fixed and the small end was free to move. Mathematically, it was proven that a triangular shaped cantilever exhibits higher strain and maximum deflection for an applied load when its base and height dimensions are equal to the rectangular beam's base and length dimensions. This higher strain and deflection values are converted into higher power outputs by piezoelectric material; thus, large power per unit area is generated by a triangular cantilever beam as compared to the rectangular beam (Mateu & Moll, 2005).

Further, Roundy et al. (2005) presented that, trapezoidal shaped cantilever has more even distribution of strain more through the structure; on the other hand, rectangular beam achieves a non-uniform strain distribution through its structure. It was also stated that a trapezoidal shaped cantilever produces more than two times the energy produced by that of a rectangular beam for a particular volume of piezoelectric. The capability of combining any number of physics, user-defined physics and expressions straight into a model makes COMSOL users creative beyond possibility or just a lot easier compared to other simulation softwares.

In this article, PVDF material based rectangular, trapezoidal and triangular cantilever beams have been utilized to study its resonant frequency using COMSOL multiphysics Software theoretically. All these structures have further been employed to verify its resonant frequency using laser doppler vibrometer (LDV). PVDF thin sheets incorporated with KNN have been prepared using spin casting method.

2 Experimental Details

Polyvinyledene Fluoride (PVDF) (purity $\geq$ 99.99%) purchased commercially from Sigma Aldrich was used as a precursor material and N,N Dimethylformamide (DMF) was used as solvent. Potassium Sodium Niobate ($K_{0.5}Na_{0.5}NbO_3$) was prepared using solid state synthesis reaction using ball milling technique. Sodium Carbonate [Na_2CO_3; 99.95% pure], Potassium Carbonate [K_2CO_3; 99.90% pure], and Niobium Oxide [Nb_2O_5; 99.90% pure] powders were utilized as raw precursors and mixed together for the synthesis of stoichiometric $K_{0.5}Na_{0.5}NbO_3$ (KNN) powder. In order to avoid the deficiency of 'K' and 'Na' ions at high sintering temperature as both 'K' and 'Na' are volatile in nature, 1% excess of Na_2CO_3 and K_2CO_3 powders were taken before ball milling. A high-speed ball milling system [Make: Metrex] was used to prepare the powder and ethanol was used as a solvent at a speed of 250 rpm for 30 h. The prepared powder was then dried and calcinated in a muffle furnace at 550 °C [Make: Metrex] for 4 h.

Solution-cast method has been employed to prepare PVDF flexible thin films. 1.25 g of PVDF and 10% KNN was mixed with 12.5 ml of DMF and was constantly stirred with 700 rpm at 60 °C till the powder get completely dissolved. Then the prepared homogeneous solution was transferred to a petridish and kept overnight at 65 °C to get dried completely. Finally, the film was peeled off from the petri dish. The prepared thin sheet and the process flow is shown in Fig. 1.

These prepared PVDF-KNN sheets were further utilized for the preparation of cantilevers of different shapes (rectangular, triangular and trapezoidal). Laser Doppler Vibrometer (LDV) was used to compute the resonant frequency. Schematic of the employed LDV set-up shown in Fig. 2.

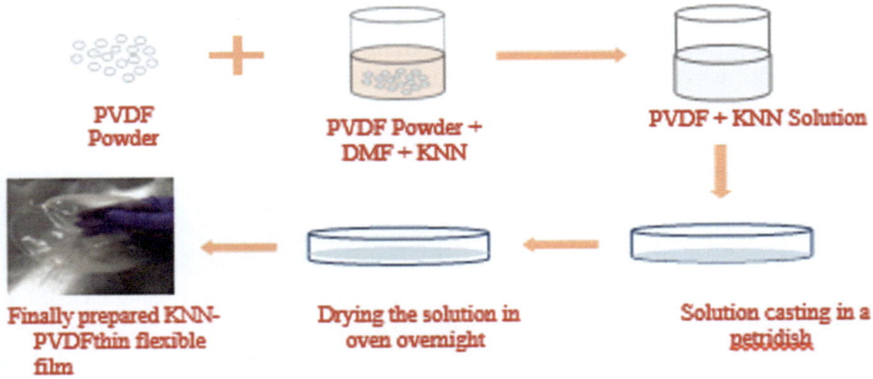

Fig. 1 Preparation of the PVDF-KNN flexible thin film using solution casting technique

Fig. 2 Laser doppler vibrometer (LDV) set up

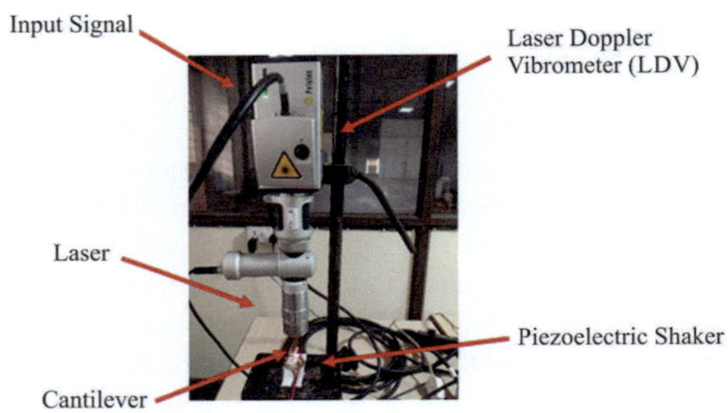

3 Results and Discussion

Structural Characterization

Figure 3 depicts the XRD pattern of the fabricated PVDF-KNN composite thin sheet. Analysis of the XRD pattern shows that the crystallographic peaks observed correspond to the orthorhombic phase of KNN (JCPDs File No. 00-079-7689). Peaks at 26.50 and 39.50 indicate the formation of α-phase of PVDF and the peak observed at 36.10 corresponds to the β-phase of PVDF (Singh et al., 2014). It is credible from the XRD pattern that PVDF peaks are dominated by the peaks of KNN due to high crystalline nature of the prepared KNN powder. Lattice parameters of KNN in the prepared sheet are found to be a = 5.73 Å, b = 4.00 Å, and c = 5.67 Å.

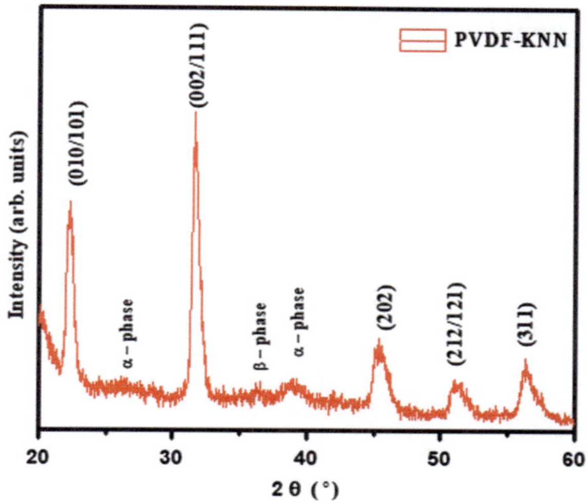

Fig. 3 The XRD pattern of the prepared PVDF-KNN thin sheets

4 Results and Discussion

Piezoelectric energy harvesters (PEH) can be used to power electronic devices that need low power input and thus it is crucial to produce high power at lower frequency. Here, natural or resonant frequency of the cantilevers were characterized by using Laser Doppler Vibrometer (LDV) (Model: Polytech Vibroflex).

LDV is a simple tool that measure the frequency difference between the incident and the reflected beam using Doppler Effect. Resonant frequency is the frequency at which the cantilever or any object vibrates with the maximum amplitude. A piezocoin has been utilized to excite the PVDF-KNN cantilevers at its natural frequency. A power supply of 2 V and frequency sweep of 1-100 kHz has been given by the function generator to make the piezocoin vibrates. It has been observed that with the change in the dimension of the cantilever, the resonant frequency of the cantilever also changes.

We investigated three shapes of the cantilever beams, namely, rectangle, triangular and trapezoidal. The observed resonant frequency of the prepared cantilevers was found to be 28.46, 61.19, 41.80 Hz for rectangle, triangle and trapezoidal respectively. Figure 4 shows the maximum amplitude v/s frequency curve for all the prepared cantilevers.

In an attempt to validate the experimental results produced by LDV, eigen frequency calculation of the prepared cantilevers with the same dimension was done utilizing COMSOL Multiphysics Software. Simulation is an effective tool for evaluating the enactment of a proposed or preferred system. Simulation is basically needed before altering an existing system or building a new one, so as to reduce the chances of fiasco, to eradicate unforeseen bottlenecks, to optimize the system's performance and to prevent under or over-utilization of resources. The capability of combining any number of physics, user-defined physics and expressions straight into a model

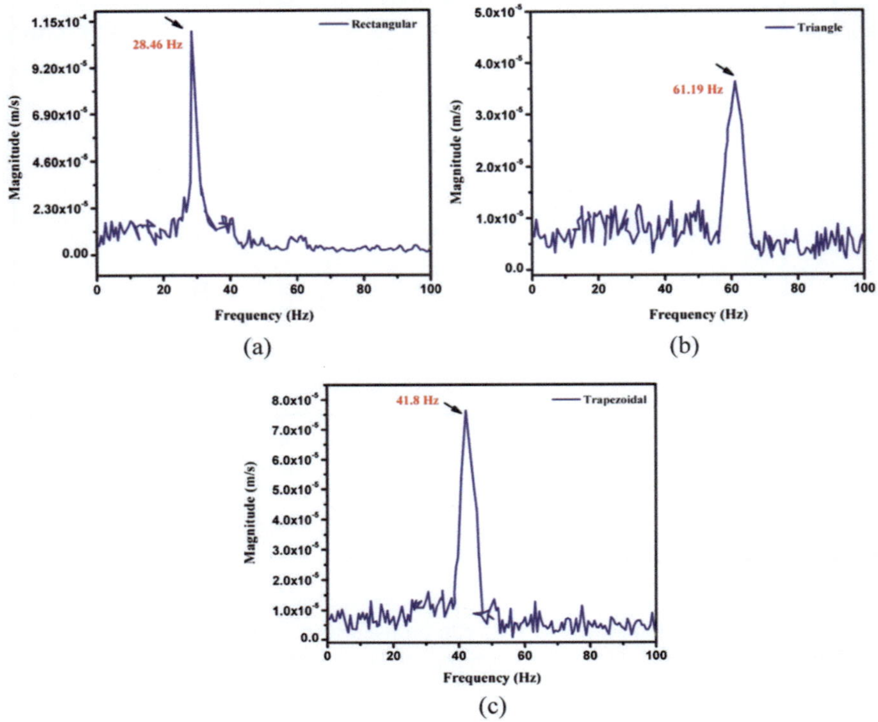

Fig. 4 Resonant frequency of the prepared cantilevers of different shapes using LDV: (a) rectangle, (b) triangle and (c) trapezoidal

makes COMSOL users creative beyond possibility or just a lot easier compared to other simulation software.

COMSOL Multiphysics utilizes finite elemental method to solve the field problems using boundary conditions. It is a technique for computing that divides an entity into tiny but distinct components. One element's physics can be roughly described by a limited number of degrees of freedom (DOFs). Each element is given a collection of characteristic equations that describe its physical characteristics, boundary conditions, and applied forces. These equations are then solved as a group of simultaneous equations to predict the behavior of the element.

A 3D unit of the piezoelectric device has been shown in Figure 5. Before monitoring the resonating frequency of the fabricated flexible cantilevers, flexibility test has been performed. The sample is observed to be cracks free even after bending and twisting as shown in Figure 6.

Three shapes of the cantilever were investigated for the calculation of their natural frequency. It can be seen from Figure 5 that the piezoelectric cantilever vibrates at a certain resonating frequency. The normal mode of vibration for all the prepared

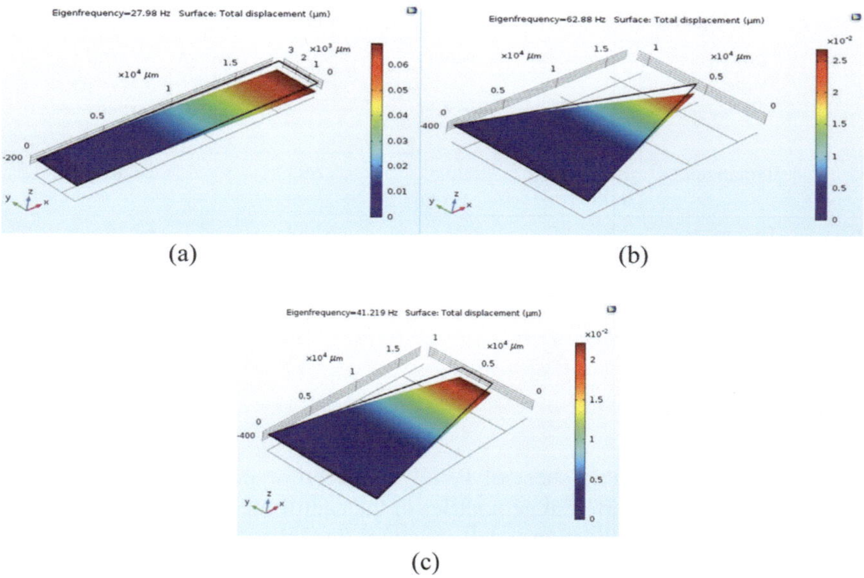

Fig. 5 Theoretically calculated resonant frequency of the prepared cantilevers of different shapes; (a) rectangle, (b) triangle and (c) trapezoidal

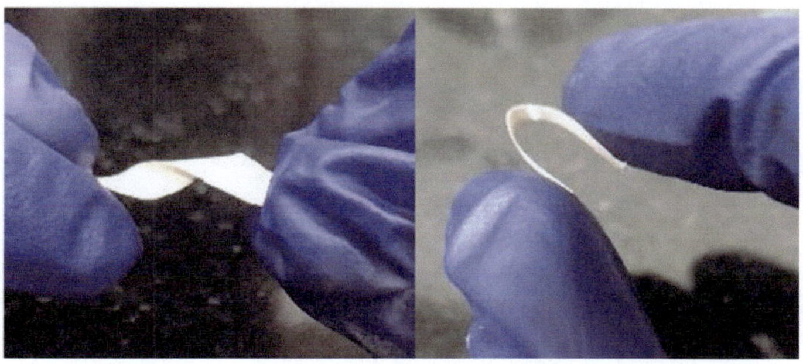

Fig. 6 Bending and twisting test for prepared cantilevers

cantilever has been shown in Figure 5(a)-(c). The resonating frequency was found to be 27.98, 62.88, and 41.22 Hz, respectively, for rectangle, triangle and trapezoidal.

A comparison for the prepared cantilevers of different shapes is given with their dimensions, in Table 1 and it is evident from the table that the experimentally obtained values are in close agreement with the theoretical results. Therefore, the fabricated cantilevers can be exploited for harvesting mechanical energy by utilizing its piezoelectric properties.

Table. 1 Comparison of the prepared cantilevers along with their dimension

S. No	*Cantilever Shapes*	*Dimension*	*Volume (*109 μm^3)*	*Natural frequency (theoretical) Hz*	*Natural frequency (experimental) Hz*
1	Rectangular	1.8 × 0.3 × 0.0045	2.43	27.98	28.46
2	Triangular	1.8 × 1.0 × 0.0045	4.24	62.88	61.19
3	Trapezoidal	1.8 × 0.3 × 0.0045	5.16	41.22	41.80

5 Conclusion

In this article, thin flexible sheet of PVDF has been prepared by using Solution casting method. Three different geometries of the cantilever beams were considered for energy harvesting application and their natural frequency has been calculated. The theoretical and experimental values of natural frequency for the first mode are found to be comparable. The values for the natural frequency computed theoretically are 29.98, 62.88, and 41.22 Hz and the values for the natural frequency observed experimentally using LDV are 28.46, 61.19, and 41.80 Hz for rectangle, triangular and trapezoidal, respectively. These match within experimental errors showing potential of using them as an energy harvester in future devices.

References

Alluri, N. R., Chandrasekhar, A., Jeong, J. H. & Kim, S.-J. (2017). Enhanced electroactive β-phase of the sonication-process-derived PVDF-activated carbon composite film for efficient energy conversion and a battery-free acceleration sensor. *Journal of Materials Chemistry C*, **5(20):** 4833–4844. https://doi.org/10.1039/C7TC00568G

Choi, W. J., Jeon, Y., Jeong, J.-H., Sood, R. & Kim, S. G. (2006). Energy harvesting MEMS device based on thin film piezoelectric cantilevers. *Journal of Electroceramics*, **17(2-4):** 543–548. https://doi.org/10.1007/s10832-006-6287-3

Elvin, N., & Erturk, A. (2013). Introduction and Methods of Mechanical Energy Harvesting. In Advances in Energy Harvesting Methods (pp. 3–14). Springer New York. https://doi.org/10.1007/978-1-4614-5705-3_1

Friswell, M. I. & Adhikari, S. (2010). Sensor shape design for piezoelectric cantilever beams to harvest vibration energy. *Journal of Applied Physics*, **108(1):** 014901. https://doi.org/10.1063/1.3457330

Hashim, A. A., Mahmoud, K. I. & Ridha, H. M. (2021). Geometry and shape optimization of piezoelectric cantilever energy harvester using COMSOL multiphysics software. *International Review of Applied Sciences and Engineering*, **12(2):** 103–110. https://doi.org/10.1556/1848.2021.00170

Hosseini, R. & Hamedi, M. (2015). Improvements in energy harvesting capabilities by using different shapes of piezoelectric bimorphs. *Journal of Micromechanics and Microengineering*, **25(12):** 125008. https://doi.org/10.1088/0960-1317/25/12/125008

Huang, H.-H., & Chen, K.-S. (2016). Design, analysis, and experimental studies of a novel PVDF-based piezoelectric energy harvester with beating mechanisms. *Sensors and Actuators A: Physical*, **238:** 317–328. https://doi.org/10.1016/j.sna.2015.11.036

Kang, M. G., Jung, W. S., Kang, C. Y. & Yoon, S. J. (2016). Recent progress on PZT based piezoelectric energy harvesting technologies. *Actuators*, *5*(1). https://doi.org/10.3390/act5010005

Kanno, I., Ichida, T., Adachi, K., Kotera, H., Shibata, K. & Mishima, T. (2012). Power-generation performance of lead-free (K,Na)NbO3 piezoelectric thin-film energy harvesters. *Sensors and Actuators A: Physical*, **179:** 132–136. https://doi.org/10.1016/j.sna.2012.03.003

Kumar, B. & Kim, S.-W. (2012). Energy harvesting based on semiconducting piezoelectric ZnO nanostructures. *Nano Energy*, **1(3):** 342–355. https://doi.org/10.1016/j.nanoen.2012.02.001

Mateu, L. & Moll, F. (2005). Optimum Piezoelectric Bending Beam Structures for Energy Harvesting using Shoe Inserts. *Journal of Intelligent Material Systems and Structures*, **16(10):** 835–845. https://doi.org/10.1177/1045389X05055280

Minh, L. V, Horikiri, F., Shibata, K., Mishima, T. & Kuwano, H. (2011). A new lead-free (K, Na) NbO_3 piezoelectric material and its microfabrication for micro energy harvester. December 2014, 4–7.

Roundy, S., Leland, E. S., Baker, J., Carleton, E., Reilly, E., Lai, E., Otis, B., Rabaey, J. M., Sundararajan, V. & Wright, P. K. (2005). Improving Power Output for Vibration-Based Energy Scavengers. *IEEE Pervasive Computing*, **4(1):** 28–36. https://doi.org/10.1109/MPRV.2005.14

Singh, P., Borkar, H., Singh, B. P., Singh, V. N. & Kumar, A. (2014). Ferroelectric polymer-ceramic composite thick films for energy storage applications. *AIP Advances*, **4(8):** 0–11. https://doi.org/10.1063/1.4892961

Wang, L., Tong, X., Yang, H., Wei, Y. & Miao, Y. (2019). Design and analysis of a hollow triangular piezoelectric cantilever beam harvester for vibration energy collection. *International Journal of Pavement Research and Technology*, **12(3):** 259–268. https://doi.org/10.1007/s42947-019-0032-1

Wang, B., Zhang, C., Lai, L., Dong, X. & Li, Y. (2021). Design, Manufacture and Test of Piezoelectric Cantilever-Beam Energy Harvesters with Hollow Structures. *Micromachines*, **12(9):** 1090. https://doi.org/10.3390/mi12091090

Xiong, Y., Song, F. & Leng, X. (2020). A piezoelectric cantilever-beam energy harvester (PCEH) with a rectangular hole in the metal substrate. *Microsystem Technologies*, **26(3):** 801–810. https://doi.org/10.1007/s00542-019-04608-8

Xu, Q., Gao, A., Li, Y. & Jin, Y. (2022). Design and Optimization of Piezoelectric Cantilever Beam Vibration Energy Harvester. *Micromachines*, **13(5):** 675. https://doi.org/10.3390/mi13050675

Zhao, X., Shang, Z., Luo, G. & Deng, L. (2015). A vibration energy harvester using AlN piezoelectric cantilever array. *Microelectronic Engineering*, **142:** 47–51. https://doi.org/10.1016/j.mee.2015.07.006

Part II
Materials for Environmental Remediation

Chapter 6
Synthesis and Characterization of Cu-MOF (Copper-Metal Organic Framework) for Gas Sensor and Electron Emission Devices

Mohd Sarvar, Shah Masheerul Aalam, Mohd Sadiq, Mohd Shahid Khan, and Javid Alia

1 Introduction

Research on metal organic frames (MOFs) has gained much attention recently because of its promise for creating frameworks that integrate organic and inorganic chemistry (Almeida et al., 2012) for a range of applications, including energy (Bobbitt et al., 2017; Chidambaram & Stylianou, 2018), catalysis, sensing, separation, and purification. Due to its high design ability, huge surface area, excellent compatibility with both organic and aqueous media, and controlled porosity, MOFs, which are porous crystalline materials, can be regarded as affordable materials depending on the metal source utilised in their creation. As gases and biomolecules can attach to and interact with the functional groups in the organic ligand of MOFs through hydrogen bonding, hydrophobic (stacking), electron donor/acceptor interactions, or the formation of a dative covalent bond, the involvement and application of MOFs in the field of chemical sensors is progressively expanding (Alegre et al., 2016; Carrasco, 2018; Karmakar et al., 2017). The interactions between the gas molecules and the structural metal or organic linker give MOFs their high degree of selectivity (Zhu & Xu, 2014). In addition, hollow and porous metal oxides and composites created by calcining MOFs have become more well-liked as components for the production of sensors due to their abundance of reaction sites and ease of access to gas (Stavila et al., 2014).

M. Sarvar · S. M. Aalam · M. Sadiq · J. Alia (✉)
Material Science Lab, Department of Physics Jamia Millia Islamia, New Delhi 110025, India
e-mail: jali1@jmi.ac.in

M. Sadiq
ARSD College, University of Delhi, New Delhi 110007, India

M. S. Khan
Laser Spectroscopy Lab, Department of Physics Jamia Millia Islamia, New Delhi 110025, India

R. K. Kotnala et al. (eds.), *Advanced Functional Materials for Sustainable Environments*, https://doi.org/10.1007/978-3-031-62620-3_6

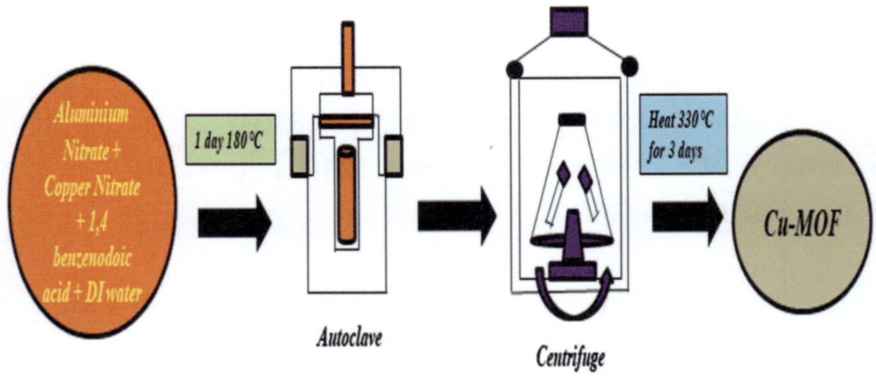

Fig. 1 Preparation of Cu-MOF

These MOF-made porous transition metal oxides have properties that are at par with or even superior to those of the corresponding metal oxides produced by more conventional techniques like hydrothermal, chemical precipitation and solid-state reaction procedures (Salunkhe et al., 2017). As a result, MOF-based devices are gaining more and more attention for the detection and field emission of various gases and vapours. This article concentrates on acetone, a gaseous species that can be toxic to humans even at very low concentrations. The biomedical and environmental industries have consistently worked to create highly sensitive and specific acetone gas sensors. One of the most common organic solvents utilised in businesses, medical facilities, and research labs is acetone, the most basic ketone. Short-term exposure to high levels of acetone can irritate nose, throat, lungs, and eyes; long-term exposure can result in a severe headache and narcosis. Acetone is utilised as a biomarker for diabetes since it is a substantial volatile organic compound (VOC) found in exhaled human breath and frequently exceeds 1.8 ppm in diabetic patients (Karmaoui et al., 2016; Salunkhe et al., 2017). Nowadays, the medical field is primarily concerned with non-invasive glucose monitoring for diabetic patients both at home and in hospitals. In this study, Cu nanoparticles (MOF) are incorporated into Al-MOF metal-organic frameworks to form Cu-MOF nanostructures for field emitters and gas sensors. The field emitter and sensor based on nanostructures have been shown to perform better (Fig. 1).

2 Experimental Details

Aluminum nitrate nonahydrate (0.45 g), copper nitrate hexahydrate (1.3 g), and 1,4-benzene-dioic acid (0.288 g) were combined and mixed in 30 ml deionized water to create the Al-based Cu-MOF. The resulting solution was transferred to a high-pressure autoclave in a 50 ml Teflon-lined container. After being sealed and maintained at 180 °C for a day, the autoclave was left to naturally cool down to ambient temperature. Centrifugation was used to separate the precipitate, and it was repeatedly rinsed in deionized water. Subsequently, the product was exposed to an air temperature of 330 °C for three days in order to eliminate the foreign molecules from its pores.

3 Result and Discussion

The as-prepared Cu-MOF nanostructure's morphologies were examined using SEM examination. Figure 2 shows common SEM images of the samples. The homogenous distribution of C, O, Al, and Cu in Cu-MOF is shown by the EDS elemental mappings (inset of Fig. 3), and the measured values are given in the accompanying table.

In order to ascertain the structures of the Cu-MOF as it was synthesised, Raman spectra were obtained (Fig. 4). The sharp peak, located between 531,980 and 1785 cm^{-1}, is attributed to band S_1 due to disorder activation, while the acute peak, located at 253,467 cm^{-1}, is linked to the E_g mode of Cu and Al nanoparticles. Cu-MOF is in charge of the characteristic peaks at 252, 461, 531, 980 and 1785 cm^{-1} in the spectrum. This peak, which is assigned to 253 cm^{-1} and 531 cm^{-1} is Cu stretching mode, confirm the coexistence of Cu-MOF. Al stretching mode is assigned to 461 cm^{-1}, C-H stretching to 980 cm^{-1} and the C=C stretching bond to 1785 cm^{-1} (Rebollo Plata et al., 2014).

Figure 4 showed the FTIR spectra of Cu-MOF, which is a useful analytical approach to validate the existence of functional groups. Cu-MOF showed two distinct peaks of high intensity at 1396 cm^{-1} that was attributable to the coordinated carboxylic acid's asymmetric and symmetric stretching modes, respectively. Weak and thin bands at 923 and 634 cm^{-1} were attributable to the vibration of the aromatic ring systems (C-H) and (C-H), respectively. As a result, the existence of aromatic ring indicates that the finished good contains the organic ligand (linker).

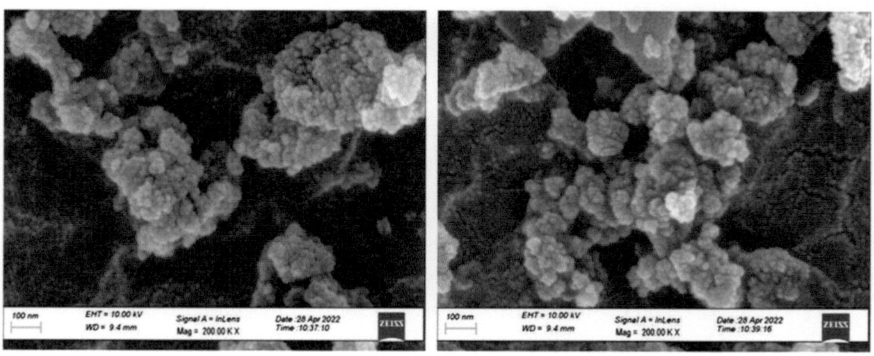

Fig. 2 SEM micrograph of as-grown Cu-MOF

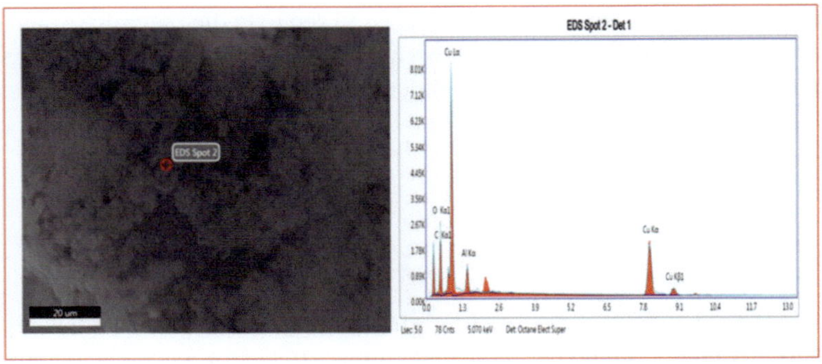

Element	Weight %	Atomic %
C K	30.65	54.09
O K	20.90	27.69
AlK	4.55	3.57
CuK	43.91	14.65

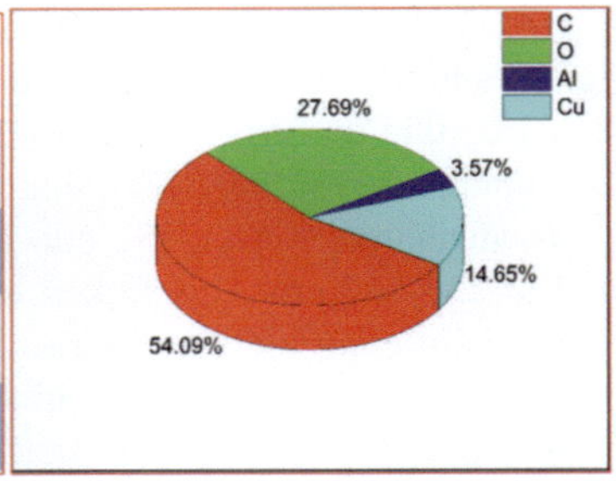

Fig. 3 EDX graphs of as grown Cu-MOF and chemical composition weight % from EDX

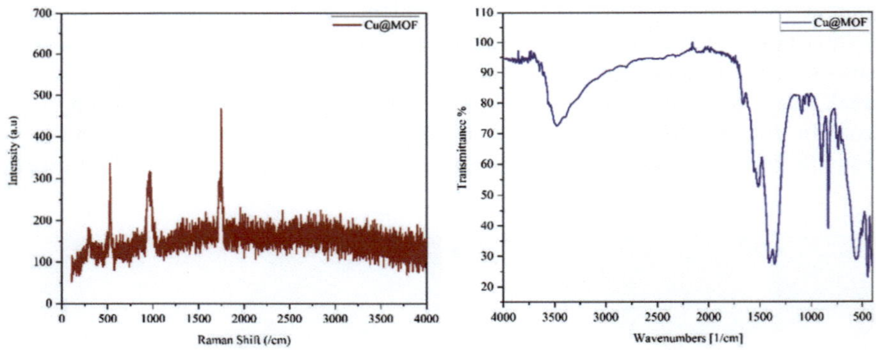

Fig. 4 Raman and FTIR Spectroscopy graphs of as grown Cu-MOF

The big peaks in the spectral area between 3400 and 3600 cm^{-1} may be caused by crystalline water or the acidic OH of carboxylic groups.

UV-visible (UV-vis) absorption spectroscopy is used to qualitatively assess each object's optical absorption characteristics. The wavelengths of absorption peaks at 395 nm are shown in Fig. 5. The traditional Tauc's relation is used to generate the Tauc plot from UV-vis spectra. Cu-MOF's direct optical band gaps were determined at 2.45 eV by extrapolating the Tauc plot's tangent.

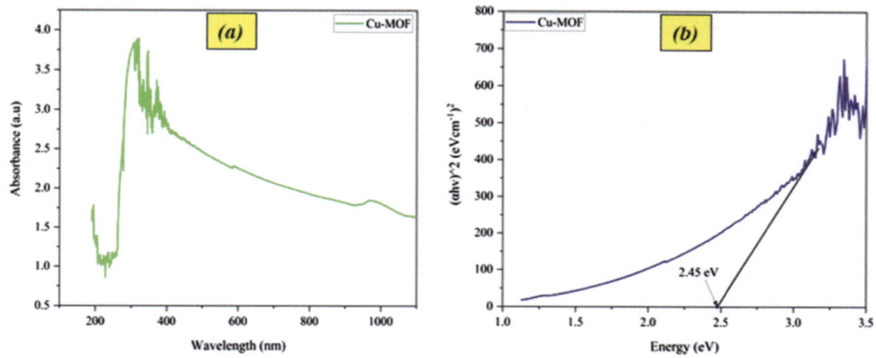

Fig. 5 UV-visible spectroscopy and direct band-gap Tauc plot image of as-grown Cu-MOF

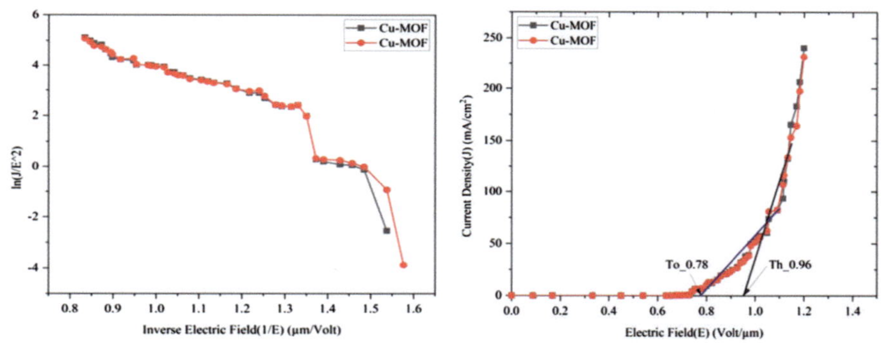

Fig. 6 Field emission and FN plot images of as-grown Cu-MOF

Through a strong electric field and a high vacuum, electrons migrate from a cathode to an anode through a process known as tunnelling. Field emission is produced through tunnelling in the quantum tunnelling mechanism. Field emission measurements of the synthesised Cu-MOF emitter were conducted with a diode set-up. A higher voltage was maintained for the anode than the cathode. As-grown samples with a 125 m gap were utilised as the cathode in the field emission setup. The diode set-up was held in place by a vacuum of around 10^{-6} bar. The anode was connected to a high-voltage DC source prior to the emission current being measured with a Keithley multimeter. According to the plot, Cu-MOF nanostructures have a turn-on voltage of 0.78 V/μm and a current density of 241 mA/cm^2 at 1.1 V/μm as shown in Fig. 6.

The sensors were put to the test by being exposed to the target gas acetone at room temperature. The measurements were made at 100, 90, and 80 ppm. Figure 7 shows that of resistance variation verifies the quick response and quick recovery.

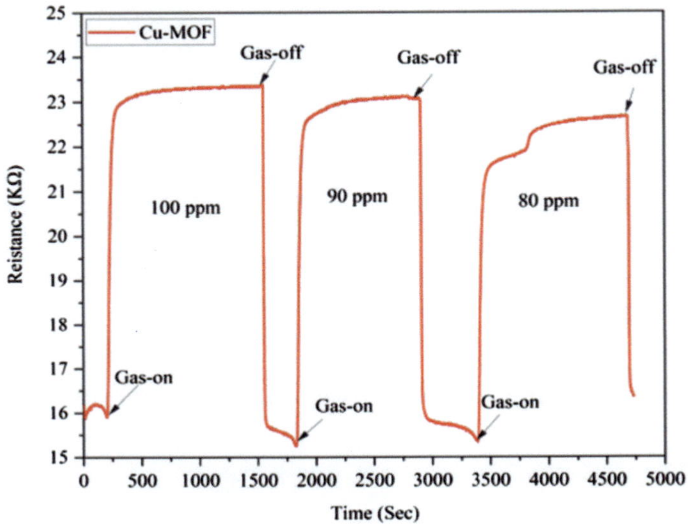

Fig. 7 Gas sensing performance images of as-grown Cu-MOF

Acknowledgements Mohd Sarvar thanks the DST, New Delhi, for the DST-INSPIRE fellowship (Ref No. DST/INSPIRE FELLOWSHIP/2019/IF-190286).

References

Alegre, R.J.V., Marqués, L.E., Herrera, R.P. and Díaz, R.P. (2016). Metal-organic frameworks (MOFs) bring new life to hydrogen-bonding organocatalysts in confined spaces. *Cryst Eng Comm,* **18:** 3985–3995.

Almeida, P.F.A., Klinowski, J., Vilela, S.M.F., Tome, J.P.C. et. al. (2012). Ligand design for functional metal-organic frameworks. *Chemical Society Review,* **41:** 1088–1110.

Bobbitt, N.S., Mendonca, M.L., Howarth, A.J., Islamoglu, T., Hupp, J.T. et.al. (2017). Metal-organic frameworks for the removal of toxic industrial chemicals and chemical warfare agents. *Chemical Society Review,* **46:** 3357–3385.

Carrasco, S. (2018). Metal-organic frameworks for the development of biosensors: A current overview. *Biosensors,* **8:** 92–121.

Chidambaram, A. and Stylianou, K.C. (2018). Electronic metal-organic framework sensors. *Inorganic Chemistry Frontiers,* **5:** 979–998.

Karmakar, A., Samanta, P., Desai, A.V. and Ghosh, S.K. (2017). Guest-responsive metal-organic frameworks as scaffolds for separation and sensing applications. *Accounts of Chemical Research,* **50:** 2457–2469.

Karmaoui, M., Leonardi, S.G., Latino, M., Tobaldi, D.M. et.al. (2016). Pt-decorated In_2O_3 nanoparticles and their ability as a highly sensitive (<10 ppb) acetone sensor for biomedical applications. *Sensors & Actuators B Chemical,* **23:** 697–705.

Rebollo, P.B., Sampedro, M.P., Gallardo, G., Ortega, M.N., et. al. (2014). Growth of metal micro and/or nanoparticles utilizing arc-discharge immersed in liquid. *Revista mexicana de física,* **60:** 227–232.

Salunkhe, R.R., Kaneti, Y.V. and Yamauchi, Y. (2017). Metal-Organic Framework-Derived Nanoporous Metal Oxides toward Supercapacitor Applications: *Progress and Prospects. ACS Nano*, **11:** 5293–5308.

Shooto, N.D. and Dikio, E.D. (2015). Synthesis, Morphology and Lead Ion Adsorption Properties of metal organic frameworks of copper and cobalt. *Chemical Sciences,* **6**: 1–9.

Stavila, V., Talin, A.A. and Allendorf, M.D. (2014). MOF-based electronic and opto-electronic Devices. *Chemical Society Review*, **43:** 5994–6010.

Yan, B. (2019). Photofunctional MOF-based hybrid materials for the chemical sensing of Biomarkers. *Journal of Material Chemistry C,* **7:** 8155–8175.

Zhu, Q.L. and Xu, Q. (2014). Metal-organic framework composites. *Chemical Society Review*, **43:** 5468–5512.

Chapter 7
Ammonia Gas Sensing Characteristics of MWCNT and Bi-MWCNT Operating at Room Temperature

Shah Masheerul Aalam, Mohd Sarvar, Mohd Sadiq, Md Taj Alam, and Javid Ali

1 Introduction

Since few years, there has been deep interest in novel materials and processing methods for gas sensing applications that can distinguish between gaseous molecules. Gas sensing research aims to develop an electronic nose with adequate sensitivity, selectivity, and reproducibility to be able to detect every type of gas that is present in ambient atmosphere at low concentration levels. Alcohol, H_2, NH_3, NO_2, liquefied petroleum gas (LPG), and CO_2 are a few examples of gases that are lethal, hazardous, or flammable. In food preparation, environmental clean-up, agriculture, and medical diagnostics, NH_3 is the most widely used of these gases.

Based on semiconductor metal oxides, NH_3 gas sensors have been created, but they must operate at high temperatures (Chen et al., 2009; Chopra et al., 2002). Researches are searching for new materials to replace these in order to conserve energy and lower the working temperature. Carbon nanotubes (CNTs) have a lot of benefits, which makes them very promising in this regard. CNTs, for example, have a high area to weight ratio and operate efficiently at room temperature. Many CNT gas sensor models have been created, including gas ionization sensors (Devi et al., 2006), capacitive sensors (Huang et al., 2009). CNT-FET, resonant-circuit sensor (Paolo et al., 2009), and resistive sensor. These models were used to apply CNTs to NH_3 sensing applications. In comparison to other models, used to construct and setting the measurement systems, the resistance sensor model is an easier method.

S. M. Aalam · M. Sarvar · M. Sadiq · M. T. Alam · J. Ali (✉)
Material Science Lab, Department of Physics Jamia Millia Islamia, New Delhi 110025, India
e-mail: jali1@jmi.ac.in

M. Sadiq
A.R.S.D. College, University of Delhi, New Delhi 110021, India

R. K. Kotnala et al. (eds.), *Advanced Functional Materials for Sustainable Environments*, https://doi.org/10.1007/978-3-031-62620-3_7

Multi-walled carbon nanotube (MWCNT) films were produced on a silicon substrate in this study using the low-pressure chemical vapour deposition (LPCVD) method. Fe was used as a catalyst. FESEM was used to analyze the MWCNTs' diameters and structural details. A static measurement system was used to examine the MWCNT films' NH_3 sensitivity at room temperature. The MWCNTs were embellished with Bi-nanoparticles to increase CNT sensitivity and facilitate NH_3 gas contact.

2 Experimental Set-Up

The MWCNT film was created using low pressure chemical vapour deposition for the resistance sensing model. On the surface, SiO_2 layer has been formed by keeping the Si substrate in the furnace and increasing the temperature to 900°C for 10 h. Fe catalyst layer was deposited before the MWCNT film was grown. The reaction container was then filled with the substrate. Using the LPCVD technique, Acetylene (C_2H_2) was used as the gas source to generate the MWCNT film at 750°C for 30 min. The MWCNTs were shielded from atmospheric oxygen using inert N_2 gas. Following the growth phase, the MWCNTs were naturally cooled in the reaction compartment while always being surrounded by an inert N_2 gas flow. By using FESEM, the acquired MWCNTs' morphology was identified. A static measurement device with a Keithley 6514 source metre was then used to assess the MWCNT-based sensor's NH_3 resistance variation and sensitivity. To improve the MWCNT film's ability to detect NH_3 gas, a Bi-layer coating on MWCNT was applied. The sensitivity and response of the MWCNT-based sensor were improved using the metal nanoparticle catalyst.

3 Results and Discussion

The FESEM micrographs shown in Fig. 1 provides evidence that the morphological tests were accurate. The MWCNT created on the Si substrate is visible in the micrograph, which is dense, long, and highly magnified. The micrograph B, which clearly demonstrates the decoration of NPs (nanoparticles) on the surface with diameters of only a few nanometres, confirms that the Bi-decorated MWCNT exists.

The diameter spread of the MWCNTs was verified by the Renishaw Raman Spectrometer wavelength of 532 nm. In this article, the G-band and D-band are found at 1583 and 1347 cm^{-1}, respectively. In contrast to D peak at 1347 cm^{-1}, the graphitic G peak in our work appears at 1583 cm^{-1} in the purest MWCNT and has a high intensity. In graphitic materials, the G band is created by the stretching and fracturing of E_{2g} by sp^2 hybridised C-C bonds. The D band, which is frequently related to A_{1g}, develops as a result of the disorder of graphitic planes, defects in nanotubes,

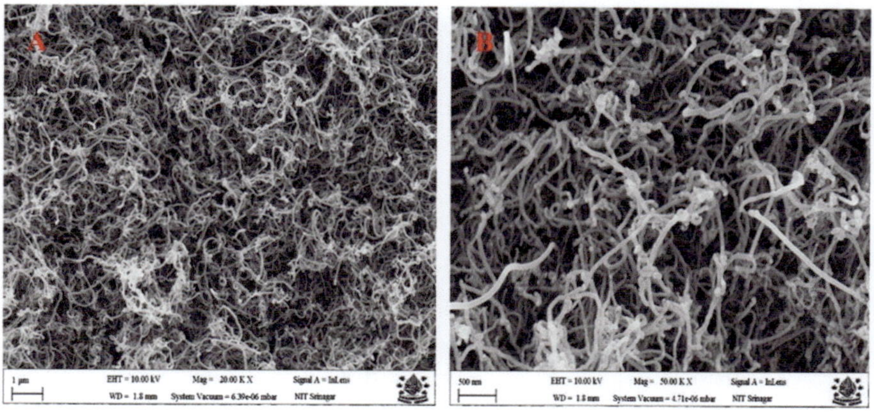

Fig. 1 FESEM micrographs: (A) Bare CNT and (B) Bi-decorated CNT

and other impurities. In decorated the graphitic nature of the material decreases with the surface decoration of the Bi nanoparticles as shown in Fig. 2.

The same NH_3 experiments were carried out on the MWCNT sensor that had been coated with Bi nanoparticles. The sensor reaction magnitude was significantly improved in the range of 1 to 5 secs, from pristine 82 to about 120%. When the target gas reached saturation, it was taken out of the chamber. Highly sensitive, quick responses, and complete recovery were observed in decorated sensors when compared to the reaction, sensitivity, and recovery qualities of a pristine MWCNT

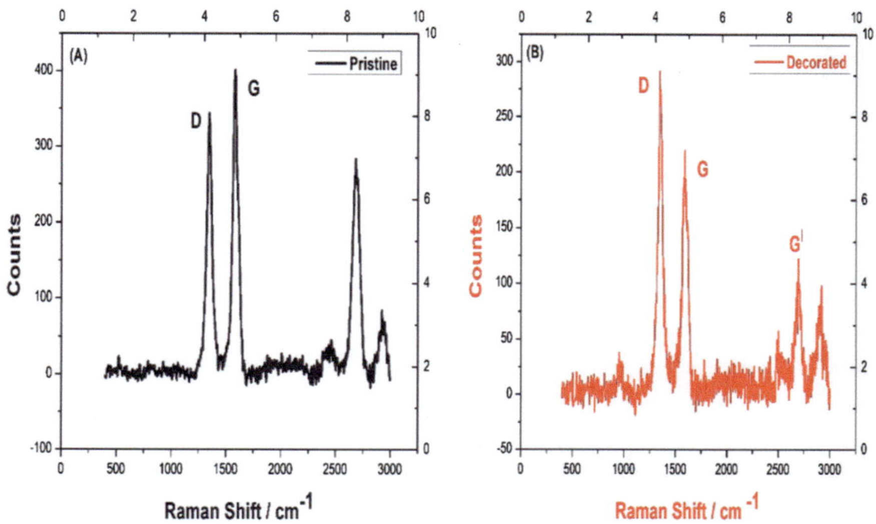

Fig. 2 Raman spectra of: (A) Pristine and (B) Bi-decorated

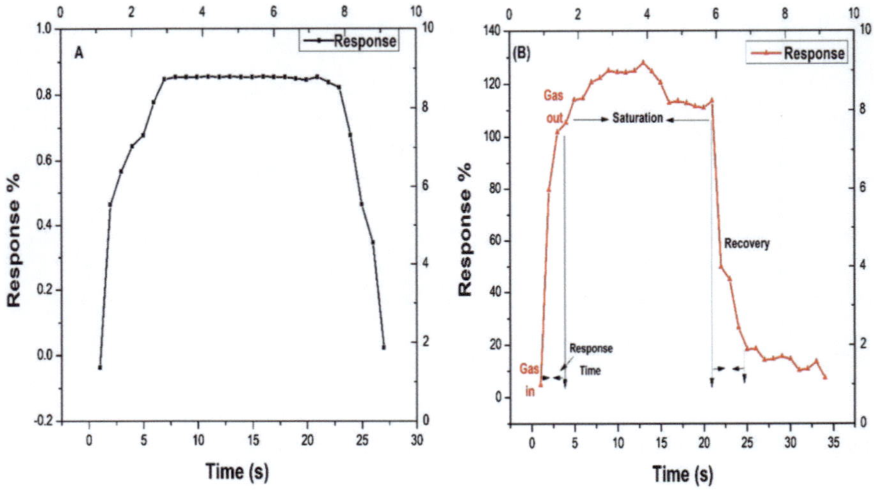

Fig. 3 Graph of: (A) Pristine and (B) Bi-decorated MWCNT sensor response towards NH_3

sensor. The improved response as shown in Fig. 3 shows that the Bi-nanoparticles create areas on the MWCNT surface where gas molecules can interact. An almost one-and-a-half-fold reaction was observed in comparison to a bare MWCNT sensor. This graph clearly shows how significantly better responsiveness has been observed with Bi-coated nanoparticles (Suehiro et al., 2003).

The response time for the pristine and the decorated sensors are 1–8 s and 1–5 s, respectively. The response for the two set of sensors is 82% for the bare MWCNT and 120% for the Bi-embellished MWCNT sensor. The sensitivity of the as-prepared samples is calculated for the reducing gas and it comes out to be 0.15 (15%) for bare MWCNT and 0.39 (39%) for decorated MWCNT sensor.

4 Conclusion

This article analysed the detection capabilities of the as-prepared carbon nanotube-based ammonia gas sensors. Using the LPCVD process at 750°C, properly formed, randomly oriented MWCNTs were successfully created on a Si substrate with a Fe coating on the surface. According to a morphological analysis, MWCNTs have grown exceptionally well and (Bi)-nanoparticles have been correctly deposited on their surface. The decoration of MWCNT sensor by Bi-nanoparticles has shown tremendous increase in sensor response, sensitivity, and for NH_3 detection, in comparison to pristine MWCNT sensor. An increase in the sensor reaction was observed as the concentration of the target gas increased. The Bi-nanoparticles decorated MWCNT sensor might be a step-up in terms of its potential for use in different gas sensing applications.

Acknowledgements We all the authors are grateful to the National Institute of Technology, Srinagar for providing FESEM and Raman characterizations. The Miranda House College (University of Delhi) provided the sensing set-up and cooperation, for which we are also grateful. One of the authors (S.M. Aalam) is very thankful to the Ministry of Higher Education (Govt. of India) for awarding the PMRF under the lateral entry program in the May cycle of 2022 with PMRF-ID 3302523.

References

Chen, Y. et al. (2009). Novel capacitive sensor: Fabrication from carbon nanotube arrays and sensing property characterization. *Sens. Actuators B*, **140:** 396–401.

Chopra, S. et al. (2002). Carbon-nanotube-based resonant-circuit sensor for ammonia. *Appl. Phys. Lett*, **80:** 4632–4634.

Devi, G.S. et al. (2006). NH_3 gas sensing properties of nanocrystalline ZnO based thick films. *Anal. Chem. Acta*, **568:** 41–46.

Huang, J.R. et al. (2009). A novel highly sensitive gas ionization sensor for ammonia detection. *Sens. Actuators A*, **150:** 218–223.

Paolo, B. et al. (2009). Carbon nanotubes-based transistors as gas sensors: State of the art and critical review. *Sens. Actuators B*, **140:** 304–318.

Radouane, L. et al. (2010). Gas sensors based on multiwall carbon nanotubes decorated with tin oxide nanoclusters. *Sens. Actuators B*, **145:** 411–416.

Suehiro, J. et al. (2003). Fabrication of a carbon nanotube-based gas sensor using Di electrophoresis and its application for ammonia detection by impedance spectroscopy. *J. Phys. D Appl. Phys*, **36:** L109–L114.

Thong, L.V. et al. (2010). Comparative study of gas sensor performance of SnO_2 nanowires and their hierarchical nanostructures. *Sens. Actuators B*, **150:** 112–119.

Chapter 8
Air Pollution Monitoring and Gas Leakage Detection with In-built Alert System

Aftab Alam, Nithya Priya, Raviranjan Singh, Akansha, Inderbir Kaur, and Vandana Batra

1 Introduction

In current times, air pollution is a major issue for the environment. Numerous steps are being taken to ensure the reduction in air pollution. In general, air pollution is caused by many pollutants like CO, CO_2, SO_2 etc. and also by natural gases. The increasing pollution caused by such pollutants can lead to drastic effects on the environment. Further, exposure to these harmful gases leads to many health issues such as Bronchial reactivity, reduced oxygen availability for tissues and an increase in bacterial infection (Al-Ali et al., 2010).

Currently, in India, environmental pollution has aggravated because of various causes like the release of pollutants from industries, the climate of certain regions, geography and culture. However, currently, it has been observed that there is some air quality deterioration in urban and rural India due to the usage of fuel (like coal, diesel, petrol, and fuel oil) and toxic air pollution emissions from chemical Industries, especially located in clusters (Gurjar and Nagpure, 2016) (Fig. 1).

As per WHO (World Health Organization), 2.4 billion of the world's population use fuel for domestic purposes, such as biomass (animal dung, crop wastage and wood), coal and kerosene oil, which further leads to the generation of harmful air pollution. As per available data, 3.2 million casualties per year were reported in 2020 due to domestic air pollution (Puthumana et al., 2021; IEA, 2022).

A. Alam · R. Singh · Akansha · I. Kaur
Department of Electronic Science, Bhaskaracharya College of Applied Sciences, University of Delhi, New Delhi, India

N. Priya · V. Batra (✉)
Department of Physics, Bhaskaracharya College of Applied Sciences, University of Delhi, New Delhi, India
e-mail: vandana.batra@bcas.du.ac.in

R. K. Kotnala et al. (eds.), *Advanced Functional Materials for Sustainable Environments*, https://doi.org/10.1007/978-3-031-62620-3_8

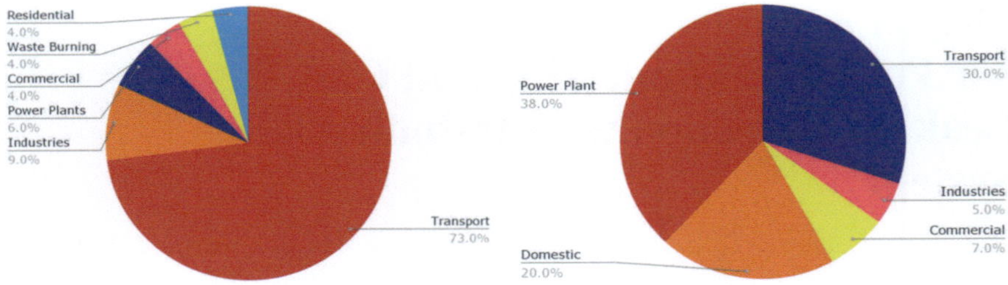

Fig. 1 Sector-wise emissions of: (a) carbon monoxide (CO) and (b) carbon dioxide (CO_2) in Delhi (Gurjar and Nagpure, 2016)

Natural gases are considered to be denser than air and they do not diffuse quickly which can cause suffocation on inhalation and also possibly trigger explosions. Natural gases like liquid petroleum gas (LPG), are primarily composed of butane and propane both of which are extremely flammable and also odorless. Therefore, a strong odorant, ethanethiol, is used enabling easy and fast detection of LPG leakage. However, some people might not be able to sense this odorant at low concentrations (Shrivastava et al., 2013). Hence there is a dire need to have a detection system which can detect various harmful gases present in the environment (Ghosh and Dhar, 2019).

In particular, more air quality measurement devices are required that can provide proper information regarding the level of pollution in the air. Various other sensors for providing data on air quality have been introduced by researchers while developing air pollution monitoring systems (Ghazali et al., 2012; Bangali and Shaligram, 2013; Okokpujie et al., 2018; Mariselvam and Dharshini, 2020; Zahra et al., 2020). Furthermore, IoT-based systems provide real-time monitoring of air quality in a given region and also provide information regarding pollution concentration and congestion instantaneously (Cynthia et al., 2019). The IoT-based technology can also be used as portable pollution monitoring systems, which can be transported anywhere and set up nearby, used to detect not only presence of conventional pollutants like NO, SO_2, etc. but also greenhouse and other gases (Yang, 2022). Portable wireless devices have also been developed that can detect VOC (volatile organic compounds) in real time and notify their high concentrations (Tsow et al., 2009). A system was also proposed to detect gas leakage, using different gas sensors like MQ-2, MQ-7 and MQ-4. These sensors transmit a signal to the microcontroller, which triggers externally connected devices. Further, this system is reported to be fast compared to other manual detection systems (Pandey et al., 2017).

Pollution measurement systems are designed and used in many countries across the globe to cater this need but manufacturing and installation of these systems are very expensive (Park et al., 2019). This article propose a prototype of a portable, cost-effective, compact, air quality monitoring and gas leakage detection in one design only. Further, this prototype has the capability of user alert and power cut to prevent major accidents because of LPG gas leakage. The prototype also informs the user about the ambient air quality.

The present article comprises four sections apart from the section which covers the need for air pollution detection systems. Section 2 covers the proposed design including the working principle and the algorithm, followed by the implementation of the design in Sect. 3. Section 4 covers the results and discussions. Conclusions and future scope are covered in Sect. 5.

2 Proposed Design of the Prototype with Working Principle

There are four levels in the proposed design of the prototype. The first level is collecting information of measuring parameters for an ambient gas level and environment, through government survey results. The features and characteristics of the sensors to be used are considered in the second level. The third level is about threshold value fixation, detection, making decisions, and output display on LCD. Fourth level is the acquisition of sensor data, operated by the Arduino microcontroller, and sent to the user's phone for information through the GSM module. Users can track sensor readings and also get an alert on the leakage of LPG gas or the presence of harmful gases in the environment and power of area automatically cut (Fig. 2).

Figure 3 shows the flow chart based upon above mentioned design.

Figure 4 shows the design of prototype, both Sensors (MQ-2 and MQ-135) are connected to Arduino uno.

The Arduino analyzes the information and, based on the code threshold value, displays it on the LCD and alerts the GSM. When the air quality of the surroundings is low or if there is a gas leakage then a message to alert the consumers is sent via

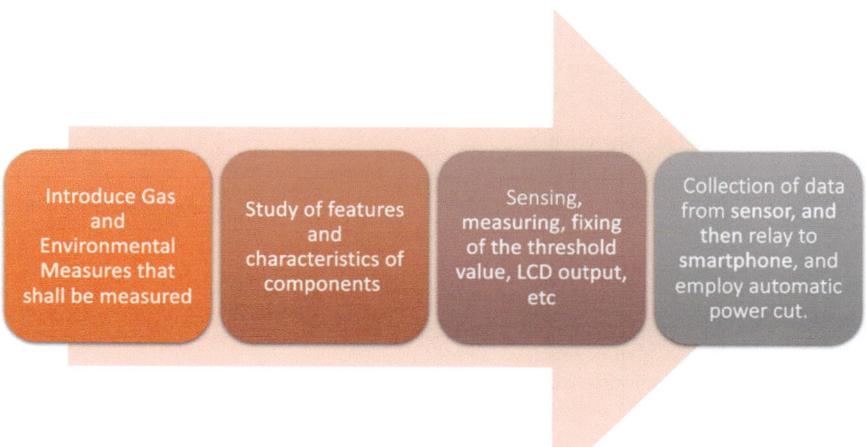

Fig. 2 System overall procedure

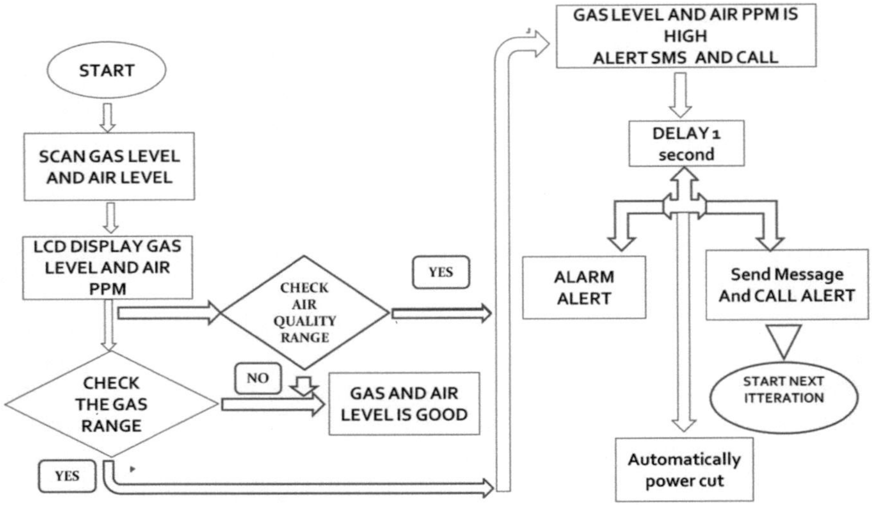

Fig. 3 The working of prototype

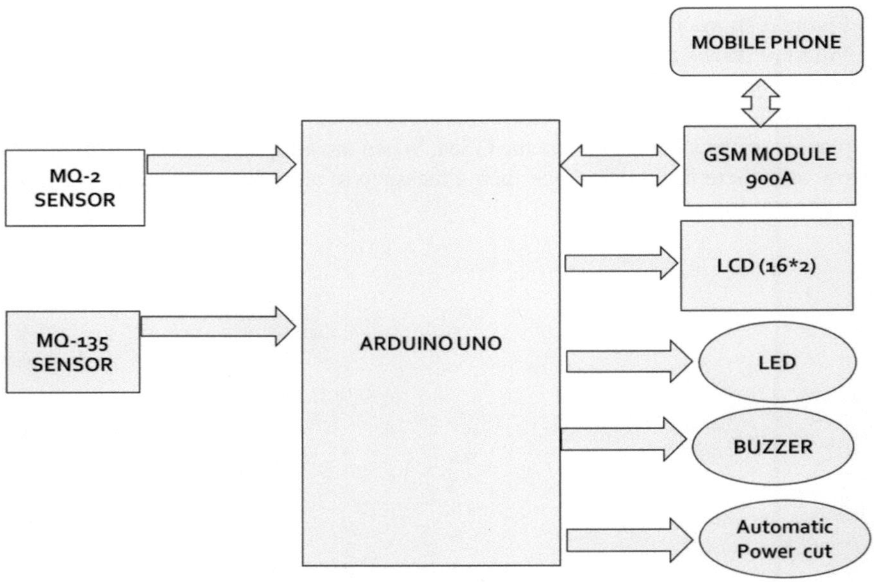

Fig. 4 Air pollution monitoring and gas leakage detection system

SMS on the mobile phone, with help of GSM900a Module. An alert through the buzzer is also incorporated in the design. To avoid any incidents of fire, the power in that region will be switched off automatically within a time interval of a few seconds.

3 Implementation of Proposed Design

Based upon Fig. 4 a prototype design was developed to test it. The proposed design was tested on Proteus software successfully (Fig. 5).

Proteus is an electronic prototyping board, which runs on Mac, Windows and Linux. In this proposed system, the Arduino is connected to the sensor MQ-2 and MQ-135, LED, buzzer, GSM Module, Relay and LCD display.

The function of the MQ2 sensor is to detect the leakage of gas and transmit a signal to the controller. It can be used to detect gases like propane, hydrogen and LPG concentration with threshold value greater than 300 ppm (300 to 1000 ppm). The conductivity of the sensor is increased on detection of gas concentration beyond a certain threshold value (Shrivastava et al,. 2013). Additionally, the MQ-135 sensor is used for air pollution monitoring. They detect the presence of gases and send signals to the controller. GSM900a module is connected with the Arduino Uno in order to send alert SMS and calls. When the gas sensor gives input to the Arduino, the GSM will send SMS and call alerts, on the registered mobile number as an output.

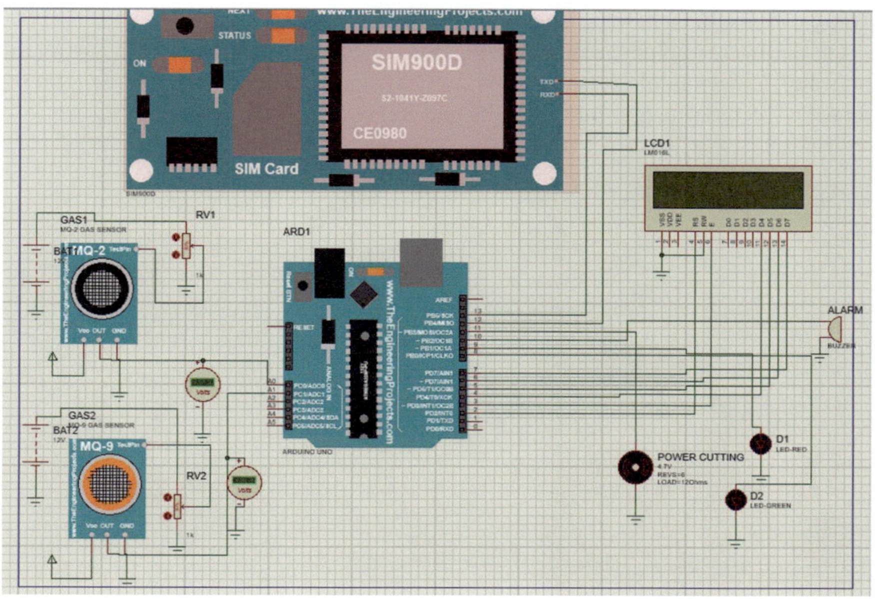

Fig. 5 Software Circuit on Proteus

Fig. 6 Components used in design of air quality monitoring and LPG detection system: (a) Arduino, (b) relay module, (c) GSM module, (d) LCD display, (e) MQ-2 sensor and (f) MQ-135 sensor (Mahzan et al., 2018; Kanti et al., 2020)

A relay, a kind of electrical switch or key that opens or closes the electrical circuit, is used to isolate the main circuit. The objective of the relay is to disconnect power to the main circuit if the leakage value of gas exceeds the threshold value. The LCD is connected with the Arduino to display the gas leakage status as well as the air quality of the atmosphere (as shown in Fig. 6).

The hardware implementation (see Fig. 5) was achieved using Arduino Uno micro-controller, GSM module 900A, a 16 by 2 LCD (liquid crystal display) Screen, MQ-2 and MQ-135 gas sensors, Buzzer, LED and Relay (Fig. 7).

Fig. 7 Implementation of proposed design using hardware circuit

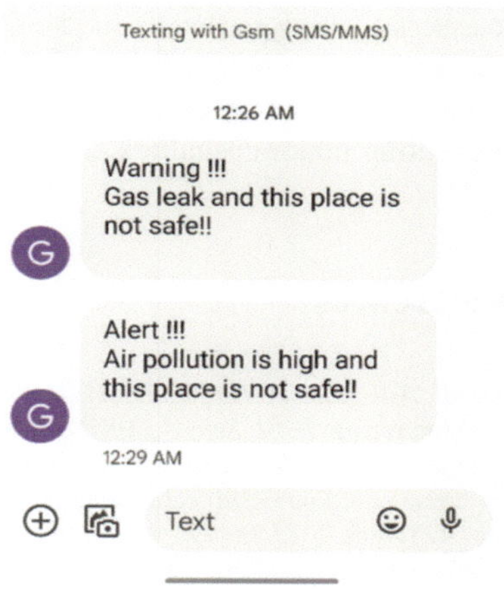

Fig. 8 Screen shot of received SMS message on detection of poor air quality and gas leakage

4 Results and Discussion

The hardware circuit (Fig. 7) was running successfully and was able to detect ambient air quality and LPG gas level. The design has been evaluated by exposing the prototype to a small amount of LPG. When the air quality is found to be satisfactory (AQI: 0 to 200) and the LPG gas level is within safe limits (< 300 ppm), the LCD displays "the area is safe" whereas if the LPG gas level is found to be high (300–1000 ppm), LCD displays "the gas level is high, this place is not safe" (Fig. 8).

Further, the Arduino notifies GSM, and consequently, GSM sends an SMS as well as a call to the user's phone. Additionally, in this situation, the power of that area is cut-off automatically within a time interval of a few seconds, to avoid any incidents of fire. Our prototype is also designed to notify the user in case of poor air quality. The display on the LCD shows "Poor Air Quality" if the AQI level is high (AQI >200). Further, an alert message is also delivered via SMS with the help of GSM900a Module.

The design of the proposed system incurs a total cost of around 1000 INR. Our portable, cost-effective, compact proposed prototype for air quality and leakage detection system was satisfactorily achieved.

5 Conclusions and Future Scope

A cost-effective system to monitor air pollution and gas leak detection with an in-built alert system, on real time basis and at any preferred site, has been designed successfully. This user interaction system informs the consumers about the ambient air quality and also warns in case of leakage of LPG gas. It can be further customized to assess specific gas levels and air quality for the usage in industrial settings, residential complexes and laboratories. The prototype can be further improved by inserting a feature of sending an alert to the ambulance and fire control services to help in preventing major casualties.

References

Al-Ali, A.R., Zualkernan, I. and Aloul, F. (2010). A Mobile GPRS-Sensors Array for Air Pollution Monitoring. *IEEE Sens J.*, **10(10):** 1666–1671. https://doi.org/10.1109/JSEN.2010.2045890.

Bangali, J. and Shaligram, A. (2013). Design and implementation of security systems for smart home based on GSM technology. *International Journal of Smart Home* **7(6)**. https://doi.org/10.14257/ijsh.2013.7.6.19.

Cynthia, J, Saroja, M.N., Sultana, P. and Senthil, J. (2019). IoT-based real time air pollution monitoring system. *International Journal of Grid and High Performance Computing* **11(4)**. https://doi.org/10.4018/IJGHPC.2019100103.

Ghazali, M.Z., Noor, N.M., Ramly, N. and Putit, S. (2012). Development of microcontroller based mobile gas sensing robot. *In:* Procedia Engineering.

Ghosh, P. and Dhar, P.K. (2019). GSM Based Low-cost Gas Leakage, Explosion and Fire Alert System with Advanced Security. *In:* 2nd International Conference on Electrical, Computer and Communication Engineering, ECCE 2019.

Gurjar, B.R. and Nagpure, A.S. (2016). Indian megacities as localities of environmental vulnerability from air quality perspective. *Journal of Smart Cities* **1(1)**. https://doi.org/10.18063/jsc.2015.01.003.

IEA I (2022). The Energy Progress Report. Washington DC. © World Bank.

Kanti, A., Jagdish, P.M. and Soumya (2020). Air Pollution Monitoring System using Arduino with MQ135 Sensor. *International Research Journal of Modernization in Engineering Technology and Science* **2(8):** 412–421.

Mahzan, N.N., Enzai, N.I.M., Zin, N.M. and Noh, KSSKM (2018). Design of an Arduino-based home fire alarm system with GSM module. *In*: *Journal of Physics*: Conference Series.

Mariselvam, V. and Dharshini, M.S. (2020). IoT based level detection of gas for booking management using integrated sensor. *In*: Materials Today: Proceedings.

Okokpujie, K., Noma-Osaghae, E., Modupe, O., John, S. and Oluwatosin, O. (2018). A smart air pollution monitoring system. *International Journal of Civil Engineering and Technology,* **9(9)**.

Pandey, R.C., Verma, M. and Sahu, L.K. (2017). Internet of Things (IOT) Based Gas Leakage Monitoring and Alerting System with MQ-2 Sensor. *International Journal of Engineering Development and Research,* **5(2):** 2321–9939.

Park, J., Oh, Y., Byun, H.H. and Kim, C.K. (2019). Low cost fine-grained air quality monitoring system using lora wan. *In*: International Conference on Information Networking.

Puthumana, J.S., Ngaage, L.M., Borrelli, M.R., Rada, E.M., Caffrey, J and Rasko, Y. (2021). Risk factors for cooking-related burn injuries in children, WHO Global Burn Registry. *Bull World Health Organ,* **99(6):** 439–445. https://doi.org/10.2471/BLT.20.279786.

Shrivastava, A., Prabhaker, R., Kumar, R. and Verma, R. (2013). GSM based gas leakage detection system. *International Journal of Technical Research and Applications,* **1(2):** 42–45.

Tsow, F., Forzani, E., Rai, A., Wang, R., Tao, N.J., Tsui, R., Mastroianni, S. and Knobbe, C. and Gandolfi, J. (2009). A Wearable and Wireless Sensor System for Real-Time Monitoring of Toxic Environmental Volatile Organic Compounds. *IEEE Sens J*, **9(12)**. https://doi.org/10.1109/JSEN.2009.2030747.

Yang, Y. (2022). IoT-based air pollution monitoring system. Highlights in Science, Engineering and Technology MSESCS, **17:** 299–307. https://doi.org/10.54097/hset.v17i.2619.

Zahra, M.M.A., Mohsin, M.J. and Abdul-Rahaim, L.A. (2020). Artificial intelligent smart home automation with secured camera management-based GSM, cloud computing and arduino. *Periodicals of Engineering and Natural Sciences,* **8(4)**.

Chapter 9
Bacterial Cellulose as Filter Membranes for Waste Water Treatment: Recent Trends and Applications

Samriddh Srivastava and Garima Mathur

1 Introduction

With increase industrialization, the utilization as well as release of heavy metals in environment have increased in past few decades. These heavy metals are considered as peculiar pollutants of environment. Research have sparked several exploratory studies in treatment of heavy metals in waste water through various chemical processes involving oxidation and reduction reactions as well as techniques like filtration, ion exchange chromatography, osmosis and flocculation. All these techniques are cost effective and environment friendly. Some heavy metals found in water and their respective safe limits according to U.S. Environmental Protection Agency (USEPA) [https://www.epa.gov/caddis-vol2/metals] are listed in Table 1. Apart from chemical contaminants, biological pollutants have also been observed in polluting the environment (Parveen S. et al., 2022). In lieu of utilizing skilled labor, high cost equipment and chemicals, membrane-based filter membranes have gained a great interest. Bacteria synthesized membranes are more employed for different applications as compared to chemically synthesized membranes, as the former are devoid of contaminants and exhibit enhanced purity (Nunes et al., 2020). In this context, bacteria from family AAB (acetic acid producing bacteria) have been utilized greatly for production of bacterial cellulose films which are further used as membrane filters. The classification of acetic acid producing bacteria (AAB) encompasses several commonly encountered genera, namely *Kozakia*, *Swaminathania*, *Acetobacter*, *Acidomonas*, *Asaia*, *Gluconacetobacter* and *Saccharibacter*. While various species such as *Komagataeibacter*, *Aerobacter*, *Agrobacterium*, *Acetobacter*, *Rhizobium*, *Sarine*, *Achromobacter*, *Alcaligenes*, *Azotobacter* and others are known for their ability to produce cellulose, the predominant cellulose producer

S. Srivastava · G. Mathur (✉)
Plant and Microbial Biotechnology Group, Department of Biotechnology Jaypee Institute of Information Technology, A-10, Sector-62, Noida, Uttar Pradesh, India
e-mail: garimacity@gmail.com

R. K. Kotnala et al. (eds.), *Advanced Functional Materials for Sustainable Environments*, https://doi.org/10.1007/978-3-031-62620-3_9

Table 1 Hazardous metals found in water and their respective safe limit [https://www.epa.gov/cad dis-vol2/metals]

Metal	*Hazardous Quantity (μg/L)*	*USEPA Safe Limit (μg/L)*
Arsenic (As)	>10	10
Cadmium (Cd)	>3	3
Chromium (Cr)	>50	50
Lead (Pb)	>10	10
Mercury (Hg)	>1	1

is typically identified as *Komagataeibacter xylinus* (Gorgieva et al., 2019). The *Komagataeibacter* genus is primarily associated with the family *Acetobacteraceae.* Through the analysis of their genetic sequences, members of this genus have been classified into two groups: the *Komagataeibacter xylinus* and the *Komagataeibacter liquefaciens* (Jung et al., 2005; Urbina et al., 2021; Pogorelova et al., 2020). The bacterial cellulose (BC) films exhibit higher crystallinity, water holding capacity as well as mechanical strength. These properties make BC an interesting biopolymer for applications in waste water treatment (Swingler et al., 2021). One such study demonstrates the preparation of BC composite- carboxymethylated BC which was utilized in removal of chemical pollutants like copper and lead from waste water. In addition to this, desorption of chemical pollutants onto the BC membrane filters was successfully achieved using ethylene diamine tetra acetic acid (EDTA) treatment (Dutt et al., 2020). Further, the BC films were also modified into composites with phthalocyanine which was used in removing contaminants in the form of dyes from waste waters (Campano et al., 2016). This chapter discusses the structure of BC and its suitability for developing membrane-based filters. Further, the effect of different BC composites as well as its applications as membrane filters in waste water treatment has been elaboratively discussed. Development and successful applications of novel BC composites-based filter membranes have several challenges. This chapter also comprehensively discusses various challenges and its associated solutions in development of membrane filters.

2 Structural Properties of BC

Bacterial cellulose (BC) and plant cellulose share the same chemical composition, consisting of a linear chain homopolymer of glucose (Glc) monomers or residues linked by (β-1,4) bonds, with a chemical formula of $(C_6H_{10}O_5)_n$, BC does not acquire the same macromolecular organization of chemical structure and properties like plant cellulose (Krystynowicz et al., 2002) Table 2. Comparing the power of polymerization of plant and BC both have different ranges of polymerization; plant cellulose comes under the range of 1300–1400 and BC ranges between 300–1000. Bacterial cellulose has the purest form of cellulose, glucose residues attached with glyosidic bond with alternating units rotated at 180 degrees, chains, and cellulose sheets are strongly connected with hydrogen bonding both intra- and intermolecular results as translucent structure (Xu et al., 2021). The structural density of cellulose ribbons like structure ranges from species to species besides the culture pathology. For example, BC fibers synthesis by *Komagataeibacter xylinum* be made up of microfibril like structure that ranges 3–4 $\times$ 10^{-6} mm (3–4 nm) thick and 70–80 nm wide (Srivastava et al., 2021).

The array of cellulose thread and the structural confirmation of the porous matrix structure of bacterial cellulose depends upon the microbial strain, methodology of synthesis, contrast in culture schedule and condition, chemical composition of culture medium, number of inoculums, and carbon sources. Different types of carbon sources like sucrose, glucose, mannitol, and many other suitable carbon sources polymerized by microbial cell and translocate it through pore known as terminal complexes, translocated glucan chain form structure of protofibrils, bundles and finally form ribbon-shaped microfibrils (80 $\times$ 4 nm) (Srivastava et al., 2021). The terminal complexes like structure controlled the structural properties of glucan chains and their assembly in microfibrils, bundles, and strip or ribbon (Chen et al., 2021). In *Komagataeibacter xylinus,* based on the arrangement of fibrils cellulose is divided into two types, viz. lace-like cellulose type-I and nebulous cellulose type-II. In cellulose type-I, the β-1, 4 glucose chains are aligned or arranged parallelly and uniaxially. On the other hand, in cellulose type-II, the positioning of the chains is irregular or arbitrary, lacking the same parallel and uniaxial arrangement found in type-I cellulose, and microfibrillar ordering of cellulose fibers and their utmost purity mainly contribute tensile strength and crystallinity of BC (Lin et al., 2020). The thin ultrafine fibrous structure of bacterial cellulose provides the substratum for its substantial applications. The arrangement of fibrils and their unique purity provides a special path to bacterial cellulose for the production of applications loaded BC-based composite. High water holding capacity, nano porosity, cell adhesion, flexibility like properties

Table 2 Difference between plant and bacterial cellulose [Srivastava and Mathur, 2021]

Properties	*Bacterial cellulose*	*Plant cellulose*
Tensile strength	70–300 MPa	<1 GPa
Young's modulus	5–17 GPa	130 GPa
Water holding capacity (WHC) (%)	>95	25–35
Size of fibres 54 (nm)	20–100	2–10 nm
Crystallinity (%)	74–96	40–85
Relative hydrophilicity (%)	40–50	20–30
Purity (%)	>99	<80
Degree of polymerization	14,000–16,000	300–10,000
Porosity (%)	>85	<75
Total surface area (m^2/g)	>150	<10

of bacterial cellulose is directly influenced by the structure, alignment and orientation of cellulose fibrils which provide a wide application for microbial cellulose (Table 3). The unique structural composition and useful properties of bacterial cellulose become a point of attraction for many researchers and industries. Biocompatible and biodegradable feature of BC provides a special place to BC in medical research and bioengineering (Kosseva et al., 2022).

BC is known to possess nanofibril arrangement and enhanced purity as well as porosity due to its unique structure. Due to these distinctive features, BC is known as potential alternative to plant based cellulose (with hemicellulose and lignin). This also improved crystallinity, mechanical capabilities, better water holding capacity

Table 3 Properties of Bacterial Cellulose (Bae and Shoda, 2004)

Properties	*Parameter*	*Bacterial cellulose*		
Morphology	Width (nm)	Nanofibril: 1.5–4		
		Fibril: 40		
	Length (μm)	1–9		
	Aspect ratio (L/w)	667–6000		
Physical properties	Crystallinity	84–89%		
	Polymerization degree (DP)	300–10,000		
	Structure	SC: 3D interwoven mesh		
		AC: amorphous agglomeration		
	Porosity	Micro/nanopores		
Thermal properties	Degradation stages	1st	2nd	3rd
	Decomposition temperature range (°C)	250–399	399–500	500–561
	Maximum weight loss temperature (°C)	350	496	525
	Activation energy (kJ/mol)	96	167	794
	Total activation energy (kJ/mol)	1058		
Mechanical properties	Tensile strength (MPa)	200–2000		
	Young's modulus (GPa)	15–138		
Water absorption properties	Water holding capacity	100–400		
	Rehydration ability	Low		
Optical properties	Light transmittance	~90%		
Acoustic properties	Sonic velocity	High (~ aluminum)		
	Dynamic loss	Low		
Other properties	Biocompatibility	Non-Toxic		
	Biodegradability	High		

(WHC), and effective antibacterial activity. However, before BC can be widely adopted at an industrial scale, it needs to overcome certain technological challenges, including hydrophobicity, limited solubility, and relatively high production costs. Researchers have conducted several studies to address these challenges associated with lower productivity as well as high cost of raw materials utilized for BC production (Kosseva et al., 2022; Bae and Shoda, 2004).

These production process difficulties may be resolved by using the fed-batch operation mode and repeated batch culture techniques. In fed-batch operation mode nutrients are added incrementally during the fermentation process. Instead of providing all the nutrients at once, they are introduced gradually as the bacteria consume them. This approach allows for better control over nutrient availability, sustains bacterial growth, and maximizes cellulose production. It offers improved process control, sustained growth conditions, and increased cellulose yield, making it valuable for optimizing and scaling up industrial bacterial cellulose production (Bae and Shoda, 2004). Consequently, repeated-batch culture is a technique employed in the production of bacterial cellulose, wherein a portion of a previous culture is reused to initiate subsequent fermentation cycles. By implementing the repeated-batch culture technique in bacterial cellulose production, significant time and resource savings can be achieved as there is no requirement for a fresh culture for each cycle. Moreover, through successive cycles, the bacteria adapt and become increasingly efficient, resulting in enhanced cellulose yields. Careful monitoring and control of growth conditions are necessary for consistent performance. While challenges may arise, repeated-batch culture is a cost-effective and efficient method for scaling up bacterial cellulose production. Both approaches have demonstrated higher yields in bacterial cellulose production. By incorporating a low-cost nutrient medium and employing genetically modified bacteria, the process holds potential for broader implementation. However, to actualize this possibility, further extensive research and investigation are necessary (Bae and Shoda, 2004; Li et al., 2022).

3 New BC Composite Based Filter Membranes

Magnificent properties of BC films, makes it an excellent component as filter membranes. The BC membranes are known for exhibiting hydrophilic and oleophobic properties which means during the filter process of waste water, only certain water-based molecules are allowed to pass through BC membranes (Bandehali et al., 2021). These BC based sheets are known to be semi-permeable to water and does not allow the passage of oil-based components through it. The modifications of BC membranes with certain composites make it more permeable to oil-based components and less permeable to water (Zhu et al., 2021). These modifications make BC more suitable to applications related to oil/water emulsions. Further, the modifications of BC membranes also make it suitable for removal of heavy metals from the water during waste water treatment (Liu et al., 2019). However, the modifications related to this content, are often based on enhancing the mechanical strength as well

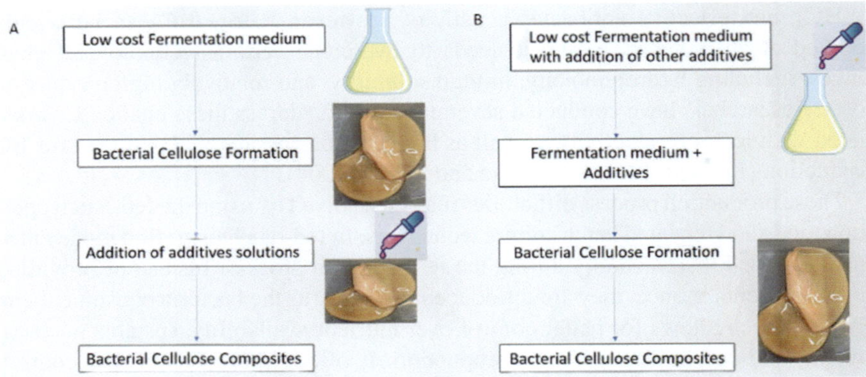

Fig. 1 Modification techniques of bacterial cellulose: (A) ex situ and (B) in situ method (Li and Cheng, 2019)

as flexibility of the membranes. Figure 1 highlights the techniques for modification of BC.

The membranes with enhanced mechanical properties are converted into reusable membranes which are further used for several applications in distinct fields (Li et al., 2019). In addition, grafting technique has been widely applied while modifying the BC membranes. During grafting, the important functional groups, especially carboxyl and thiol groups, are often added to the BC membranes which results in enhancement of its adsorption capacities (Karim et al., 2016). Hence, the modified BC membranes exhibit higher absorption of heavy metals or oil-based components during filtration of waste waters. Pore size plays an important role in filtration processes. The components having larger diameter as compared to pore size of filter membranes, are non-selective to the membranes. Hence, the modifications of BC membranes are also done with respect to modifications in pore size. It is known that denser fibrils of BC membranes can result in blockage which could further impair the filtration process. Hence, the pore size of the BC membranes is often modified with respect to its utilization in filtration applications (Alves et al., 2020).

Several studies have been conducted which explores the undiscovered potential of BC membranes in filtration processes. On study demonstrated the utilization of BC membranes with enhanced antimicrobial properties as well as water contaminants filter membranes. In this study, the modified BC films were not only utilized for removal of contaminants from waste water but also utilized for removal of pathogenic viable cells from the waste water too (Sjahro et al., 2021; Norfarhana et al., 2022). This was made successful by addition of functional groups like amines to the BC membranes during modifications. Further, the addition of chitosan, aerogels and phosphate based functional groups were also added during the modifications of BC membranes. These modifications resulted in enhanced filtration of larger contaminants from the water. In addition, BC films have also been modified using the polyethyleneimine groups for enhanced capacity to remove the complex ions like copper and lead from waste water (Jin et al., 2017). The modification of the BC membranes using polyethyleneimine was done using the technique of reductive amination. This modification enhanced the bio-adsorption ability of the membranes which resulted in more removal of the contaminants from the waste water (Zhuang et al., 2019). Copper ions were also removed using modified BC membranes. In the study related to removal of copper ions, the membranes were modified with ex situ methods of addition of chitosan on to the BC membranes. The produced BC membranes were treated with chitosan solution at different concentration. This modification enhanced the adsorption capacity of BC membranes against copper ions. The phosphate group-based modifications have been also utilized in enhancing the filtration capacity by the BC membranes. The BC membranes modified with phosphate groups are kept in the solution of dimethylacetamide and urea. The treatment of the cellulose membranes with these solutions confirmed the successful addition of phosphate groups and also enhanced the electrostatic interactions and hence resulted in easy attraction of charged groups from waste water, resulting in successful removal (Galdino et al., 2020). Further, these membranes have been successfully utilized in removal of viable cells of E. coli from the waste water (Sjahro et al., 2021). In one recent study, the recovered membranes were washed with the solution of NaOH and were fully purified. Later these membranes were used for removal of microbial contaminants from the waste water effluents (Liu et al., 2021).

Organic pollutants have been known to damage the environment and their removal is a crucial step. Latest research findings have explored that the modifications of BC membranes with the cyclodextrin oligomer composites have led to the successful removal of organic pollutants. This study includes the preparation of thin BC membranes with higher porosity and high mechanical strength which were modified by using solution of cyclodextrin and oligomer. The BC membranes were observed to exhibit higher resistance towards the passage of organic pollutants like phenols, glyphosate and 2,4-dichlorophenol from the waste water (Urbina et al., 2021). Several dyes are often released from industries into the water streams which contaminate water flora and fauna. The treatment of water to remove the toxic dyes has been successfully done with the application of BC membranes. In latest study, the chitosan blend BC based membranes were used to remove the toxic dyes from waste water. The results revealed the removal of dyes with higher rejection rate and stable pressure. During the filtration process of waste water, the fouling is one of the major issues which could restrict the overall process and results in the damage of BC membranes. During this study, the modification of BC films with chitosan improved the antifouling properties towards the oil and protein-based components present in the waste water. In addition, the removal of dyes was also made successful using modification of polyaniline using ex situ oxidative polymerization (Fig. 1). The modified films were then utilized in the removal of methyl orange dye as well as microbial contaminants from the potable water. The findings also claimed the removal of microbial contaminants by four time using the modified BC membranes (Hassan et al., 2021).

The treatment of waste water has also been done without any modification of BC membranes. In one such study, the removal of oil-based components have been successfully done using the change in filtration rate and efflux pressure rate. Moreover, the removal of contaminants from waste water were also investigated in stabilized and non-stabilized wet cellulose solutions, which resulted in enhanced removal rate with 92%. Oil-water emulsions were also removed using the modified BC membranes using the hydrolysis of alkoxysilanes. These modifications helped in restoration of 3D structure of membranes during dying process and also increased the wettability (Sai et al., 2020). The enhancement of physico-chemical properties, resulted in more enhanced removal of pollutants from waste water. 100% removal rate of oil emulsions from waste water was also achieved from modified BC membranes. The membranes used in this study were modified using BC membranes which were produced using the corn steep liquor medium, followed by washing in solution of NaOH (Hassan et al., 2021). 99% efficiency of oil emulsion removal from waste water was also achieved from modified BC membranes, using stainless steel mesh, which resulted in enhanced wettability and roughness of BC films (Cakar et al., 2014). The modifications of BC membranes and their suitable applications in removal of contaminants and pollutants from waste water is summarized in Table 4.

4 Opportunities and Challenges of BC Based Membranes Filters

Bacterial cellulose (BC) offers an emerging platform for development within the waste water treatment but challenges exist. These include cost-effectiveness of production, optimizing contaminant removal efficiency, ensuring regeneration and reusability and maintaining structural integrity in water. However, opportunities lie in functionalization and surface modification for enhanced adsorption and antimicrobial properties, developing composite materials for improved performance, exploring biofilm-based systems for biological treatment, and utilizing bacterial cellulose membranes in filtration and separation processes. The production of bacterial cellulose for water treatment faces challenges that need to be addressed. Cost-effectiveness is a key concern, requiring optimization of production processes and fermentation conditions to scale-up manufacturing. BC production varies significantly among various species and strains as well as the carbon source, however, necessitates greater effort in the early production stages and leaves potential for additional basic research contributions in this field. Glucose, fructose, and glycerol are the traditional carbon sources used in BC manufacturing, which dramatically raises costs and accounts for about 30% of all BC production costs (Revin et al., 2018). Recently, economical local sources for BC production have been suggested, such as industrial wastes or byproducts namely corn steep liquor (CSL), sugarcane waste, molasses and date syrup are considered as additional alternates to Hestrin-Schramm and Yamanaka media for BC production (Liu et al., 2014; Srivastava et al., 2022). Alternative carbon sources have been proposed as potential options to increase BC production and make it more

Table 4 Modified BC membranes and their suitable applications in waste water treatment

Research Study	*Research Findings*	*References*
Effects of anode functional groups (O-, N- and S-) on bacteria that are electroactive throughout the development of a biofilm	Important functional groups, especially carboxyl and thiol groups were added to the BC membranes resulting in enhancement of its adsorption capacities	Karim, et al., 2016
Vacuum-filtration of cellulose nanofiber suspensions was used to create multi-layered nanocellulose membranes, which were then coated with cellulose nanocrystals containing sulphate or carboxyl surface groups	Modifications in pore size of BC filter membranes which prevented the blockage of membranes during filtration process	Alves et al., 2021
Making bacterial cellulose membranes and using them as filters for water remediation and industrial wastewater treatment	100% removal rate of oil emulsions was achieved from modified BC membranes, produced using the corn steep liquor medium, followed by washing in solution of NaOH	Sjahro et al., 2021
Numerous approaches to using cellulose derivatives in the treatment of home and industrial wastewater, including nano, micro, and natural forms	BC films were utilized for removal of contaminants and pathogenic viable cells from waste water too	Norfarhana et al., 2022
By reductively amine dialdehyde with polyethyleneimine, polyethyleneimine-bacterial cellulose (PEI-BC) was created	Addition of polyethyleneimine groups enhanced the removal of copper (Cu) and lead (Pb) from waste water	Zhuang et al., 2019
Preparation, characterization, and use of phosphate functionalized bacterial cellulose with micro-fibrous structure for U(VI) adsorption	BC membranes modified with phosphate (PO_4^{3-}) groups, dimethylacetamide (C^4H^9NO) and urea resulted in enhanced electrostatic interactions and attraction of charged groups from waste water	Galdino et al., 2020
Cyclodextrin (CD) oligomer was added to the ultrafine nanofibers of 3D bacterial cellulose to create a film-like water filter	BC membranes with the cyclodextrin oligomer composites resulted in successful removal of organic pollutants	Urbina et al., 2018
Create new, ecologically friendly membranes using in situ and ex situ methods using chitosan (Ch) as the functional component and bacterial cellulose (BC) as a template for wastewaters treatment	BC films with chitosan improved the anti-fouling properties towards the oil and protein-based components present in the waste water	Zhijiang et al., 2020
Grafting, electrospinning, and surface coating were used in tandem to create a nanofiber composite nano-filtration membrane	BC membranes with modification of polyaniline using ex situ oxidative polymerization resulted in removal of methyl orange dye as well as microbial contaminants from the potable water	Hassan et al., 2017

(continued)

Table 4 (continued)

Research Study	*Research Findings*	*References*
Crosslinked cellulose nanofibers (CNF) and never-dried bacterial cellulose (BC) were utilized to remove oil from stabilized and unstabilized oil-in-water emulsions with droplet sizes smaller than 1 m	Oil-based components resulted in enhanced removal rate with 92% of contaminants from waste water	Sai et al., 2020
For effective oil-water separation, a simple and environmentally friendly approach of fabricating super wettable bacterial cellulose membranes (BCMs) through a one-step filtration process	99% efficiency of oil emulsion removal from waste water was achieved from modification using stainless steel mesh, which resulted in enhanced wettability and roughness	Sai et al., 2014

affordable. By repurposing waste materials and incorporating them into the production process, it not only reduces costs but also contributes to sustainable and eco-friendly practices in BC manufacturing. Due to the limited research and development in the area of finished filter products based on bacterial cellulose (BC), there is a need to explore less complex, affordable and readily applicable alterations beyond conventional post-synthetic oxidation and grafting processes (Srivastava et al., 2022). These alternative approaches aim to simplify the production and modification of BC-based filters while maintaining their effectiveness in water treatment. By seeking out these modifications, it becomes possible to enhance the practicality and cost-effectiveness of BC filters, making them more accessible for various applications. These alterations may involve novel surface modifications, functionalization techniques, or innovative manufacturing processes that can optimize BC's filtration capabilities and improve its overall performance as a water treatment material. Further exploratory research conducted in this sector are needed as a part of future prospects in order to unlock and further explore the full potential of BC-based filter products.

5 Conclusion

Due to their exceptional binding and adsorbing abilities, composites have proven to be a potential agent to remove different contaminants from wastewater. They can be changed in a number of ways to improve their performance. Industrial wastewater contamination by hazardous effluents causes deadly environmental problems. Membrane filtration and chemical precipitation are considered as two most commonly utilized techniques for getting rid of dangerous contaminants. These techniques are often very costly and are known to produce toxic waste products which are bad for mankind and the environment. This circumstance concentrates attention on biological methods for eliminating dangerous pollutants from waste water. It is also known that employment of biomass obtained from agricultural and microbial sources offer several advantages of biosorption. As these raw materials naturally known as adsorbents, have been effectively utilized for waste water treatment. This includes cellulose derived from various sources, such as microorganisms and plants. As BC exhibits high purity and possess three-dimensional functional nanostructure, it has superior applications as membrane filters as compared to plant-based cellulosic material. This characteristic of exhibiting high purity is often associated to the presence of functional group near hydroxyl group in the structure of BC, hence, boosting its capability as filter membrane. This creates a lot of opportunity for the quick development of filters that are highly effective at removing impurities while also being cost-effective. Therefore, there will be need for technology advancement and a greater understanding of BC, its composites, and it is as-yet-undiscovered metal absorption mechanism.

Statements and Declarations **Contributions:** Both authors have equally contributed. The submitted work is part of the doctoral research of Mr. Samriddh Srivastava, under the guidance of Dr. Garima Mathur.

Conflict of interest: The authors declare that they have no conflict of interest in the publication.

Funding: The authors did not receive support from any organization for the submitted work.

Competing interests: The authors have no financial and nonfinancial competing interests to declare that are relevant to the content of this article.

Acknowledgements The authors are grateful to the Department of Biotechnology, Jaypee Institute of Information Technology, Noida, Uttar Pradesh, India, for providing necessary facilities to execute this work.

References

Alves, A.A., Silva, W.E., Belian, M.F., Lins, L.S.G. and Galembeck, A. (2020). Bacterial cellulose membranes for environmental water remediation and industrial wastewater treatment. *International Journal of Environmental Science and Technology*, **17:** 3997–4008.

Bae, S. and Shoda, M. (2004). Bacterial Cellulose production by fed batch fermentation in molasses medium. *Biotechnology Progress,* **20:** 1366–1371.

Bandehali, S., Sanaeepur, H., Amooghin, A.E., Shirazian, S. and Ramakrishna, S. (2021). Biodegradable polymers for membrane separation. *Separation and Purification Technology*, **269:** 118731.

Cakar, F., Katı, A., Özer, I., Demirbağ, D.D., Şahin, F. and Aytekin, A.Ö. (2014). Newly developed medium and strategy for bacterial cellulose production. *Biochemical Engineering Journal*, **92:** 35–40.

Campano, C., Balea, A., Blanco, A. and Negro, C. (2016). Enhancement of the fermentation process and properties of bacterial cellulose: A review. *Cellulose*, **23(1):** 57–91.

Chen, S.Q., Lopez-Sanchez, P., Wang, D., Mikkelsen, D. and Gidley, M.J. (2018). Mechanical properties of bacterial cellulose synthesized by diverse strains of the genus *Komagataeibacter*. *Food Hydrocolloids*, **81:** 87–95.

Dutt, M.A., Hanif, M.A., Nadeem, F. and Bhatti, H.N. (2020). A review of advances in engineered composite materials popular for wastewater treatment. *Journal of Environmental Chemical Engineering*, **8(5):** 104073.

Galdino, C.J.S., Maia, A.D.M., Meira, H.M., Souza, T.C., Amorim, J.D.P., Almeida, F.C.G., Costa, A.F.S. and Sarubbo, L.A. (2020). Use of a bacterial cellulose filter for the removal of oil from wastewater. *Process Biochemistry*, **91:** 288–296.

Gorgieva, S. and Trček, J. (2019). Bacterial cellulose: Production, modification and perspectives in biomedical applications. *Nanomaterials*, **9(10):** 1352.

Hassan, E., Hassan, M., Abou-Zeid, R., Berglund, L. and Oksman, K. (2017). Use of bacterial cellulose and crosslinked cellulose nanofibers membranes for removal of oil from oil-in-water emulsions. *Polymers*, **9:** 388.

Jin, X., Xiang, Z., Liu, Q., Chen, Y. and Lu, F. (2017). Polyethyleneimine-bacterial cellulose bioadsorbent for effective removal of copper and lead ions from aqueous solution. *Bioresource Technology*, **244:** 844–849.

Jung, J.Y., Park, J.K. and Chang, H.N. (2005). Bacterial cellulose production by *Gluconacetobacter hansenii* in an agitated culture without living non-cellulose producing cells. *Enzyme and Microbial Technology*, **37(3):** 347–354.

Karim, Z., Claudpierre, S., Grahn, M., Oksman, K. and Mathew, A.P. (2016). Nanocellulose based functional membranes for water cleaning: Tailoring of mechanical properties, porosity and metal ion capture. *Journal of Membrane Science*, **514:** 418–428.

Kosseva, M.R., Zhong, S., Li, M., Zhang, J. and Tjutju, N.A. (2020). Biopolymers produced from food wastes: a case study on biosynthesis of bacterial cellulose from fruit juices. In *Food industry wastes*, Academic Press, p. 254.

Krystynowicz, A., Czaja, W., Wiktorowska-Jezierska, A., Gonçalves-Miśkiewicz, M., Turkiewicz, M. and Bielecki, S. (2002). Factors affecting the yield and properties of bacterial cellulose. *Journal of Industrial Microbiology and Biotechnology*, **29(4):** 89–195.

Li, C. and Cheng, S. (2019). Functional group surface modifications for enhancing the formation and performance of exoelectrogenic biofilms on the anode of a bioelectrochemical system. *Critical Reviews in Biotechnology,* **39(8):** 1015–1030.

Li, G., Wang, L., Deng, Y. and Wei, Q. (2022). Research progress of the biosynthetic strains and pathways of bacterial cellulose. *Journal of Industrial Microbiology and Biotechnology*, **49(1):** 71.

Lin, D., Liu, Z., Shen, R., Chen, S. and Yang, X. (2020). Bacterial cellulose in food industry: Current research and future prospects. *International Journal of Biological Macromolecules*, **158:** 1007–1019.

Liu, J., Chen, L., Cui, H., Zhang, J., Zhang, L. and Su, C.Y. (2014). Applications of metal–organic frameworks in heterogeneous supramolecular catalysis. *Chemical Society Reviews*, **43(16):** 6011–6061.

Liu, Z., Qin, D., Zhao, J., Feng, Q., Li, Z., Bai, H. and Sun, D.D. (2019). Efficient oil/water separation membrane derived from super-flexible and superhydrophilic core–shell organic/inorganic nanofibrous architectures. *Polymers*, **11(6):** 974.

Liu, F., Chen, C., Qian, J. (2021). Film-like bacterial cellulose/cyclodextrin oligomer composites with controllable structure for the removal of various persistent organic pollutants from water. *Journal of Hazardous Materials*, **405:** 124122.

Norfarhana, A.S., Ilyas, R.A. and Ngadi, N. (2022). A review of nanocellulose adsorptive membrane as multifunctional wastewater treatment. *Carbohydrate Polymers* **291:** 119563.

Nunes, S.P., Culfaz-Emecen, P.Z., Ramon, G.Z., Visser, T., Koops, G.H., Jin, W. and Ulbricht, M. (2020). Thinking the future of membranes: Perspectives for advanced and new membrane materials and manufacturing processes. *Journal of Membrane Science*, **598:** 117761.

Parveen, S., Bhat, I.U., Khanam, Z., Rak, A.E., Yusoff, H.M. and Akhter, M.S. (2022). Phytoremediation: In situ Alternative for pollutant removal from contaminated natural media: a brief review. *Biointerface Research in Applied Chemistry*, **12(4):** 4945–4960.

Pogorelova, N., Rogachev, E., Digel, I., Chernigova, S. and Nardin, D. (2020). Bacterial cellulose nanocomposites: Morphology and mechanical properties. *Materials*, **13(12):** 2849.

Revin, V., Liyaskina, E., Nazarkina, M., Bogatyreva, A. and Shchankin, M. (2018). Cost-effective production of bacterial cellulose using acidic food industry by-products. *Brazilian Journal of Microbiology*, **49:** 151–159.

Sai, H.; Jin, Z.; Wang, Y.; Fu, R.; Wang, Y. and Ma, L. (2020). Facile and green route to fabricate bacterial cellulose membrane with superwettability for oil–water separation. *Advanced Sustainable Systems*, **4:** 2000042.

Sjahro, N., Yunus, R., Abdullah, L.C., Rashid, S.A., Asis, A.J. and Akhlisah, Z.N. (2021). Recent advances in the application of cellulose derivatives for removal of contaminants from aquatic environments. *Cellulose*, **28(12):** 7521–7557.

Srivastava, S. and Mathur, G. (2021). Bacterial Cellulose: A Versatile Biopolymer. *Recent Trends in Biosciences and Biotechnology*, Delhi, pp. 76–96.

Srivastava, S. and Mathur, G. (2022) *Komagataeibacter saccharivorans* strain BC-G1: an alternative strain for production of bacterial cellulose. *Biologia* **77:** 3657–3668.

Srivastava, S. and Mathur, G. (2022). Bacterial Cellulose: A Multipurpose Biomaterial for Manmade World. *Current Applied Science and Technology* **23(3):** 55003.

Swingler, S., Gupta, A., Gibson, H., Kowalczuk, M., Heaselgrave, W. and Radecka, I. (2021). Recent advances and applications of bacterial cellulose in biomedicine. *Polymers*, **13(3):** 412.

Urbina, L., Guaresti, O., Requies, J., Gabilondo, N., Eceiza, A., Corcuera, M.A. and Retegi, A. (2018) Design of reusable novel membranes based on bacterial cellulose and chitosan for the filtration of copper in wastewaters. *Carbohydrate Polymers*, **193:** 362–372.

Urbina, L., Corcuera, M.Á., Gabilondo, N., Eceiza, A. and Retegi, A. (2021). A review of bacterial cellulose: sustainable production from agricultural waste and applications in various fields. *Cellulose*, **28(13):** 8229–8253.

Xu, Y., Liu, X., Jiang, Q., Yu, D., Xu, Y., Wang, B. and Xia, W. (2021). Development and properties of bacterial cellulose, curcumin, and chitosan composite biodegradable films for active packaging materials. *Carbohydrate Polymers*, **260:** 117778.

Zhijiang, C., Ping, X., Cong, Z., Tingting, Z., Jie, G. and Kongyin, Z. (2018). Preparation and characterization of a bi-layered nano-filtration membrane from a chitosan hydrogel and bacterial cellulose nanofiber for dye removal. *Cellulose*, **25:** 5123–5137.

Zhu, F., Zheng, Y.M., Zhang, B.G. and Dai, Y.R. (2021). A critical review on the electrospun nanofibrous membranes for the adsorption of heavy metals in water treatment. *Journal of Hazardous Materials*, **401:** 123608.

Zhuang, S. and Wang, J. (2019). Removal of U (VI) from aqueous solution using phosphate functionalized bacterial cellulose as efficient adsorbent. *Radiochimica Acta*, **107:** 459–467.

Chapter 10
Impact of Sintering on Specific Heat, Ion Conducting Channels and Electrical Properties of $Na_{0.5}Bi_{0.5}TiO_3$ SOFC Electrolyte

Pragati Singh, Raghvendra Pandey, and Prabhakar Singh

1 Introduction

SOFCs are electrochemical devices that generate electricity through fuel oxidation. The electrolyte material, a key component, must exhibit fast oxygen ion conduction. Ongoing research aims to increase the conductivity of oxygen ion conductors and reduce the operating temperature of fuel cells. The various electrolyte materials such as ZrO_2, $LaGaO_3$, CeO_2, LaMOX, apatite, hexagonal perovskite structured electrolytes possess the conductivity of the order ~ 9.74×10^{-4} S/cm, 0.03 S/cm, 2.5×10^{-4} S/cm, 0.06 S/cm, 0.01 S/cm and, 10^{-2} S/cm respectively, in the intermediate temperature range (IT-SOFC) i.e., 550–700°C (Fan et al., 2016). Sodium bismuth titanate ($Na_{0.5}Bi_{0.5}TiO_3$ or NBT), a well-known ferroelectric material (Khatua et al., 2017), simultaneously attracted the SOFC electrolyte research community (Li et al., 2013; Yang et al., 2016). It is a promising ionic conductor with a transference number >0.9. The structural debate surrounding NBT continues in the research community, with arguments for both rhombohedral (R3c) and monoclinic (Cc) structures at room temperature (Garg et al., 2008). However, in the temperature range suitable for intermediate-temperature solid oxide fuel cells (IT-SOFCs) (550–700 °C), NBT transitions to a non-ferroelectric cubic structure, which is preferable for electrolyte

P. Singh · P. Singh (✉)
Department of Physics, Indian Institute of Technology (Banaras Hindu University) Varanasi, Varanasi 221005, India
e-mail: psingh.app@iitbhu.ac.in

P. Singh
Department of Sustainable Energy Engineering, Indian Institute of Technology Kanpur, Kanpur, Uttar Pradesh 208016, India

R. Pandey
Department of Physics, A.R.S.D. College, University of Delhi, New Delhi 110021, India

R. K. Kotnala et al. (eds.), *Advanced Functional Materials for Sustainable Environments*, https://doi.org/10.1007/978-3-031-62620-3_10

applications. Interestingly, both ferroelectric and electrolyte applications require highly dense samples at the microstructure level, despite their contrasting material requirements. For SOFC electrolyte materials, the sample needs to exhibit good ionic conductivity, whereas ferroelectric materials have different considerations. This fact goes against the selection of ferroelectric materials.

Ion conduction in electrolyte materials for SOFCs occurs through the formation of ion-conducting channels (Adams, 2006). The existence of these ion channels has been confirmed both theoretically and experimentally (S Adams 2006). In $LaGaO_3$, there is the formation of saddle point around AAB critical triangle. In one of our papers on $LaAlO_3$, it was predicted that the angle of the critical isosceles triangles ABB/AAB plays a crucial role in ion conduction (Verma et al., 2020). An extensive literature survey reveals that in NBT ceramics, Bi-O is helpful in the migration of ions through AAB critical triangle termed as saddle point and ion conduction occurs through volatilization of Bi_2O_3 (Yang & Li, 2017). The theoretical density functional theory simulations also predicted the critical triangle for NBT (Zhang et al., 2011). There are possibilities of three critical triangles whose areas are proportional to the activation energy, i.e. Na-Na-Ti 1.0–1.3 eV, Bi-Bi-Ti 0.22 eV and Na-Bi-Ti 0.6–0.85 eV (Yang et al., 2018). Moreover, it is also reported that the grain growth affected through processing, oxygen vacancy concentration and activation energy for ion migration are correlated with each other (Khatua et al., 2017). There are several reports on the synthesis of NBT ceramics through various methods such as hydrothermal method, different morphologies etc. (Aksel, & Jones, 2010; Pookmanee et al., 2004; Ma et al., 2013; Zhou et al., 2016, Lu et al., 2013; Ramana et al., 2011). The literature concludes that grain size plays a crucial role in altering properties such as domain size, octahedral tilting, conductivity etc. Hence, it opens up the research for the grain/domain structure tuning for improving the electrical properties.

Present study, attempts the formation of ion-conducting channels through critical triangle/parallelogram and its variation with sintering temperature. The correlation between the formation of the critical triangle with conductivity and specific heat is also established. For the exploration of ionic contribution, the conductivity obtained from Nyquist plots were fitted with the space charge model. Further, the change in morphology has been studied with the variation of sintering temperature.

2 Preparation and Characterization of Samples

2.1 Sample Preparation

The ceramic samples of $Na_{0.5}Bi_{0.5}TiO_3$ were synthesized through the traditional solid-state reaction method. Precisely measured amounts of Na_2CO_3, Bi_2O_3 and TiO_2 powders were thoroughly mixed to achieve stoichiometric proportions. Prior to calcination, the sample was preheated at 300 °C for 2 h. The mixture was then calcined at 820 °C for 2 h. After calcination, the mixture was grinded and mixed with 2% polyvinyl alcohol (PVA) as a binder. The samples were then sintered in air at various temperatures ranging from 1000 °C to 1150 °C in 50 °C increments. Additionally, one sample was sintered at 1180 °C for 2 h at a heating rate of 5 °C/min. The samples sintered at different temperatures were labeled T_1, T_2, T_3, T_4 and T_5, corresponding to 1000 °C, 1050 °C, 1100 °C, 1150 °C, and 1180 °C, respectively.

2.2 *Characterization Techniques*

The phase formation of the samples was analyzed using a Rigaku Miniflex desktop X-ray Diffractometer (XRD) equipped with CuKα radiation ($\lambda = 1.5405$ Å). The XRD scans were conducted in the 2θ range of approximately 20–70° with a step size of 0.01°. Rietveld refinement method implemented in the FULLPROF suite software package was employed to refine the diffraction patterns. Thermo gravimetric and differential scanning calorimetric analysis were performed on the sintered powders using a Metler DSC/TGA system under a continuous nitrogen purging atmosphere. Raman spectra of the sintered powders were obtained using a Thermo SCIENTIFIC DXR2 Micro-Raman spectrometer with an excitation wavelength of 532 nm. The sintered samples were analyzed for microstructures using a scanning electron microscope (EVO-Scanning Electron Microscope MA15/18) and the density of the pellets were determined utilizing an Archimedes density kit (DENVER SI-234). Additionally, the sintered pellets underwent a process of platinum paste coating and were cured at 700 °C for 30 min. Conductivity measurements were conducted using a Wayne Kerr LCR meter (6500 P Series) at various temperatures from room temperature to 700 °C with a step size of 5 °C.

3 Results

3.1 *Thermo Gravimetric Analysis*

Figure 1 shows the variation of mass loss and dm/dT for the sintered samples (T_1, T_2, T_3, T_4 and T_5). It is observed that there is continuous mass loss along with the multiple kinks showing Type-E isotherm for the samples sintered up-to 1150 °C. The multiple kinks in mass loss curves indicate multistage decomposition reaction. However, for T_5 sample, mass gain appears with the temperature and is nearly 1.7% (i.e. less than 4%). This mass gain indicates non-stoichiometry, impure phase and oxidation of the T_5 sample. To verify the formation temperatures, the derivative of mass, dm/dT has been plotted with temperature. It is observed that dm/dT decreases after 600 °C for T_1, T_2 and T_4 samples. While there is no change observed in dm/dT curve for T_3 sample. Whereas, in T_5 sample, there is a hump in dm/dT curve below

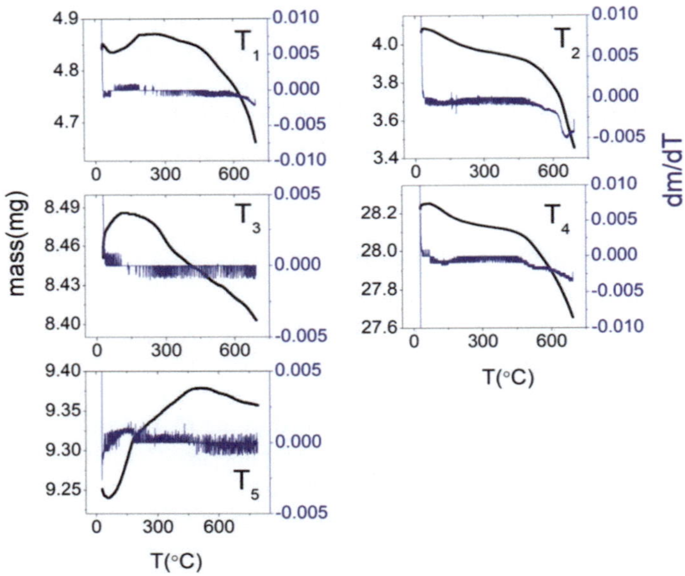

Fig. 1 Variation of mass loss and dm/dT for the sintered powders (T_1, T_2, T_3, T_4 and T_5). A sudden increase in mass is observed for T_5 sample

200 °C and no change is observed after that is correlated with the mass gain observed in the TGA spectrum.

3.2 Differential Scanning Calorimetric Analysis

For the thermodynamical stability of the samples, free energy and entropy were calculated. Figure 2 (a, b and c) shows variation of specific heat (C_v), Gibb's free energy (ΔG) and equilibrium constant (K) with temperature. For T_5 sample, values of $C_v<0$, $\Delta G>0$ (positive) and $K<1$ (Taylor & Floyd, 1952). Thus, the reaction is not favored in T_5 case. Moreover, the value of ΔG is negative and $K>1$ in all other samples. Hence, the samples sintered at all sintering temperatures except 1180 °C are thermodynamically stable to work at high temperatures. The higher negative value of ΔG for the sample sintered at 1050 °C indicates that forward reaction for the formation of product is more favored. Figure 2(d) shows the variation of C_v with the sintering temperature at RT and 700 °C. It is observed in Fig. 2(d) that the sample sintered at 1150 °C has lower value and the sample sintered at 1050 °C has higher value of specific heat. This shows that T_2 possesses higher electronic and phononic contributions in comparison to T_4 sample.

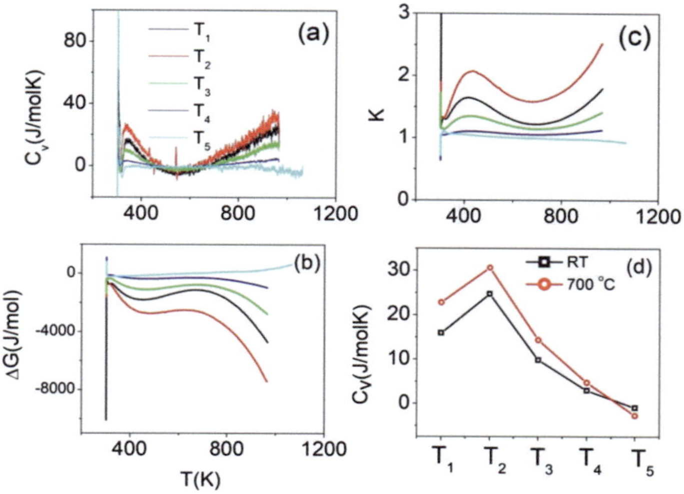

Fig. 2 Variation of: (a) specific heat, (b) Gibb's free energy (c) equilibrium constant K with temperature and (d) variation of specific heat with the sintering temperature

3.3 X-ray Diffraction and Microstructure Analysis

Figure 3 shows X-ray diffractograms of the prepared samples. The samples are observed to be in single phase up to 1150 °C. At 1180 °C, a secondary phase has grown and that has become major phase. On matching with standard JCPDS files, it is matching well with the rutile phase of TiO_2 (PCPDF No. 898304). It is observed that there is shifting of XRD peaks at $2\theta \approx 32°$, but no definite trend is observed. First, the peak shifts towards lower angle at 1050 °C, and then towards right at the higher angle for 1100 °C. Again left shifting is observed for 1150 °C i.e. 'S' shaped variation with sintering temperature.

To observe this observation very critically, we have done Rietveld refinement using Full Prof Suite Package using R3c symmetry (Singh et al., 2020a, b, c). It is found that the R-factors for all the samples refinement lie within the appreciable range (Pragati Singh et al., 2020a, b, c). Even after applying the zero correction, no definite trend was obtained in the shifting of the peaks and thus lattice parameters obtained after refinement are in accordance with the left shifting and right shifting of the XRD peaks (i.e. again 'S' shaped curve) (Fig. 3(b)). In addition to it, the Rietveld refinement of the sample sintered at 1180 °C (T_5) could not be achieved with single symmetry as the major phase observed is of rutile TiO_2 on matching with the JCPDS file No. 898304.

For T_5 sample (Fig. 4), Rietveld refinement has been achieved with P42/mnm + R3c symmetry and the goodness of factor lies within the appreciable range. As observed earlier, for T_5, C_v<0 and ΔG>0, proving that this sample is non-stoichiometric and phase impure. Hence, T_5 sample is discarded from further

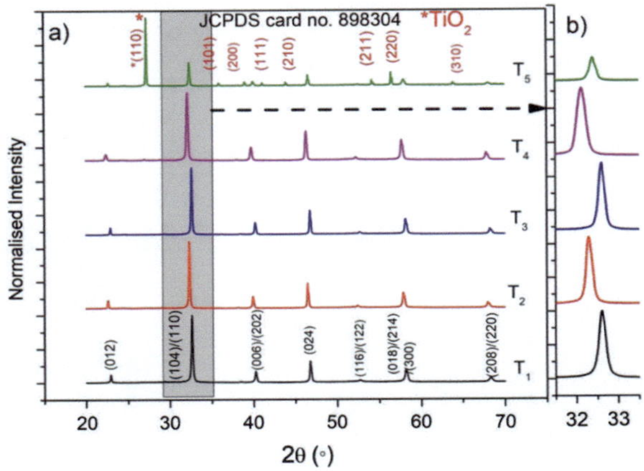

Fig. 3 (a) X-ray diffractograms of the samples sintered at different sintering temperatures, 1000 °C, 105 °C, 1100 °C, 1150 °C and 1180 °C labelled as T_1, T_2, T_3, T_4 and T_5 and (b) Enlarged peak (104)/(110) at $2\theta \approx 32°$ showing the left and right shifting with the variation of sintering temperature

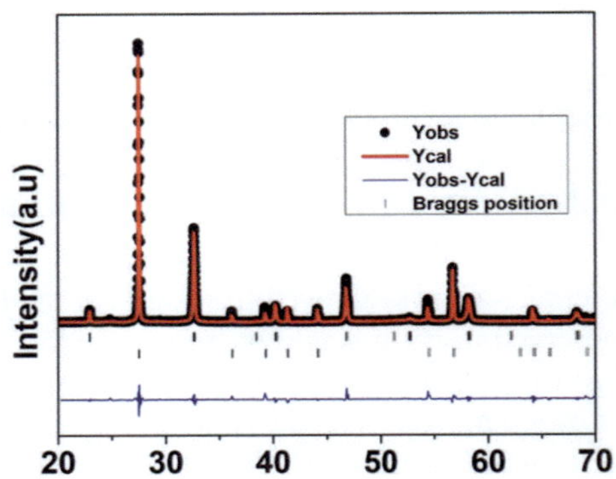

Fig. 4 Rietveld refinements of the samples sintered at 1180 °C

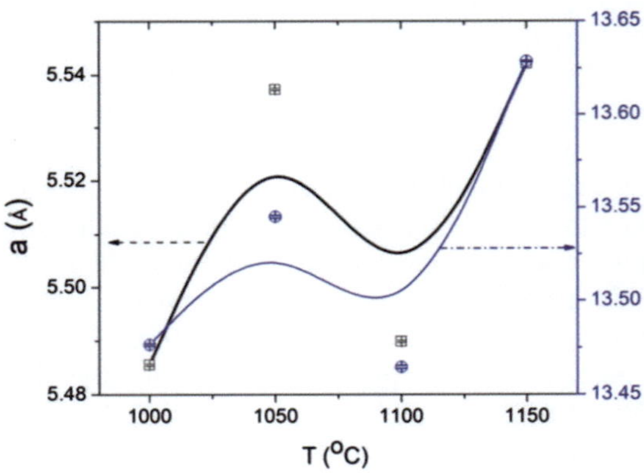

Fig. 5 Lattice parameters obtained after Rietveld refinement of the studied samples

measurements. Lattice parameters obtained after Rietveld refinement of all the investigated samples are shown in Fig. 5.

3.4 Raman Studies

In order to elucidate our observation of rhombohedral structure of NBT and phononic contribution observed in specific heat, Raman measurement was done. Figure 6 shows the Raman spectra of the samples sintered at 1000 °C (T_1), 1050 °C (T_2) and 1150 °C (T_3). According to Bilbao crystallographic server, the number of peaks for R3c and Cc symmetry is 13 and 9, respectively. The Raman spectra were convoluted using 10 peaks i.e. between 9 and 13 (Fig. 6). The increase in the number of peaks from Cc symmetry clearly violates the existence of Cc symmetry. The fitting of the Raman spectrum with less number of peaks supports our observation of R3c symmetry.

Figure 6 depicts four regimes, regime I comprise of three modes E (TO1), A1 (TO1) and A1 (LO2) at ~ 109 cm-1, 134 cm-1 and 178 cm-1, respectively belonging to Bi-O and Na-O bonds. Regime II consists of E (TO2) and A1 (TO4) at ~ 246 cm-1 and 276 cm-1 assigned to TiO_6 octahedra. Regime III and IV are due to oxygen vibrations of NBT assigned as E(TO8), E(LO8), A1(LO8) and A1(LO9) at ~ 549 cm^{-1}, 591 cm^{-1}, 737 cm^{-1} and 830 cm^{-1} (Singh et al., 2020a, b, c; Ruth & Sundarakannan, 2016; Singh et al., 2022, Singh et al., 2021; Jackson et al., 2014). To analyze the

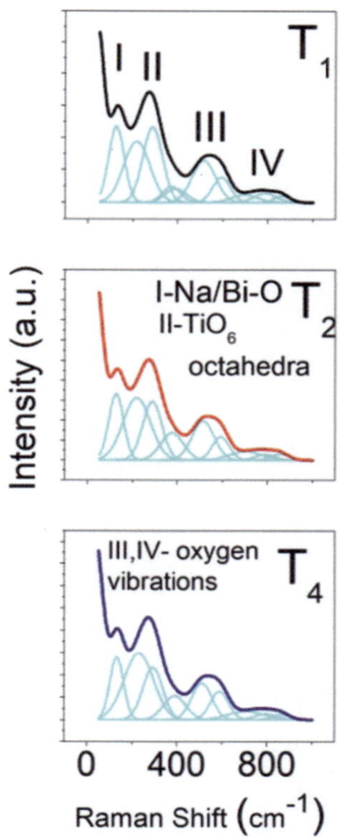

Fig. 6 Deconvoluted Raman spectra of the samples sintered at 1000 °C (T_1), 1050 °C (T_2) and 1150 °C (T_3). Data has been deconvoluted with 10 peaks showing the existence of R3c symmetry. Black, red and blue lines show the experimental data and cyan coloured peaks show the deconvoluted peak positions

phononic contributions, FWHM, intensity and wavenumber are studied with the variation of sintering temperature.

With the increase in sintering temperature, first, it is observed that there is a decrease in wave number shift of $A1(LO_9)$ up to 1050 °C followed with the increase in wave-number shift of $A1(LO_9)$ at 1150 °C (regime IV) (Fig. 7(a)) (i.e. a 'U' shaped curve). Simultaneously, with the increase in sintering temperature, there is an increase in wave-number shift of $A1(TO_4)$ (regime II). While the intensity and FWHM of $A1(LO_9)$ followed the reverse trend in comparison to the trend of the corresponding wave number shift. Whereas, the intensity and FWHM follow the inverse trend with each other, i.e. maxima in intensity and minima in FWHM or vice versa (Fig. 7(b)). Hence, the interplay between the oxygen vibrations of NBT and TiO_6 octahedron is appearing which may affect the conductivity of the specimens with the sintering temperature.

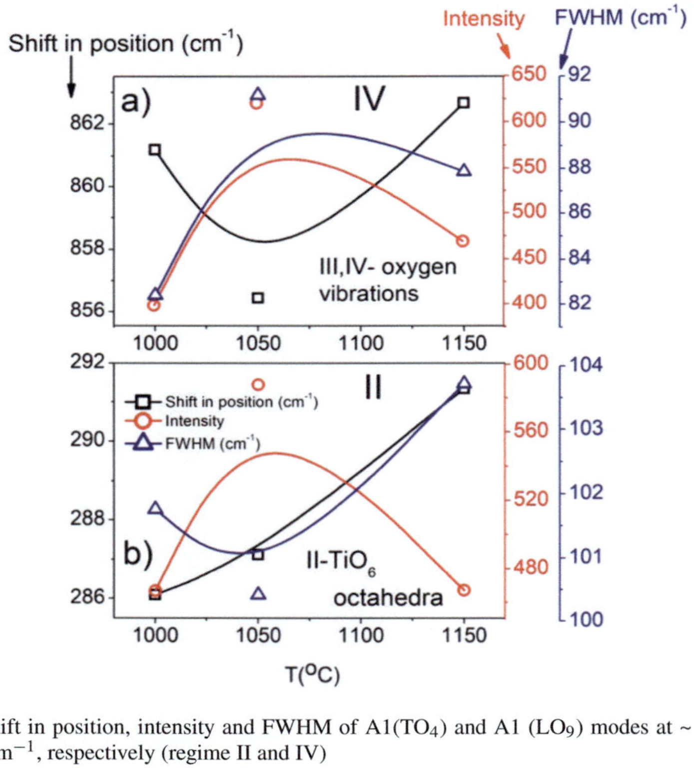

Fig. 7 Shift in position, intensity and FWHM of $A1(TO_4)$ and A1 (LO_9) modes at ~ 276 cm^{-1} and 830 cm^{-1}, respectively (regime II and IV)

3.5 Microstructural Studies

The correlation between sintering temperature and grain growth is widely recognized. The SEM micrographs (Fig. 8) of the samples sintered at different sintering temperatures 1000 °C, 1050 °C, 1100 °C and 1150 °C. It is discerned from the micrographs that for T_1, spherical grains are observed. As the grain size increases, firstly there is a change in morphology from spherical for T_1 to the mixture of spheres and cubes for T_2 and finally to cube-shaped grains for T_3 and T_4. Here, the sintering has occurred in three stages from phase formation at 1000 °C to recrystallization at 1050 °C and grain growth at 1150 °C. The change in morphology indicates different internal energy of the compounds as denser grains represent lower specific heat in the sample sintered at 1150 °C. This change in morphology should also affect the octahedral tilting and conductivity of the studied specimens.

Before analyzing conductivity data, we will study the oxygen content from SEM micrographs as Na and Bi both are volatile. Therefore, the elemental composition with the variation of sintering temperature was investigated from EDX measurements. Figure 9(a) depicts the variation of difference of stoichiometric and experimental Na and Bi content and (b) variation of difference of stoichiometric and experimental Ti and O content. It is observed that Bi and Ti possess minima at 1050 °C and there after increases with the increase in sintering temperature. Moreover, it is also observed in Fig. 9(a), that there is a maximum at 1050 °C, there after decreases with the increase in sintering temperature (i.e. 'U' shaped curve) for Na and O content. This 'U' shaped curve with the sintering temperature is similar to the trend observed in specific heat observed in Fig. 2(d). Moreover, it is also observed that the

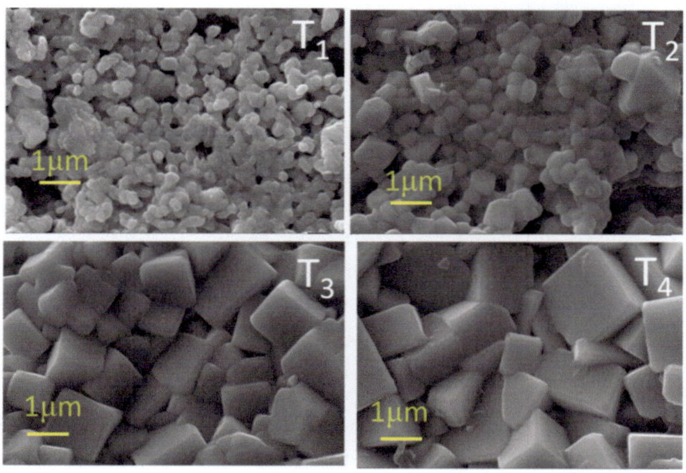

Fig. 8 SEM images depict the samples sintered at varying temperatures: 1000 °C, 1050 °C, 1100 °C and 1150 °C

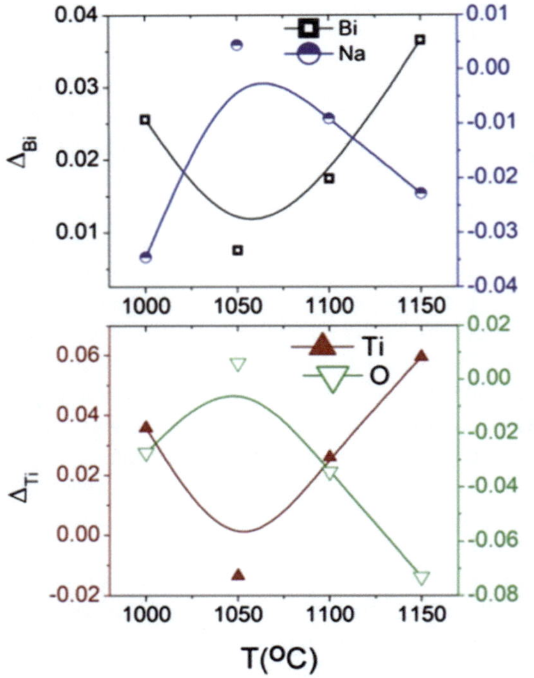

Fig. 9 Variation of difference of stoichiometric and experimental content: Na and Bi and Ti and O

sample sintered at 1150 °C, the sample is highly Na and O deficient in comparison to Bi deficiency.

3.6 Electrical Characterization

Figure 10(a) illustrates the changes in Impedance Nyquist plots recorded at 700 °C for samples sintered at various temperatures. Notably, a larger depressed semi-circle is succeeded by a nearly flat, depressed semi-circle. The positioning of the semi-circle's center below the X-axis signifies the existence of Non-Debye type behaviour. Figure 9 (inset) shows the variation of bulk conductivity (σ_{bulk}) with sintering temperature. It is observed that there is a minima for T2 followed with the increase in conductivity with the sintering temperature. Figure 10(b) shows the variation of R_{bulk} and E_a. It is observed that there is a maxima in the value of resistance at 1050 °C followed with 'S' shaped curve. Whereas, there is a maxima in activation energy plot followed with an 'S' shaped curve with the increase in sintering temperature.

As seen from results section that lattice parameters, conductivity and activation energy follow an 'S' shaped curve with the increase in sintering temperature. Whereas, the specific heat and conductivity are inversely proportional to each other, i.e. for T_2, the specific heat possesses maxima and conductivity possesses minima.

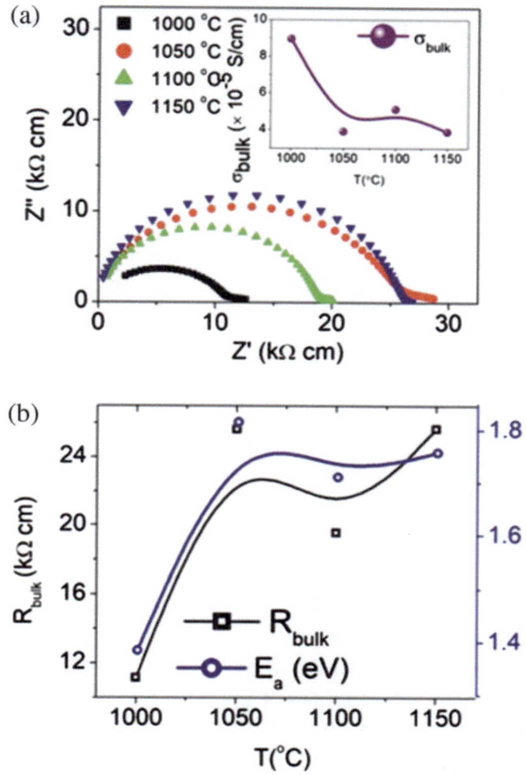

Fig. 10 Impedance Nyquist plots with the change of sintering temperature Variation of: (a) (inset) σ_{bulk} with sintering temperature and (b) R_{bulk} and E_a with sintering temperature

The specific heat and conductivity are lower for T_4 sample. Also, from Raman spectra and O content, it is observed that at 1050 °C, either a maxima or minima are observed in FWHM and intensity (i.e. a 'U' shaped curve).

4 Discussions

We have two peculiar observations which need to be clarified, i.e. observation of 'S' shaped curve and observation of 'U' shaped curve in specific heat and conductivity. In NBT sample, rhombohedral phase possesses a-a-a- local in-phase octahedral tilting which affects the oxygen ion motion in these samples (Khatua et al., 2017). Figure 11 depicts the 'U' shaped variation of tetragonal strain and 'S' shaped curve for octahedral tilting with sintering temperature. The tetragonal strain is basically due to the stretching of TiO_6 octahedra. The octahedral tilting calculated using relation (1) is due to the bending of TiO_6 octahedra:

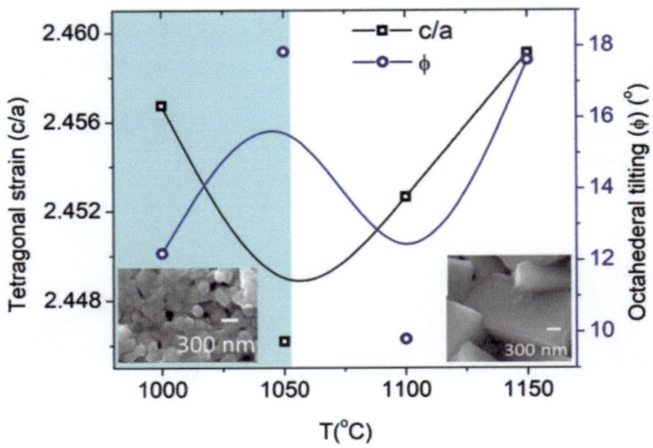

Fig. 11 Variation of tetragonal strain (c/a) and octahedral tilting (θ) with sintering temperature

$$\cos^2 \theta = \frac{2 - 2\cos\theta}{\cos\theta + 5} \quad (1)$$

where θ is the bond angle Ti-O-Ti and φ the octahedral tilting. Thus, competitive bending and stretching of TiO_6 octahedra suggests 'S' and 'U' shaped curve. At 1000 °C and 1050 °C, spherical grains are observed and there is competitive bending and stretching while in the samples sintered at 1100 °C and 1150 °C, the bending and stretching cooperate to form cubic grains. The specific heat comprises electronic and phononic contributions. The phononic contributions from Raman spectra reveal the 'U' shaped curve in the TiO_6 octahedral vibrations and 'U' shaped curve from oxygen vibrations of NBT. Simultaneously, our samples indicate 'U' shaped curve in Na and O deficiency.

Hence, the sample sintered at 1150 °C is highly oxygen deficient with higher sized cubic grains and octahedral tilting and lower specific heat suggesting more phononic contribution in comparison to the electronic. The change in morphology along with the increase in grain size with the sintering temperature suggests the change in octahedral tilting (see Fig. 12) which alters the oxygen ion movement. The electronic contribution of the specimen has ruled out as the activation energy of the samples is greater than 1.5 eV.

The activation energy of ~ 1.75 eV again indicates the formation of critical triangle using Na, Bi and Ti. For electrolyte materials, there is a need to study the formation of ion-conducting channels in the material (Adams, 2006; Singh et al., 2020a, b, c). As discussed earlier, the literature suggests the movement of oxygen ion through Bi-O bond in NBT. But, the bond dissociation energies of the possible bonds are in the order (Na-Na (73.0813 kJ/mol) < Bi-Bi (200.4 kJ/mol) < Na-O (256.1 kJ/mol) < Bi-O (337.2 kJ/mol) < O-O (498.36 kJ/mol)) (Bhosale et al., 2017); which suggest the Na-volatility is more critical than Bi and O volatility. The EDX results confirm the fact as our samples are more Na and O deficient.

Hence, to illustrate the path for the oxygen ion in the lattice, structure obtained from diamond software was studied (Fig. 13). Figure 13 illustrates the emergence

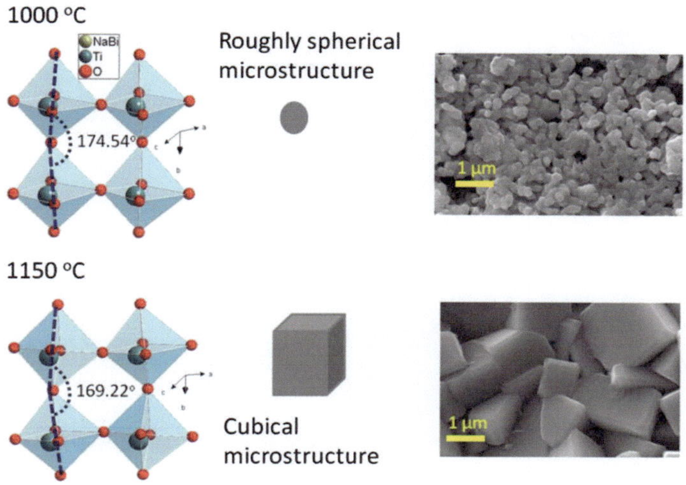

Fig. 12 The change in octahedral tilting and morphology from spherical to cubical grains with the sintering temperature

of ion channels through the application of Valence Shell Electron Repulsion Theory (VSEPR theory) following the Rietveld refinement of the XRD data for the investigated samples. It is observed that there are formation of three triangles or one trapezium along with the combinations of triangles 1, 2, 3 and 4. After studying in detail, scalene triangles 1 and 2 combine to form a parallelogram and thus distance and angles parameters are plotted for triangles (represented as 1 and 2) (Fig. 13). Here, angles α' and 2β' as mentioned in Fig. 13 of the critical triangle form a 'U' shaped curve with the increase in sintering temperature and follow well with the specific heat of the studied systems. The length of the critical triangle, s'_{Ti-Ti} and $s_{Na/Bi-Ti}$ show two trends i.e. 'S' and 'U' shaped curves and are greater than the diameter of O^{2-} ion (2.8 Å). This is clearly indicating the possibility of two triangles or one parallelogram where diagonal of the parallelogram i.e. Ti–Ti distance is affecting the lattice parameters and activation energy for the movement of ions through Ti–Ti distance (Fig. 14).

Hence, 'S' shaped curve is formed and hence affecting conductivity and activation energy. In Fig. 13, there is the formation of multiple long ion channels along critical triangle, z- and y-axis, where the electron density is minimum. These long ion channels not only felicitate the charge carriers to move but due to multiple directions of channeling, conductivity decreases. To assess the ionic contribution in the specimens, conductivity data of the studied compositions has been modeled using the Space Charge Ionic Model (Tschope, 2001). Figure 15 illustrates the fitting of the space charge ionic model and presents the corresponding fitting parameters for the investigated specimens. It suggests that the sample sintered at 1150 °C is highly ionic in comparison to other samples.

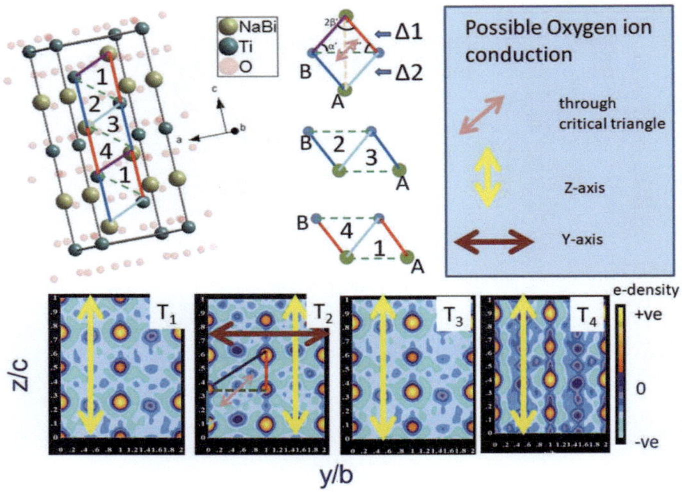

Fig. 13 The path of oxygen ion in the lattice using structure obtained from diamond

Fig. 14 Variation of: $s_{Na/Bi\text{-}Ti}$ and α' with sintering temperature and s'_{Ti-Ti} distance and 2β' with sintering temperature

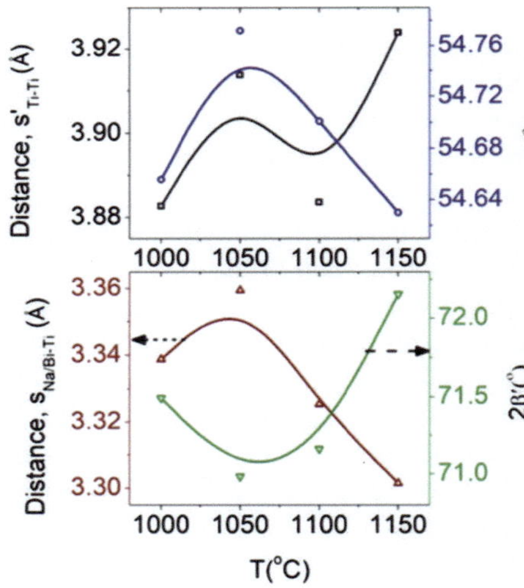

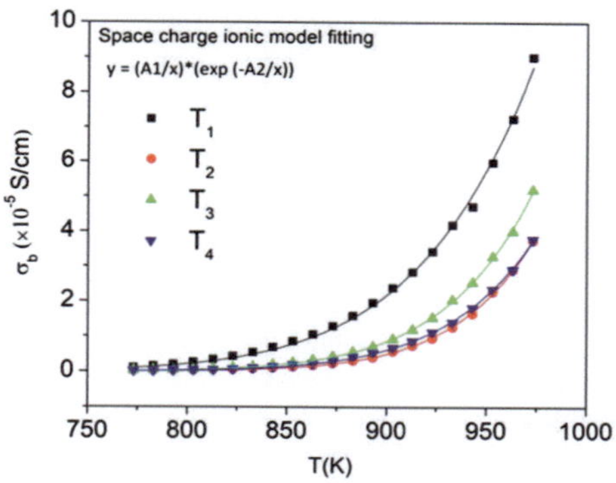

Fig. 15 Space charge ionic models fitting for the examined specimens

5 Conclusion

With the increase in sintering temperature from 1000 °C to 1150 °C, change in morphology from 'spherical' to 'cubic' grains is observed though the samples are crystallized in rhombohederal structure. Further, it seems that there is a structural and micro-structural correlation, as in the cubical grains, grain growth and tetragonal strain is proportional. The octahedral tilting is not following a trend with microstructure but it is in correlation with the lattice parameters perhaps due to structural anisotropy. This structural anisotropy increased the phononic contributions in the sample sintered at 1150 °C and forms the multiple oxygen ion conduction channels along critical triangle, z- and y-axis. These channels are also observed to vary with sintering temperature through variation of Na and O deficiency in comparison to Bi deficiency.

References

Adams S. (2006). From bond valence maps to energy landscapes for mobile ions in ion-conducting solids. *Solid State Ionics*, **177:** 1625–1630.

Aksel, E. and Jones, J.L. (2010). Phase Formation of Sodium Bismuth Titanate Perovskite During Solid-State Processing. *Journal of the American Ceramic Society,* **93:** 3012–3016.

Bhosale, D.R., Yusuf, S.M., Kumar, A., Mukadam, M.D. and Patil, S.I. (2017). High oxide ion conductivity below 500 °C in the garnets LaxY 3- x Fe 5 O12 + δ. *Physical Review* Materials., **1:** 2475–9953.

Garg, R., Senyshyn, A., Boysen, H. and Ranjan, R. (2008). Structure and phase transition of $Na_{0.5}La_{0.5}TiO_3$. *Journal of Physics: Condensed Matter,* **20:** 505215.

Jackson, D.K., Toulouse, J. and Luo, H. (2014). Phonon modes and central peaks in an A -site relaxor: A low-frequency Raman study of sodium bismuth titanate. *Physical Review B,* **90:** 1098–0121.

Khatua, D.K., Mehrotra, T., Mishra, A., Majumdar, B., Senyshyn, A. and Ranjan, R., (2017). Anomalous influence of grain size on the global structure, ferroelectric and piezoelectric response of $Na_{0.5}$ $Bi_{0.5}$ TiO_3. *Acta Materialia*, **134:** 177–187.

Li, M., Pietrowski, M., Souza, R., Zhang, H., Reaney, I.M., Cook, S.N., Kilner, J. A. and Sinclair, D.C. (2013). A family of oxide ion conductors based on the ferroelectric perovskite $Na_{0.5}$ $Bi_{0.5}$ TiO_3. *Nature Material*, **13:** 31–35.

Lu, R., Yuan, J., Shi, H., Li, B. and Wang, W. (2013). Morphology-controlled synthesis and growth mechanism of lead-free bismuth sodium titanate nanostructures via the hydrothermal routenanostructures via the hydrothermal route. *CrystEngComm.,* **15:** 3984–3991.

Ma, X., Zhang, W. and Xue, L., (2013). Hydrothermal synthesis of bismuth sodium titanate particles with different morphologies. *Journal of Materials Science,* **48:** 6878–6884.

Pookmaneea, P., Rujijanagulb, G., Ananta, S., Heimann, R.B. and Phanichphanta, S. (2004). Effect of sintering temperature on microstructure of hydrothermally prepared bismuth sodium titanate ceramics. *Journal of the European Ceramic Society,* **24:** 517–520.

Ramana, M.V., Kiran, S.R., Reddy, N.R., Kumar, K.V.S., Murthy, V.R.K. and Murty, B.S. (2011). Synthesis of lead free sodium bismuth titanate (NBT) ceramic by conventional and microwave sintering methods. *Journal of Advanced Dielectrics*, **1:** 71–77.

Ruth, D.E.J. and Sundarakannan, B. (2016). Structural and Raman spectroscopic studies of poled lead-free piezoelectric sodium bismuth titanate ceramics. *Ceramic International*, **42:** 4775–4778.

Singh, P., Jha, P.K., Jha, P.A. and Singh, P. (2020a). Influence of sintering temperature on ion dynamics of $Na_{0.5}Bi_{0.5}TiO_{3-\delta}$: Suitability as an electrolyte material for SOFC. *International Journal of Hydrogen Energy*, **45:** 17006–17016.

Singh, P., Jha, P.K., Sinha, A.S.K., Jha, P.A., and Singh, P. (2020b). Ion dynamics of non-stoichiometric $Na_{0.5+x}Bi_{0.5-x}TiO_{3-\delta}$: A degradation study. *Solid State Ionics,* **345:** 115158.

Singh, P., Pandey, R., Miruszewski, T., Dzierzgowski, K., Mielewczyk-Gryn, A. and Singh, P. (2020c). Signature of oxide ion conduction in alkaline earth metal doped Y_3GaO_6. *ACS Omega.* **5:** 30395–30404.

Singh, P., Pandey, R. and Singh, P. (2021). Polyol-mediated synthesis of Bi-deficient Mg^{2+}-doped sodium bismuth titanate and study of oxide ion migration behaviour with functional properties. Journal of Alloys and Compounds, **860:** 158492.

Singh, P., Pandey, R. and Singh, P. (2022). Insight into structural and electrical properties of potassium and lithium doped non-stoichiometric sodium bismuth titanate ($Na_{0.54}Bi_{0.46}TiO_{3-\delta}$). *Springer Proceedings in Materials,* **14:** 171–185.

Taylor, A. and Floyd, R.W. (1952). The constitution of nickel rich alloys of the nickel-chromium-aluminum system. *Journal* of the *Institute* of *Metals.* **81:** 456.

Tschope, A. (2001). Grain size-dependent electrical conductivity of polycrystalline cerium oxide II: Space charge model. *Solid State Ionics*, **139:** 267–280.

Verma, O.N., Jha, P.A., Singh, P., Jha, P.K. and Singh, P. (2020). Influence of iso-valent 'Sm' double substitution on the ionic conductivity of $La_{0.9}Sr_{0.1}Al_{0.9}Mg_{0.5}O_{3-\delta}$ ceramic system. *Materials Chemistry and Physics,* **241:** 122345.

Yang, F. and Li, M. (2017). Optimisation of oxide-ion conductivity in acceptor-doped $Na_{0.5}$ $Bi_{0.5}$ TiO_3 perovskite: approaching the limit. *J. Mater. Chem. A*, **5:** 21658–21662.

Yang, F., Li, M., Li, L., Wu, P., Pradal-Vel´azqueza, E. and Sinclair, D.C. (2018). Defect chemistry and electrical properties of sodium bismuth titanate perovskite. *Journal of Materials Chemistry A*, **6:** 5243.

Yang, F., Zhang, H., Li, L., Reaney, I.M. and Sinclair, D.C. (2016). High Ionic Conductivity with Low Degradation in A - Site Strontium- Doped Nonstoichiometric Sodium Bismuth Titanate Perovskite. *Chemistry of Materials,* **28:** 5269–5273.

Zhang, Y., Hu, J., Gao, F., Liu, H. and Qin, H. (2011). Ab initio calculation for vacancy-induced magnetism in ferroelectric $Na_{0.5}Bi_{0.5}TiO_3$. *Computational and Theoretical Chemistry*, **967:** 284–288.

Zhou, X., Jiang, C., Chen, C., Luo, H., Zhou, K. and Zhang, D. (2016). Morphology control and piezoelectric response of $Na_{0.5}Bi_{0.5}TiO_3$ synthesized via a hydrothermal method. *CrystEng-Comm,* **18:** 1302–1310.

Part III
Energy Generation from Water

Chapter 11
Droplet Based Triboelectric Nanogenerator (DB-TENG) by Conjunction of Photovoltaic and Triboelectric Effect

Gunjan Yadav, Kajal Jindal, Bilasini Naorem, and Monika Tomar

1 Introduction

In nano energy, the harvesting of energy from renewable energy sources has lately acquired prominence. Nanogenerators (NGs) are one of the most studied nanotechnology applications involving nanomaterials. Mostly, NGs are utilised in self-powered electrical gadgets and are categorised as piezoelectric nanogenerators (PENGs), triboelectric nanogenerators (TENGs) and thermoelectric generators (TEGs) based on key physics ideas (Wei et al., 2021; Kumar et al., 2021). Recent research has focused on developing TENG for energy harvesting. Previously, solid-state TENGs were utilised, but their endurance is poor due to material fractionation and restricted performance in situations with high humidity. These concerns do not affect liquid–solid (LS) TENGs. In comparison to solid-state TENGs, the stability and sensitivity of these generators are greater. As potential generation occurs only on the material surface, the LS-TENG operates with falling droplets. It is the ideal method for developing droplet-based (DB) TENG (He et al., 2015; Zhu et al., 2015; Cheng et al., 2019; Ouyang et al., 2022). Thus, a semiconducting layer is required for DB-TENG to generate high output voltage and power. Gallium nitride is the optimal material for this application because of its resistance at droplet interface is

G. Yadav
Department of Physics and Astrophysics, University of Delhi, New Delhi 110007, India

G. Yadav · B. Naorem · M. Tomar (✉)
Department of Physics, Miranda House, University of Delhi, New Delhi 110007, India
e-mail: monika.tomar@mirandahouse.ac.in

K. Jindal
Dpartment of Physics, Kirori Mal College, University of Delhi, New Delhi 110007, India

M. Tomar
Institute of Eminence Fellow, University of Delhi, New Delhi 110007, India

R. K. Kotnala et al. (eds.), *Advanced Functional Materials for Sustainable Environments*, https://doi.org/10.1007/978-3-031-62620-3_11

low compared to other semiconductors. GaN is robust, thin and stable under ambient conditions to preclude the formation of a channel for charge flow caused by an ionic droplet (Xia et al., 2019) triboelectric nanogenerators (TENGs). Therefore, the GaN semiconductor is advantageous for the efficiency of voltage generation with the device DB-TENG. Recently, a number of studies related to DB-TENG have been conducted which focused on the relationship between a droplet and thin film layer, and further investigations into energy harvesting applications based on substrate have also taken place. Consequently, it is important to comprehend the mobility of the droplet associated with TENG (Li et al., 2018; Chen et al., 2019; Zeng et al., 2022) an emerging high entropy energy harvesting technology, has gradually become the center of the internet-of-things (IoT). Furthermore, higher efficiency can be achieved by gathering energy from two sources. Ren et al. (2021) fabricate a p–n junction, which caused an increased voltage to 4 times and current by 13 times. Wei et al. (2021) demonstrated DB-TENG that could generate electricity under any atmospheric conditions, obtaining 52 V for a 20° inclined plane and 30 nA for rainwater and 0.8 μA for seawater. Table 1 presents a comparison of various materials based TENG devices in terms of their electricity generation capabilities.

It is inferred from Table 1 that several groups have utilized TENG, yet only a few studies have been conducted to evaluate TENG employing GaN thin films. It is evident from the table that, due to its resistance to extreme conditions such as high humidity, heat, and concentrated acids, salts and alkalis, the GaN-based TENG devices produce the greatest voltage of any other material. Zheng et al. (2021) fabricate the LS TENG by combining the photovoltaic and triboelectric effects. Their study shows that the current is produced when DI water droplets are slide across a

Table 1 Literature survey on various materials based TENG device

S. No	*Material*	*Results (V_{out} and I_{out})*	*Reference*
1	ZnO micro-balloon arrays	57 V	Deng et al. (2017)
2	MoS_2-PVDF	50 V, 30 nA	Sahatiya et al. (2018)
3	PVDF-HFP/rGO	300 V, 7 μA	Qin et al. (2019)
4	Ga-doped ZnO	89.15 V	Chen et al. (2019)
5	MoS_2	10.20 V	Karmakar et al. (2020)
6	ZnO-polystyrene composite	7 V, 99 nA	Gupta et al. (2020)
7	p-GaN NW	20 V, 23 μA	Waseem et al. (2022)
8	Cu/GaN	14.2 V	Xiao et al. (2022)
9	GaN	25 V	Chen et al. (2022)
10	GaN/Si	130 V, 21.5 μA	Wang et al. (2022)
11	GaN-Bi_2Te_3	40 V, 340 μA	Zhang et al. (2022)
12	NiO-Ti nanocomposites	62 V, 250 nA	Padhan et al. (2022)
13	NiO-Mg nanocomposites	35 V, 130 nA	Padhan et al. (2022)
14	PVC/MoS_2	398 V, 40 μA	Zhao et al. (2022)

silicon wafer, resulting in an output voltage of 4 V. Zheng et al. (2022) further demonstrated that the tribovoltaic effect on water/Si diminishes with increasing temperature. They utilized a flat surface, and dispersed a droplet at an angle of 90 onto the triboelectric layer's surface for their experiment.

The present study examines recent advances in wave energy harvesting and the impact of different substrates on TENG performance. It outlines the design philosophy of an energy harvesting device based on DB-TENG in absence and presence of UV light. Additionally, it shows how different environmental and climatic conditions in oceans affect three solutions, viz. doubly ionised water (DI), rainwater and seawater by testing their output parameters with a GaN based DB-TENG.

2 Experimental Setup

Thin films of GaN and AlN were prepared by laser molecular beam epitaxy (LMBE) using an KrF laser ($\lambda = 248$ nm). To reduce the lattice mismatch between GaN and the substrate, AlN buffer layers were used. The deposition parameters for these thin films are given in Table 2.

The thermal evaporation technique is used to deposit copper (Cu) and gold (Au) contacts. The deposition parameters are listed in Table 3. The Cu and Au electrodes are used because both metals have a high electron attraction force and a voltage gradient can occur at the two ends of the DB-TENG.

Figure 1 shows the design of TENG device. For open circuit voltage (V_{oc}) and short circuit current (I_{sc}) measurements, the semiconductor characterization system was used. In the present work, NaCl solution with a concentration of 0.6 M, which

Table 2 Deposition Parameters of GaN and AlN thin films

Parameters for Growth	*GaN*	*AlN*
Laser source, wavelength	KrF, 248 nm	KrF, 248 nm
Target	1' dia. GaN (99.9% pure)	1' dia. AlN (99.9% pure)
Ambient gas	N_2	N_2
Base pressure	1×10^{-10} mbar	1×10^{-10} mbar
Temperature during nitridation	700 °C	700 °C
Gas Flow during nitridation	8 sccm	8 sccm
Growth temperature	800 °C	800 °C
N_2 flow	1.3–4 sccm	1.4 sccm
Laser fluence	5.2 Jcm^{-2}	5.2 Jcm^{-2}
RF cell power	360 W	360 W
T-S distance	6 cm	6 cm
Film thickness	250 nm	20 nm

Table 3 Deposition parameters for the growth of metal contacts

Parameters for Electrodes	*Copper (Cu)*	*Gold (Au)*
Deposition technique	Thermal evaporation	Thermal evaporation
Source/Target	Copper flakes (99.999% pure)	Gold wire (99.999% pure)
Substrate temperature	RT	RT
Base pressure	1.6×10^{-6} Torr	1.6×10^{-6} Torr
Source to substrate distance	16 cm	16 cm
Power (W)/Current (A)	8 A	8 A
Deposition time (min)	2	3
Post deposition annealing	700°C (N_2 gas, 1 min)	700°C (N_2 gas, 1 min)
Work function (eV)	4.7 eV	5.1 eV

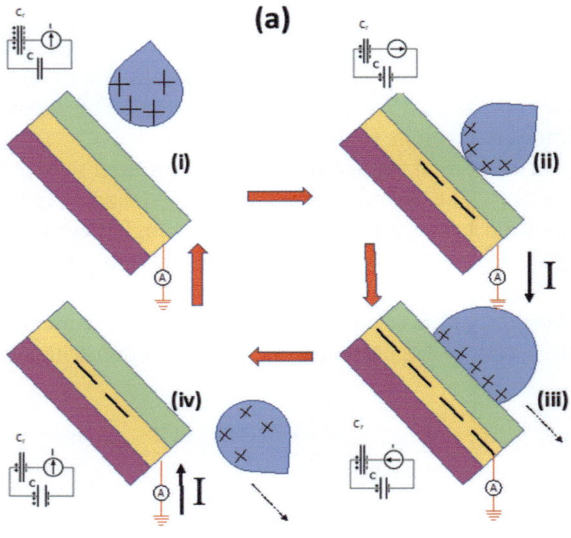

Fig. 1 Generation of current in DB-TENG

corresponds to sea water, is used as a droplet on top of GaN layer. In addition, the effect of NaCl concentration in the range of 0.1 to 1.0 M was investigated.

3 Effect of UV Radiation on DB-TENG

In the present work, two probe system is used to investigate the generation of electricity from moving NaCl droplets in the absence and presence of UV light on device. Figure 2 (a) shows the measurement setup of DB-TENG in the presence of UV light. The device DB-TENG was tilted 60° from the normal and 50 μl of NaCl with 0.6 M concentration was successively dropped onto the GaN surface. Figure 2 (b) and (c) show the V_{oc} and I_{sc} in the absence of UV light when 50 μL NaCl droplets glide over the GaN surface. The attraction of the Na^+ ions in the water droplets to the GaN contact is responsible for the positive peak in I_{sc} and V_{oc}.

Table 4 shows that generated V_{oc} increases from 18 to 42 V and the I_{sc} increases from 0.9 to 8.8 μA in the UV light compared to no UV light conditions. Measurements of V_{oc} and I_{sc} were made at fixed time intervals for seven consecutive cycles. The results obtained clearly indicate increased voltage and current generation due to the combination of photovoltaics and triboelectric effect.

4 Effect of Load Resistance on DB-TENG Under UV Radiation

To measure the power (P_{out}) generated in TENG, external load resistance (R_L) is used. Figure 3 (a) and (c) show the total output voltage (V_T) and current (I_T) in dark and light conditions. Figure 3 (b) and (d) show the P_{out} vs R_L in the absence and presence of UV light. From Fig. 3 (a) and (c), show that maximum V_T increases from 124 to 184 V at R_L of 1000 MΩ upon illumination of UV light. The maximum I_T increase from 2.3 to 22 μA at 1 kΩ were observed in dark and UV light. The maximum generated power is 64 μW in dark conditions and 824 μW in presence of UV light at a R_L of 150 MΩ for the GaN based DB-TENG.

5 Conclusion

In summary, GaN based DB-TENG is fabricated using LMBE technique and is used for effective energy harvesting from ocean wave motion by conjunction of triboelectric and photovoltaic effect. GaN based DB-TENG exhibited V_{oc} of 18 and 42 V in absence and presence of UV light. While the I_{sc} of 0.9 and 8.8 μA is observed in dark and UV conditions. The work presented here deals with the use of GaN material based TENG to extract energy in all weather conditions.

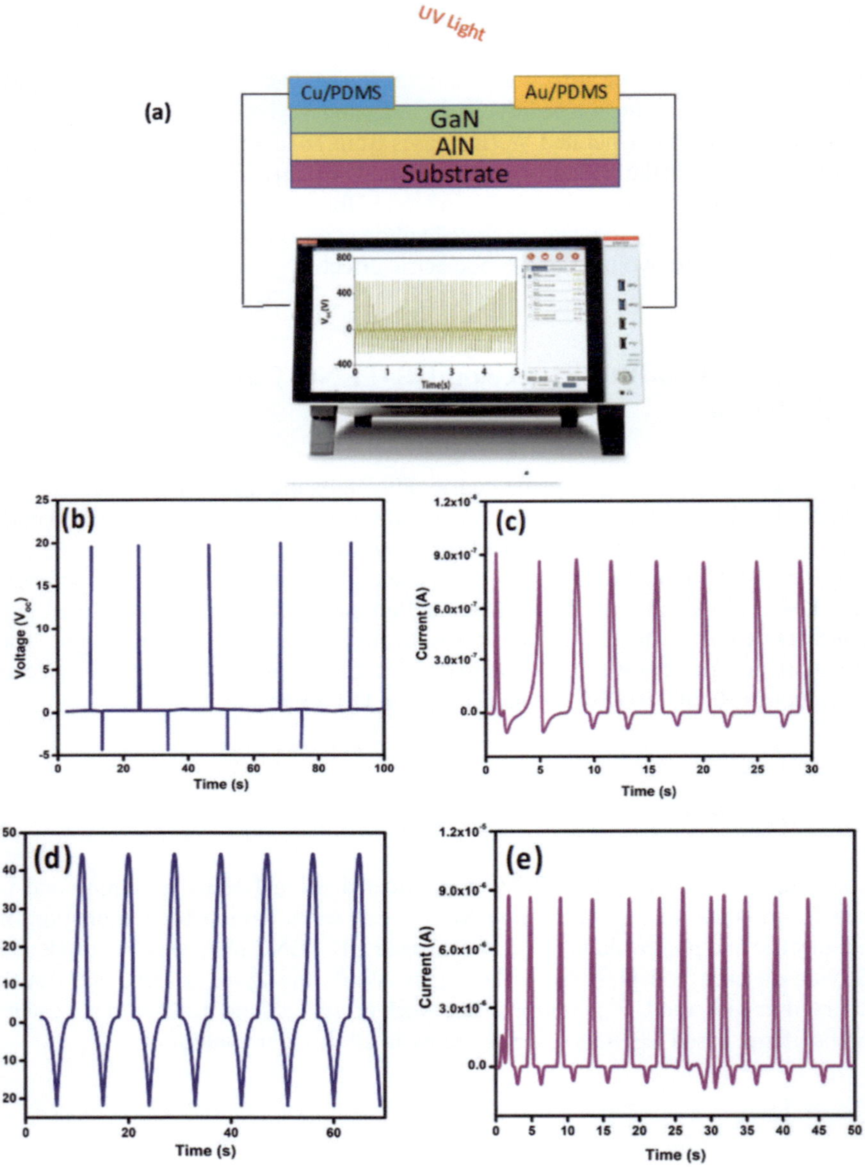

Fig. 2 (a) Schematic of DB-TENG (b) and (c) voltage and current generation in dark conditions, (d) and (e) voltage and current in light conditions

Table 4 Parameters obtained in absence and presence of UV light for DB-TENG

DB-TENG	V_{oc} *(V)*	*Isc* (μA)
No light	18	0.9
UV light	42	8.8
Ratio (UV light/NO light)	2.33	9.7

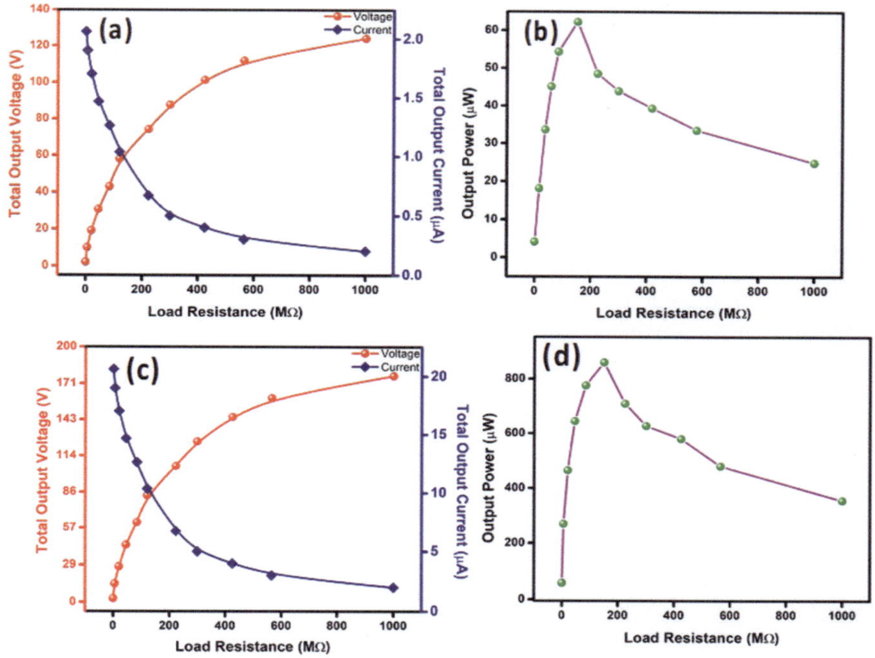

Fig. 3 (a) Variation of V_T and I_T with R_L, (b) P_{out} with R_L in the absence of UV light, (c) generated V_T and I_T with R_L and (d) P_{out} vs R_L in presence of UV light

Acknowledgements Authors are thankful to DEITY-India and Miranda House college, University of Delhi for the equipment facilities. Gunjan Yadav is thankful to UGC India for the SRF fellowship.

References

Chen, Y., Pu, X., Liu, M., Kuang, S., Zhang, P., Hua, Q., Cong, Z., Guo, W., Hu, W. and Wang, Z.L. (2019). Shape-adaptive, self-healable triboelectric nanogenerator with enhanced performances by soft solid-solid contact electrification. *ACS Nano*, **13(8):** 8936–8945.

Cheng, T., Gao, Q. and Wang, Z.L. (2019). The current development and future outlook of triboelectric nanogenerators: a survey of literature. *Adv Mater Technol,* **4(3):** 1800588.

He, X., Guo, H., Yue, X., Gao, J., Xi, Y. and Hu, C. (2015). Improving energy conversion efficiency for triboelectric nanogenerator with capacitor structure by maximizing surface charge density. *Nanoscale,* **7(5):** 1896–1903.

Kumar, S., Sharma, A., Gupta, V. and Tomar, M. (2021). Development of novel MoS2 hydrovoltaic nanogenerators for electricity generation from moving NaCl droplet. *J Alloys Compd,* **884:** 161058.

Li, X., Tao, J., Wang, X., Zhu, J., Pan, C. and Wang, Z.L. (2018). Networks of high performance triboelectric nanogenerators based on liquid–solid interface contact electrification for harvesting low-frequency blue energy. *Adv Energy Mater.,* **8(21):** 1800705.

Ouyang, R., Huang, Y., Ye, H., Zhang, Z. and Xue, H. (2022). Copper particles-PTFE tube based triboelectric nanogenerator for wave energy harvesting. *Nano Energy,* **102:** 107749.

Wei, X., Zhao, Z., Zhang, C., Yuan, W., Wu, Z., Wang, J. and Wang, Z.L. (2021). All-weather droplet-based triboelectric nanogenerator for wave energy harvesting. *ACS Nano,* **15(8):** 13200–13208.

Xia, K., Xu, Z., Zhu, Z., Zhang, H. and Nie, Y. (2019). Cost-Effective Copper‾Nickel-Based Triboelectric Nanogenerator for Corrosion-Resistant and High-Output Self-Powered Wearable Electronic Systems. Nanomater (Basel, Switzerland) **9(5)**. https://doi.org/10.3390/nano9050700

Zeng, Y., Luo, Y., Lu, Y. and Cao, X. (2022). Self-powered rain droplet sensor based on a liquid-solid triboelectric nanogenerator. *Nano Energy,* **98:** 107316. https://doi.org/10.1016/j.nanoen.2022.107316

Zhu, G., Peng, B., Chen, J., Jing, Q. and Wang, Z.L. (2015). Triboelectric nanogenerators as a new energy technology: from fundamentals, devices, to applications. *Nano Energy,* **14:** 126–138.

Chapter 12
Study of Structural, Electrical Properties of Double Perovskite Based Composites for Eco-Friendly Hydroelectric Cell Applications

Manisha, Chitralekha, S. Gaurav, and S. Shankar

1 Introduction

Due to its prospective uses and increased adaptability to produce cutting-edge microelectronic and spintronic devices, the double perovskite-based multifunctional materials have attracted growing attention in hardware advances. In double perovskite, the magnetoelectric coupling is anticipated to add another level of freedom to modify the magnetic state by an electrical field or tune the electrical qualities by a magnetic field. Due to their significant magnetodielectric effect, magnetism, and magnetoresistance close to room temperature, some well-known double perovskites such as La_2CoMnO_6, Bi_2FeMnO_6 and La_2CoMnO_6, have been vigorously investigated (Shereef et al., 2023).

This material displays a higher temperature of 225 K, along with electromagnetic coupling and magnetodielectric properties. Notably, it achieves an exceptional magnetoresistance of 84% at lower magnetic fields, the La_2CoMnO_6 framework is the most widely studied of all the double perovskites. In recent research on the La_2CoMnO_6 system's synthesis, structure, and magnetic behaviour has recently been published (Orayech et al., 2015). Many reports have been made; however the structure and such systems' magnetic behaviour continued to be discussed and is still

Manisha · Chitralekha · S. Shankar (✉)
Funtional Materials Research Laboratory, Department of Physics ARSD College, University of Delhi, New Delhi 110021, India
e-mail: sssubramaniam@arsd.du.ac.in

Manisha · S. Gaurav
Department of Applied Physics, Amity Institute of Applied Sciences, Amity University, Noida, India

Chitralekha
Department of Physics, Multanimal Modi College, Ch. Charan Singh University, Meerut, India

R. K. Kotnala et al. (eds.), *Advanced Functional Materials for Sustainable Environments*, https://doi.org/10.1007/978-3-031-62620-3_12

under investigation. La_2CoMnO_6-based double perovskites are interesting because of their remarkable characteristics and promise for various industrial applications. In the case of an ordered sub-lattice featuring a high spin Co^{2+} and Mn^{4+} pair, an observed ferromagnetic (FM) transition occurs at approximately 220 K. However, for a disordered sub-lattice with a low spin Co^{3+} and high spin Mn^{3+}, the FM transition is noted at around 150 K. Reports indicate that both low-temperature sintered samples and samples prepared at high temperatures exhibit transition temperatures around TC at approximately 220 K and TC at roughly 150 K, respectively. It is believed that the physical properties are significantly influenced by the synthesis procedures and associated conditions. Double perovskites hold promise for various applications, including aircraft turbine sensors, computer circuits and phased array antennas. The layered structure effectively inhibits the loss of electrical signals, a common issue among stacked layers (Krishna Murthy and Venimadhav, 2012; Yadav et al., 2022).

Barium titanate ($BaTiO_3$) is a material that appears white to grey in its powdered form and exhibits a perovskite structure. It dissolves in acids like sulfuric, hydrochloric and hydrofluoric acids, but remains insoluble in both alkalis and water. When in its purest form, it functions as an electrical insulator, yet it can accommodate a significant amount of scandium, yttrium, neodymium, samarium and other metals and with slight doping becomes semiconducting. As a semiconductor, $BaTiO_3$ is well-known for its excellent temperature co-green of resistivity residences. It implies that the fabric will display a rise at a warm temperature, referred to as the Curie temperature. Due to its percent residences, barium titanate is maximum observed to be used as a thermistor, e.g. in thermal switches. Barium titanate nano powder is used by various ceramic composites because of its remarkable characteristics like numerous semiconductive ceramics and ferroelectric ceramics. Barium titanate has a lot of potential for manufacturing ceramic capacitors (Cernea et al., 2006; Shankar et al., 2021).

2 Experimental

The binary composite (1-x) LCMO-xBT ($x = 0$, 0.1 and 1) was created via solid state reaction method. Initial ingredients of chemicals of lanthanum oxide, cobalt oxide, manganese dioxide (MnO_2) were used for preparation of LCMO. Barium carbonate ($BaCO_3$) and titanium dioxide (TiO_2) were used as starting materials for prepartion of $BaTiO_3$. All chemicals were used as in stoichiometric ratio. The starting oxides i.e., LCMO & BT were grinded individually for 6 h. The ground powders of LCMO and $BaTiO_3$ underwent calcination in a muffle furnace at different temperatures, with LCMO at 900 °C and $BaTiO_3$ at 1000 °C, respectively. The high temperature calcined powder was ground by combining it stoichiometrically to prepare the selected 0.9LCMO-0.1BT composition. Pellets were made using grinded powder for individual sample.

The powders were used to create pellets by blending them with a polyvinyl alcohol binder and subsequently hard-press into cylindrical shapes, with each pellet having a diameter of 10 mm and thickness of approximately 1 mm. These samples were sintered at 1050 °C for 5 h, along with the remaining powder. The high temperature sintered composite was named 0.9LCMO-0.1BT. Later hydroelectric cells were prepared using zinc plate and silver paste for making pattern and their response were calculated.

3 Characterization

The sintered powder and pellets were employed to acquire X-ray diffraction (XRD) patterns and FTIR spectra. XRD analyses were conducted using a Bruker D8-Advance X-ray Diffractometer, utilizing Cu-Kα radiations with a wavelength of 1.5418 Å. Fourier transform infrared spectroscopy (FT-IR) measurements were conducted using a Bruker FT-IR instrument operating within the wavenumber range from 400 to 1000 cm^{-1}. The sintered round pellets were coated with silver paste on parallel surface for observing dielectric properties by using Wenneker 6500 P high frequency LCR meter setup over the frequency range of 20 Hz to 5 MHz and at ambient temperature. The hydroelectric cell performance at dry and wet conditions were performed.

4 Results and Discussion

4.1 XRD Analysis

The XRD pattern attained for LCMO-BT series at ambient temperature and room pressure is shown in Fig. 1. The pure $La_{0.7}Ca_{0.3}MnO_3$ (JCPDS card no.70–2664), suggests that all LCMO samples exhibit a single-phase composition and crystallize in the orthorhombic perovskite structure with a Pnma space group (Penchal Reddy et al., 2014). The XRD pattern also shows the pure phase formation of $BaTiO_3$ barium titanate (Su et al., 2012). The presence of tetragonal-$BaTiO_3$ in this sample can be readily recognized by the peak splitting occurring around 46°C. All the composites show dual nature reflecting the LCMO (Yadav et al., 2017) and BT dominating peaks. However, merging and broadening of peaks observed in all composites indicating formation of composites (Orayech et al., 2015; Yuan et al., 2015). The crystallite size obtained for each sample is given in Table 1. LCMO exhibit smaller crystallite size than BT and their composite shows variation in crystallite size.

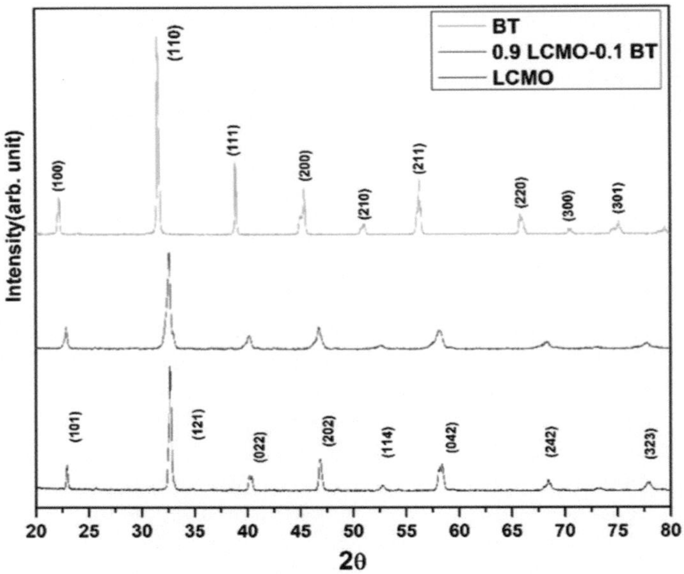

Fig. 1 The XRD pattern of LCMO, BT, LCMO-BT composite

Table 1 Crystallite size of LCMO, BT and LCMO-BT composite

Material	*Crystallite size (nm)*
LCMO	21
BT	26
0.9LCMO-0.1BT	16

4.2 *FTIR Analysis*

Figure 2 shows the room temperature FT-IR spectra of LCMO, BT and composite over wave number 400–1000 cm^{-1}. For LCMO, band in the range of 500–560 cm^{-1} are for asymmetric stretching vibration (Shereef et al. 2023). The existence of a prominent and intense band at 570 cm^{-1} in La_2CoMnO_6 confirms the formation of the double perovskite phase. The strong peak is also associated with the anharmonic asymmetric stretched mode of MnO_6, indicating the formation of MnO_6 octahedra. The symmetric stretching vibration of MnO_6 octahedra is responsible for a minor deformation in the spectra that occurs between 950 and 1000 cm-1S. In the case of pure BT, there are two prominent absorption peaks at 453 cm^{-1} and 480 cm^{-1}, corresponding to Ti–O bending and stretching vibrations, respectively (Cernea et al., 2006). O–H bond links to titanium are responsible for the bands at 1003 cm^{-1}, while metal–oxygen bonds are responsible for the sharp band at 463 cm^{-1}. The O–H stretching vibration in a moist state is attributed to the distinctive absorption band above 3500 cm^{-1}, which indicates the occurrence of water in the BT (Shankar et al.,

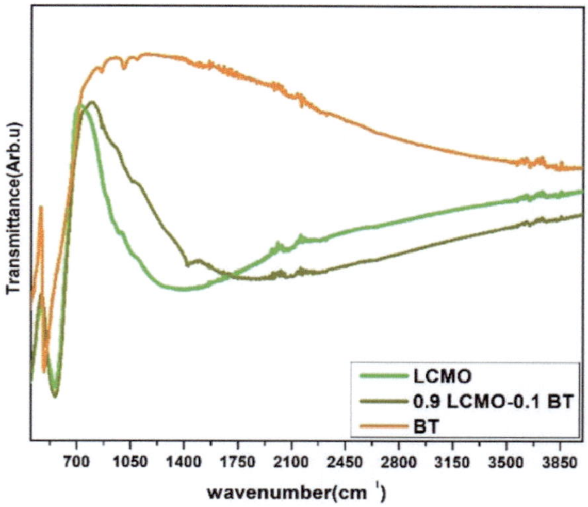

Fig. 2 FT-IR spectra of LCMO, BT, LCMO-BT Composite

2018). The Ti–O bond's vibration mode in the TiO_6 octahedra is responsible for the appearance of the band around 550 cm^{-1}. In composite 0.9LCMO-0.1BT, peak shifting and broadening 785 cm^{-1} was observed in comparison to LCMO (Bajpai et al., 2019) which confirms the formation of composite material.

4.3 Dielectric Constant vs Frequency Analysis

The dielectric constant versus frequency at room temperature is shown in Fig. 3. The plot illustrates the typical trend of a decrease in the dielectric constant with an increase in frequency. BT shows great dielectric behaviour in lower frequency region (Ahmed et al., 2018; Cernea et al., 2006). Conversely, LCMO exhibits weaker dielectric behavior. In the LCMO-BT nanocomposite, the increase in BT content leads to a significant decrease in the dielectric constant, primarily attributed to the Maxwell–Wagner-type contribution. The dielectric constant in the LCMO-BT nanocomposite experiences a sharp decrease with an increase in BT content, primarily as a result of the Maxwell–Wagner-type contribution. For composite the value of dielectric constant is greater for 0.7LCMO-0.3BT. improvement in dielectric behavior is observed in composite by addition of BT in LCMO.

4.4 Dielectric Loss vs Frequency Analysis

Figure 4 shows dielectric loss over frequency at ambient temperature. BT shows minimal loss due to its ability to have great dielectric properties. In LCMO, the

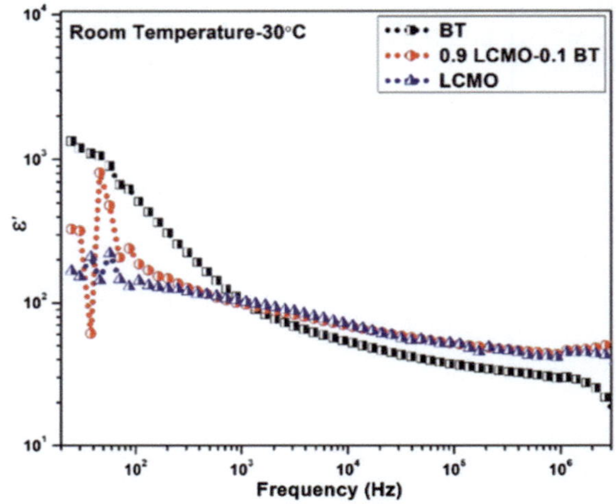

Fig. 3 Dielectric loss vs frequency of LCMO, BT, LCMO-BT Composite

dielectric loss is attributed to the existence of impurities. At lower frequencies, where the applied field's frequency and the frequencies of various ionic sites are nearly equivalent (Krishna Murthy and Venimadhav, 2012; Murthy et al., 2015), the observed loss is usually higher. In the case of barium titanate (BT), the tan δ value was lower. For 0.9LCMO-0.1BT composite, the value of tan δ is higher for 0.1% BT and decreases as BT increases.

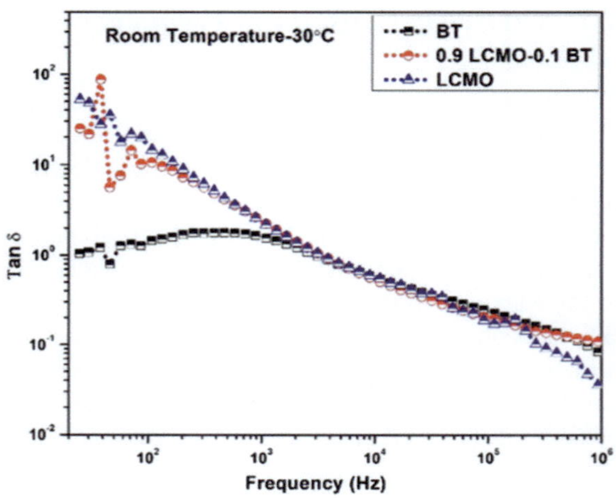

Fig. 4 Dielectric loss over frequency of LCMO, BT, LCMO-BT composite

Table 2 Hydroelectric property analysis in three conditions

Material	*Dry state (mA)*	*Apply few drops of water (mA)*	*Dip in water (mA)*
LCMO	0	0.9	2.81
BT	0	0.006	0.042
0.9LCMO-0.1BT	0	0.042	1.102

4.5 Hydroelectric Property Analysis

The functioning of the hydroelectric cells (HECs) of LCMO-BT composite were carried out and its behaviour in three states i.e., dry state, applying few drops of water and dip in water were analysed. The corresponding current noted are given in Table 2. In dry state, no current was observed. When water was sprinkled on each hydroelectric cell, some amount of current in mA was detected. In water sprinkled condition, composites of 0.9LCMO-0.1BT, shows current value 0.042 mA. All the samples performed well when HECs were fully dipped in water. The BT HEC gave current value of 0.042 mA, whereas LCMO performed better than BT by giving current of 2.81 mA when dipped in water as more dissociation of water molecules took place in HECs. The enhancement in the HEC of LCMO as compared to BT is due to larger number of oxygen vacancies present in former as compared to latter. Thereafter, the process was repeated for LCMO-BT composite, and it was observed that the HECs gave better performance as the concentration of LCMO is added to BT. In composite, 0.9LCMO-0.1BT shows current value 1.102 mA. The composite 0.9LCMO-0.1BT gave maximum current value, showing that as we increase BT concentration in LCMO, current in wet state also increases.

5 Conclusion

In summary, composite of (1-x) LCMO-xBT ($x = 0, 0.10, 1$) were successfully prepared using solid state reaction method. The structural analysis made using XRD and FT-IR spectra revealed the occurrence of both LCMO and BT phases and revealed the composite formation. The frequency variation of dielectric properties follows Maxwell–Wagner polarization and improved diffusive type relaxor behavior in the composites. Impedance measurements revealed non-Debye type thermal dependent relaxation mechanism in LCMO-BT composites. Hydroelectric cells (HECs) of LCMO-BT composite in the water sprinkled and wet states displayed ionic diffusion and stable as well as improved performance with increasing content of LCMO. Following results show improved structural, dielectric, impedance and improved HEC performance of LCMO-BT composite and demonstrate potential use in magnetoelectric and hydroelectric cell application.

Acknowledgements This work is funded by Research Scheme no. 03(1427)/18/EMR-II, CSIR, New Delhi, India and supported by UGC-DAE CSR-Indore, India for characterization facilities. The authors are also thankful to Dr. R. K. Kotnala, National Physical Laboratory, CSIR for the design and fabrication of hydroelectric cell.

References

Ahmed, K., Kanwal, F., Ramay, S., et al. (2018). Synthesis and Characterization of $BaTiO_3$/ Polypyrrole Composites with Exceptional Dielectric Behaviour. *Polymers* (Basel), **10:** 1273. https://doi.org/10.3390/polym10111273

Bajpai, N., Saleem, M., Soni, M., et al. (2019). Bismuth doped double perovskite La_2CoMnO_6: Analysis on structural, dielectric and modulus character. *In*: AIP Conference Proceedings. p 020183

Cernea, M., Monnereau, O., Llewellyn, P. et al. (2006). Sol–gel synthesis and characterization of Ce doped-$BaTiO_3$. *J Eur Ceram Soc.,* **26:** 3241–3246. https://doi.org/10.1016/j.jeurceramsoc.2005.09.039

Krishna Murthy, J. and Venimadhav, A. (2012). Magnetodielectric behavior in La_2CoMnO_6 nanoparticles. *J Appl Phys.,* **111**. https://doi.org/10.1063/1.3673862

Murthy, J.K., Chandrasekhar, K.D., Murugavel, S. and Venimadhav, A. (2015). Investigation of the intrinsic magnetodielectric effect in La_2CoMnO_6 : role of magnetic disorder. *J Mater Chem C.,* **3:** 836–843. https://doi.org/10.1039/C4TC01920B

Orayech, B., Urcelay-Olabarria, I., López, G.A. et al. (2015). Synthesis, structural, magnetic and phase-transition studies of the ferromagnetic La_2CoMnO_6 double perovskite by symmetry-adapted modes. *Dalt Trans.,* **44:** 13867–13880. https://doi.org/10.1039/C5DT01532D

Penchal Reddy, M., Zhou, X.B., Jing, L. and Huang, Q. (2014). Microwave sintering, characterization and magnetic properties of double perovskite La_2CoMnO_6 nanoparticles. *Mater Lett.,* **132:** 55–58. https://doi.org/10.1016/j.matlet.2014.06.057

Shankar, S., Kumar, M., Tuli, V. et al. (2018). Energy storage and magnetoelectric coupling in ferroelectric–ferrite composites. *J Mater Sci Mater Electron,* **29:** 18352–18357. https://doi.org/10.1007/s10854-018-9949-4

Shankar, S., Thakur, O.P. and Jayasimhadri, M. (2021). Significant improvements in dielectric, impedance, multiferroic and magnetoelectric properties of $(1-x)$ $Co_{0.5}Ni_{0.5}Fe_2O_{4-x}BaTiO_3$ bulk composites ($x = 0$, 0.10 and 0.20). *J Mater Sci Mater Electron,* **32:** 16706–16714. https://doi.org/10.1007/s10854-021-06227-4

Shereef, A., Aleena, P.A., Kunjumon, J. et al. (2023). Third order nonlinear optical properties and electrochemical performance of La_2CoMnO_6 double perovskite. *Mater Sci Eng B.,* **289:** 116262. https://doi.org/10.1016/j.mseb.2023.116262

Su, S., Zuo, R., Lv, D. and Fu, J. (2012). Synthesis and characterization of (001) oriented $BaTiO_3$ platelets through a topochemical conversion. *Powder Technol.,* **217:** 11–15. https://doi.org/10.1016/j.powtec.2011.09.045

Yadav, R., Modi, A., Para, T.A. et al. (2022). Influence of Y substitution on structural, magneto-transport, and magnetic properties of $La_{2-x}Y_xCoMnO_6$ ($0 \leq x \leq 0.3$) double perovskites. *J Magn Magn Mater.,* **554:** 169311. https://doi.org/10.1016/j.jmmm.2022.169311

Yadav, R., Para, T.A., Reshi, H.A. et al. (2017). Easy synthesis and electric, magneto-transport and magnetic properties of double perovskite La_2CoMnO_6 compound. *J Mater Sci Mater Electron,* **28:** 2970–2975. https://doi.org/10.1007/s10854-016-5882-6

Yuan, N., Liu, X., Meng, F. et al. (2015). First-principles study of La_2CoMnO_6: a promising cathode material for intermediate-temperature solid oxide fuel cells due to intrinsic Co-Mn cation disorder. *Ionics* (Kiel), **21:** 1675–1681. https://doi.org/10.1007/s11581-014-1320-z

Chapter 13
Study of Ionic Diffusion Mechanism in La^{3+} Ion Doped Spinel Ferrites Prepared Using Wet Chemical Route for Hydroelectric Cell Phenomena

Divya Bhandari, Prachi Jain, and O. P. Thakur

1 Introduction

Multiferroic materials have become a current era of research nowadays due to its extraordinary properties like high density information storage (Aslam et al., 2021), removal of toxic pollutants from degraded water bodies (Mariosi, F.R. et al., 2020) cancer therapy (Pęczkowski et al., 2023), etc. They have gained a lot of attention because they are able to exhibit more than two ferroic orderings at the same time. Multiferroics are known to be very ideal material which can show ferroelectricity and ferromagnetism at the same time or M-E coupling (Kumar et al., 2013). Cobalt ferrite is a very well known spinel ferrite having inverse spinel structure where Co^{2+} ions occupy the octahedral positions whereas Fe^{3+} ions are distributed equally among both tetrahedral and octahedral lattice positions (Jain et al., 2023a, b). The work based on the hydroelectric cell performance using multiferroics is very limited. The hydroelectric cell (HEC) is a cell which is able to produce electricity just by sprinkling few water droplets on the sample surface. Here we have studied only ionic diffusion mechanism for our material to check whether fabrication of HEC can be suggested or not (Jain et al., 2023a, b). In this article, the ionic diffusion has been done to mainly check the improvement in the hydroelectric cell upon lanthanum doping in the pure cobalt ferrites.

D. Bhandari · P. Jain · O. P. Thakur (✉)
Materials Analysis and Research Laboratory, Department of Physics, Netaji Subhas University of Technology, Dwarka, New Delhi 110078, India
e-mail: opthakur@nsut.ac.in

R. K. Kotnala et al. (eds.), *Advanced Functional Materials for Sustainable Environments*, https://doi.org/10.1007/978-3-031-62620-3_13

2 Preparation Technique

Lanthanum doped cobalt ferrites nanoparticles has been synthesized via sol–gel auto-combustion method in which the stoichiometric amounts of salts were taken and mixed in water by keeping it on magnetic stirrers for 30–30 min each. In the meanwhile prepare 50 mL citric acid solution. Add the prepared nitrate solution in the CA solution and leave it for stirring for 1 h for complete mixing. Add NH_3 solution to maintain the pH (7–8) of the solution. Then adjust the temperature of the stirrer at 80°C for 4 h until gel gets formed. After gel formation, increase the temperature to 100–150°C till the gel gets converted into blackish powdered form. The obtained powder is then grinded into fine powder and sintered at 700°C for 5 h with the heating rate of 5°C/min.

3 Results and Discussion

3.1 Crystallographic Study

Figure 1(a) shows the X-Ray diffraction patterns of pure cobalt ferrite (CFO) and lanthanum doped cobalt ferrites (LCFO) where the obtained highly intense peaks are indexed using JCPDS card number #22-1086 while Fig. 1(b) shows shifting of the highest peak at $2\theta = 35°$. The highly crystalline nature depicts the formation of nanoparticles of the La doped cobalt ferrites after sintering at 700°C. The crystallite size of the sample has been calculated using Debye Scherrer equation and found to be around 27.70 nm (Table 1):

$$D = K\lambda/\beta\cos\theta \tag{1}$$

where D is the crystallite size, the full width half maxima (FWHM), the Bragg's diffraction angle and K and are constants.

3.2 FT-IR Spectroscopy

The FTIR spectra of prepared lanthanum doped cobalt ferrite nanoparticles is plotted within the wavenumber range extending from 400–4000 cm^{-1} (Fig. 2). The stretching vibrations of Fe^{3+}-O ions at tetrahedral or A-sites corresponds to the absorption peak at 519 cm^{-1} while the vibrations around 437 cm^{-1} attributes due to octahedral sites (Jain et al., 2023a, b). Due to existence of nitrogen group, vibration of C = O bond takes place at 1204 cm^{-1}. Moreover, the peak around wavenumber 1745 cm^{-1} corresponds to the absorption of water molecule and atmospheric carbon-dioxide. The vibration of CH_2 group usually occurs near frequency 2970 cm^{-1}.

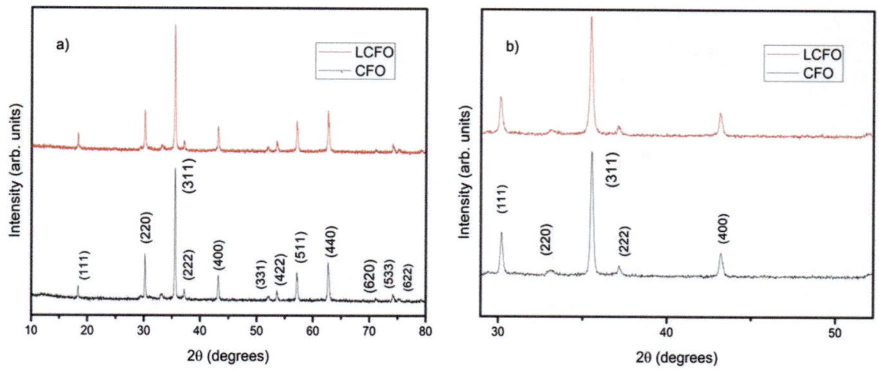

Fig. 1 (a) XRD pattern of pure and lanthanum doped cobalt ferrite ($CoLa_{0.1}Fe_{1.9}O_4$) and (b) enlarged view of XRD peaks

Table 1 The crystallite size, microstrain for the pure and lanthanum doped cobalt ferrites at 2θ = 35.7° prepared using sol–gel route

Sample name	*FWHM*	*2θ (°)*	*Crystallite size (nm)*	*Microstrain*	*Lattice parameter (Å)*	*Volume (Å^3)*
Cobalt ferrite (CFO)	0.1870	35.7913	30.49	0.303	8.3878	590.125
LCFO ($CoLa_{0.1}Fe_{1.9}O_4$)	0.1122	35.4687	27.70	0.1244	8.3145	574.788

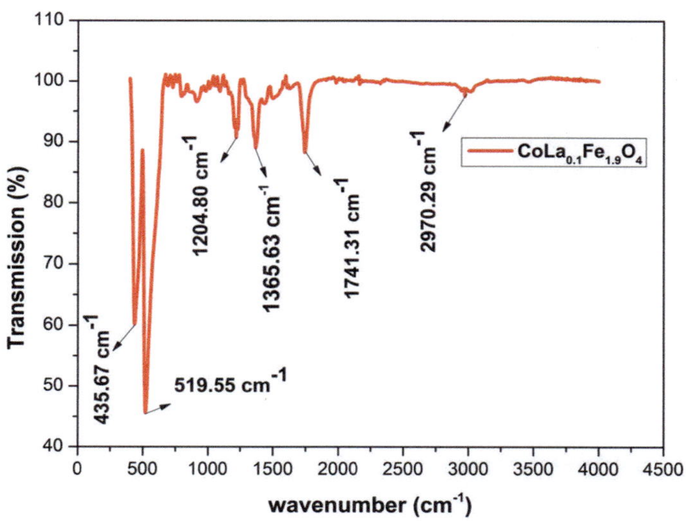

Fig. 2 FTIR pattern of lanthanum doped cobalt ferrite ($CoLa_{0.1}Fe_{1.9}O_4$)

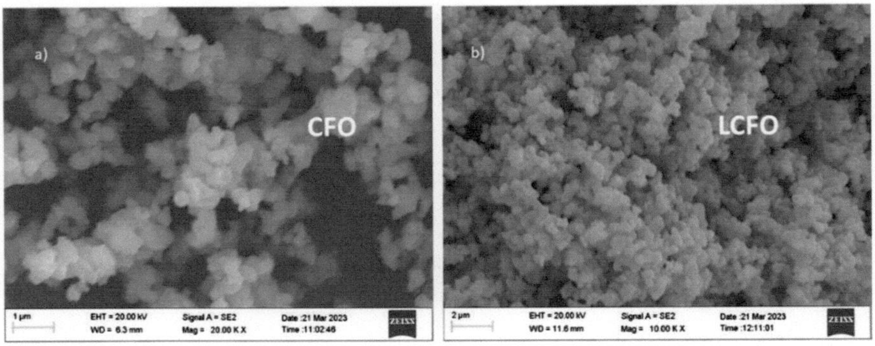

Fig. 3 (a) FESEM images of pure cobalt ferrites and (b) lanthanum doped cobalt ferrite ($CoLa_{0.1}Fe_{1.9}O_4$)

3.3 Morphological Analysis

Figure 3(a) and (b) presents the morphology of pure and La doped cobalt ferrite nanoparticles. The particle size of the nanoparticles is found to be 70 nm using ImageJ software. The obtained images show some porosity in the structure which would be helpful in the humidity sensors as well as for studying the hydroelectric cell application.

3.4 Electrical Measurements

Figure 4(a) shows the plot of ε′ against frequency from 10 Hz to 1 MHz at 100°C, 200°C and 300°C while Fig. 4(b) presents the tangent loss plots for lanthanum doped cobalt ferrite nanoparticles at different temperature values (100°C, 200°C and 300°C). It has been observed from Fig. 4(a), the steep decrement in the ε′ values with subsequent rise in frequency occurs. At high frequency, the ε′ remained constant because the effect of grains becomes more dominant as compared to the grain boundaries in the high frequency range (Kumar et al., 2010). At low frequency, the effect of grain boundary is more prominent than grain, thus ε′ is high. Similar behaviour has been observed in tangent loss plots (Fig. 4 (b)).

3.5 Ionic Diffusion Mechanism

The Cole–Cole method is best which would help in deciphering the amount of water molecules dissociating on the sample surface. The Nyquists plots of pure CFO and La^{3+} doped CFO represented in Fig. 5(a) and (b), respectively. It is quite clear from the

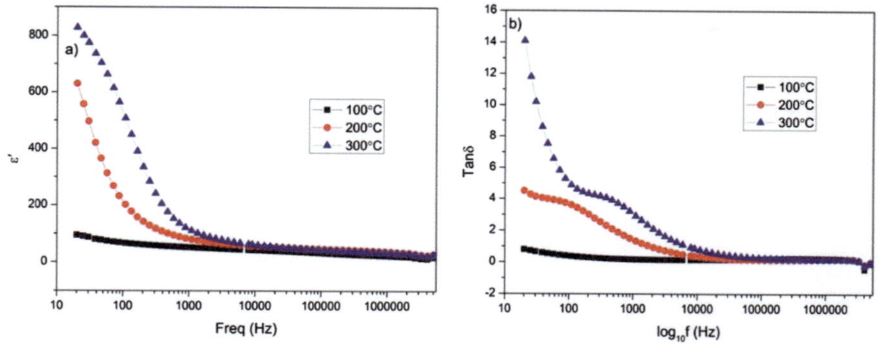

Fig. 4 (a) ε′ vs frequency for La doped cobalt ferrites and (b) Tangent loss against $\log_{10}f$ for frequency varying between 10 Hz–1 MHz

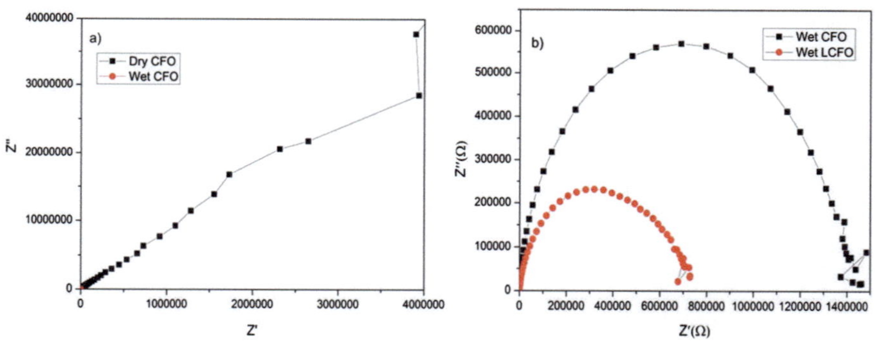

Fig. 5 (a) Ionic diffusion mechanism for pure CFO and (b) La doped cobalt ferrite ($CoLa_{0.1}Fe_{1.9}O_4$)

plots that the Cole–Cole plot show the high value of impedance in the dry sample as compared to wet one. This shows that the sample is exhibiting the conducting nature win the wet form and the diffusion of ions is taking place. Also, the improvement in the Z" values was observed on the La^{3+} doping in cobalt ferrite. Thus, it can be worthy to produce the HECs using La^{3+} doping in cobalt ferrites.

4 Conclusion

Wet chemical route has been opted for the preparation of lanthanum doped spinel ferrites nanoparticles where the XRD morphology confirmed the formation of highly crystalline spinel ferrites with the average crystallite size around 27 nm. The effects of lanthanum doping on the structural, dielectric and ferroelectric properties of pure CFO has been analyzed successfully. The decrease in the crystallite size of pure

CFO is observed on doping with lanthanum. The FESEM images has confirmed the spherical shaped grains with some irregularities present in the obtained morphology. The ε′ and tangent loss values has been increased with rise in temperature. From the Nyquist plots, ionic diffusion mechanism has been studied carefully which would further suggests their application in fabricating the hydroelectric cells.

References

Aslam, A., Razzaq, A., Naz, S., Amin, N., Arshad, M.I., Nabi, M.A.U., ... and Rehman, A.U. (2021). Impact of lanthanum-doping on the physical and electrical properties of cobalt ferrites. *Journal of Superconductivity and Novel Magnetism*, 1–10.

Jain, P., Shankar, S. and Thakur, O.P. (2023). Unveiling the impact of Ni^{2+}/Y^{3+} co-substitution on the structural, dielectric, and impedance properties of multiferroic spinel ferrite for hydroelectric cell application. *Physical Chemistry Chemical Physics*, **25(32):** 21280–21296.

Jain, P., Thakur, O. P. and Subramanian, S.S. (2023, November). Structural, Dielectric and Impedance Phenomena in Copper Ferrite Nano Powders for Hydroelectric Cell Application. In Materials Science Forum (Vol. 1099, pp. 157–162). Trans Tech Publications Ltd.

Kamran, M. and Anis-ur-Rehman, M. (2023). Influence of La^{3+} substitutions on structural, dielectric and electrical properties of spinel cobalt ferrite. *Ceramics International*, **49(4):** 7017–7029.

Kumar, M., Shankar, S., Kotnala, R.K. and Parkash, O. (2013). Evidences of magneto-electric coupling in BFO–BT solid solutions. *Journal of Alloys and Compounds*, **577:** 222–227.

Kumar, P., Sharma, S.K., Knobel, M. and Singh, M. (2010). Effect of La^{3+} doping on the electric, dielectric and magnetic properties of cobalt ferrite processed by co-precipitation technique. *Journal of Alloys and Compounds*, **508(1):** 115–118.

Mariosi, F.R., Venturini, J., da Cas Viegas, A. and Bergmann, C.P. (2020). Lanthanum-doped spinel cobalt ferrite ($CoFe_2O_4$) nanoparticles for environmental applications. *Ceramics International*, **46(3):** 2772–2779.

Pęczkowski, P., Szostak, E., Pocheć, E., Michalik, J.M., Piętosa, J., Tahraoui, T., ... and Gondek, Ł. (2023). Biocompatibility and potential functionality of lanthanum-substituted cobalt ferrite spinels. *Journal of Alloys and Compounds*, **966:** 171433.

Part IV
Advanced Materials for Energy Harvesting

Chapter 14
Antiferroelectricity: Advancements and Prospects in Future Applications

Roshan S. Thomas, Himani Saini, Shristi Chaudhary, and Shilpi Jindal

1 Introduction

We are surrounded by gadgets in todays digitalized world, from an alarm clock to transmitters. Applications including power distribution, transportation, spacecraft, medical gadgets, and weapons all depend more and more on power electronics. Capacitors are a crucial part of power conditioning in pulsed circuits because they can deliver a significant quantity of energy in a very short period of time while typically occupying the largest space and frequently weighing the most. High energy storage density dielectric materials are becoming an increasingly important and promising technology as demands for reducing the size and weight of the system coming into picture (Müller U., 2013).

Antiferroelectric (AFE) materials have garnered the most attention among all dielectrics due to their superior energy storage capabilities. Identifiable spontaneously ordered dipole moments in antiparallel arrangements give rise to zero remnant polarization in AFE materials, which are capable of transitioning to a ferroelectric (FE) phase with parallel dipole ordering in response to a sufficient electric field. This is much analogous to ferromagnetism and anti ferromagnetism (Liu et al., 2009).

Both FE and AFE materials exhibit spontaneous polarization, however they differ greatly. AFE materials will be able to demonstrate great energy-storage density and energy efficiency because the nearby polarizations of their crystals are in antiparallel orientations, creating a two-fold hysteresis loop. The application in impulsive power systems requires the good energy-storage performance.

Application of electric field can cause an antiferroelectric crystal's antiparallel dipoles to flip and become parallel, which is known as an electric field induced AFE

R. S. Thomas · H. Saini · S. Chaudhary · S. Jindal (✉)
Department of Physics, University Institute of Sciences Chandigarh University, Ludhiana-Chandigarh State Hwy, Chandigarh 140413, Punjab, India
e-mail: shilpi.uis@cumail.in

R. K. Kotnala et al. (eds.), *Advanced Functional Materials for Sustainable Environments*, https://doi.org/10.1007/978-3-031-62620-3_14

to FE phase transition. When the electric polarization P is measured as a function of the applied field E, the phase transition can be identified by the distinctive double-hysteresis loops. The phase transition is accompanied by considerable changes in other parameters, such as linear dimensions and optical characteristics, in addition to the rapid shift in polarization. Antiferroelectric materials are used in major engineering applications such as digital displacement transducers, energy storage capacitors, electrocaloric cooling devices, and flat panel displays thanks to these alterations at the phase transition. Antiferroelectric materials in the thin film form are necessary for the majority of device applications.

2 Brief History

In 2020, AFE completed 100-years of its revelation by Vlasek in Rochele salt in 1920. In his work, the origin of the polarization-electric field (P-E) hysteresis welcomed relationships to ferromagnetic materials, and thus the word "ferro" was used as a prefix to the dielectric analog. Antiferromagnetism has been recognized as peculiarity since in the 1930s, when it was crafted by Louis Néel. Following revelation of ferroelectricity in barium titanate, ($BaTiO_3$) the perovskite structure. In different areas during Second World War, researchers were working for finding some suitable dielectric material that can swap mica in various capacitors applications. As a result of this, antiferroelectricity came into picture. The majority of the studies on antiferroelectrics were conducted at Tokyo Institute of Technology (Tokyo Tech), namely on perovskite lead zirconate antiferroelectrics $PbZrO_3$ in two distributions from Shirane, Sawaguchi, Takeda (1950) and Sawagachi, Maniwa and Hoshino (1951). Kittel in 1951 gave principal hypothetical defense, recommending that antiferroelectricity ought to exist. Shirane joined Pepinsky in X-beam primary gathering. Pepinsky around then had an exceptionally progressed underlying refinement research center and utilized diffuse dispersing to help primary conclusions for polar balances and compute electron thickness maps around molecules. He utilized underlying investigation to find ferroelectric and other related materials amongst various gem structures. By 1953, second stage of antiferroelectrics had been distributed and refined in perovskite $PbHfO_3$. Around then, Newnham was working on gathering of Pepinsky. Newnham along with Shirane and Pepinsky in 1954 distributed the precious stone design of sodium niobate, ($NaNbO_3$) antiferroelectrics. In the meantime, Cross and Nicholson in UK were likewise dealing with $NaNbO_3$ precious stones and had the option to obtain twofold hysteresis curve. Simultaneous to exploratory work as well as motivated by new force of Devonshire phenomenological demonstrations of $BaTiO_3$ ferroelectric materials, Using proposes in Kittel distribution, Cross acquainted model which represent twofold hysteresis perceptions in $NaNbO_3$. Outside perovskite family, Keeling and Pepinsky in 1955 concentrated on low temperature, 148 K conduct of ammonia di hydrogen phosphate $(NH_4)H_2PO_4$ (ADP) and decided the request problem H-relocation with antiferroelectrics sublattice balance breaking. Yet another significant step accompanied Cochran following the work of Raman, Frohlich, Landau, Landauer Ginzburg, Anderson, and partners, with clarification of delicate mode hypothesis for the displacive stage changes in 1959. Although initially it was presented for ferroelastics but it was by and large relevant to numerous displacive stage advances that includes ferrodistortive and also anti-ferrodistortives. This hypothesis led to numerous significant central examinations coupling phonon scatterings and underlying transitions. In lined up with essential work, there were significant advances in the improvement of the $PbZrO_3$-$PbTiO_3$ (PZT) strong arrangement stage outline. Although many gatherings contributed but most broad as well as far reaching work was reported by Clevite Corporation in the Cleveland, USA and crafted by Jaffe, Berlincourt, Cook, Cmolik, and Jaffe in the mid-1960s (Zheludev et al., 2010). This created attention in antiferroelectric-ferroelectric (AFE -FE) limit of PZT stage outline for basic safeguard applications. Basic work at guard public research centers was additionally directed to test the actual boundaries controlling the AFE stage progress conduct. Brilliant work performed by Samara in 1970 at the Sandia National Laboratories gave an account of the hydrostatic tension consequences for the AFE transitions. The outcomes corresponded the exact perceptions inside the delicate mode hypothesis of Cochran. After 70 years of the disclosure of antiferroelectricity in $PbZrO_3$, numerous other significant commitments past initial 20 years have been taken place. Regardless, those initial 20 years supported large number of significant qualities which guide current comprehension.

3 Origin of AFE and Structural Property

3.1 Switching of P-E in Double Hysteresis Loop

Antiferroelectric character is described by capacity to incite stage progress between nonpolar AFE stage and polar FE stage, delivering momentous changes in various properties. In order to notice this exchanging, dielectric material is required to have fitting crystallographic structure having antiparallel unconstrained polarizations. Coercive fields needed for transformation should be adequately low in order to limit conductivity and in addition, dielectric breakdown strength should be high. Thus, to allow field constraining exchanging and in this manner notice the dielectric two-fold hysteresis reaction (Pan et al. 1989). Other than organization, outside conditions also influence the antiferroelectric execution. Figure 1(A) is twofold hysteresis curve in $Pb(Lu_{0.5}Nb_{0.5})O_3$ antiferroelectric material. Figure 1(B) depicts direct proportionality between EF and the temperature. Properties are also affected by the mechanical limit conditions. In this way, in the two-fold hysteresis, stresses that are applied or potentially capacity of material to go through changes in shape. During AFE-FE exchanging, and surprisingly, anode arrangement can to some degree cinch the progress and ought to be thought of. By and large, for a specific piece, EF is greater than EA while indication of EA relies upon stage. In event that testing condition is remote from stage limit, EA and EF have similar sign (exemplary twofold circle). In

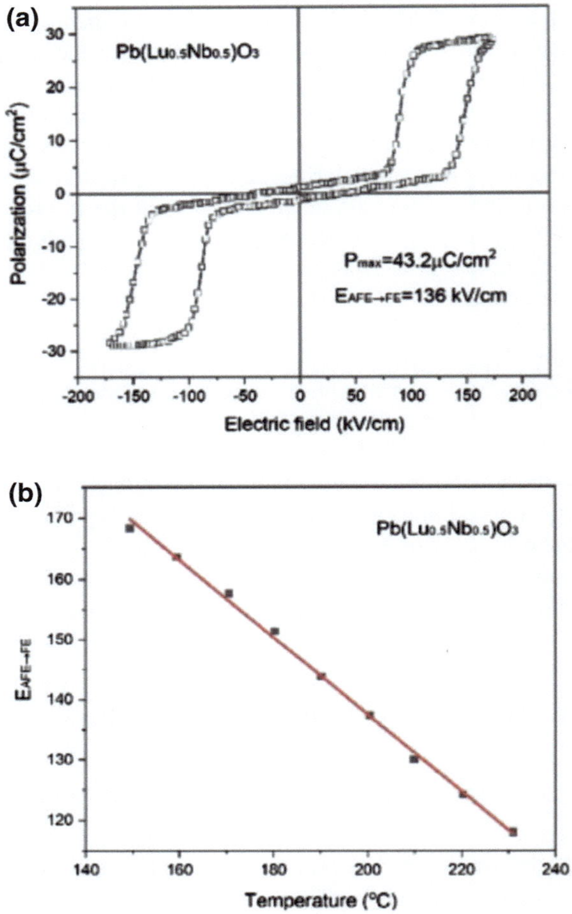

Fig. 1 Presence of antiferroelectricity in: (A) double hysteresis of $Pb(Lu_{0.5}Nb_{0.5})O_3$ and (B) dependence of critical field on temperature

any case, AFE and FE stages frequently coincide when near the stage limit. In these conditions, pinched shape P-E loop can be obtained due to the negative EA (Valant et al., 2001).

3.2 Superlattice Diffraction Between High Temperature Paraelectric and Low Temperature Antiferroelectric Phases

As momentarily portrayed over, creation of two sublattices with antiparallel polarization is part of fundamental stage transition from a nonpolar PE to an antipolar AFE. This requires modifying the translational balance of design of antiferroelectric comparative with Paraelectric to represent the need for numerous model cells to arrive at a genuine recurrent unit in antiferroelectric stage. An outcome of it is presence of extra diffraction spots or lines in antiferroelectric stage comparative with Paraelectric diffraction design. For instance, in $PbZrO_3$ (Čakare et al., 2000; Chaiyo et al., 2011), the nearby dipoles situate along <110> headings and structure an antiparallel game plan with in pairs shift; it will be depicted further underneath. Thus, net polarization is dropped to nothing, though grid boundary is quadrupled along <110>, that leads to {110} type of superlattice diffraction reflections and requires characterizing another unit cell involved eight model cells. However, it is fundamental to be exceptionally cautious while utilizing the superlattice diffraction to decide the stage. To start with, {110} spots could be undetectable on the off chance that the perception bearing is not opposite to the modulation, thus a methodical examination is required. Second, numerous other superlattice diffractions are possibly jumbling, for example, those inferable from an enemy of distortive stage change, for example, the antiphase octahedral shifting related with the AFE/FE stage that ordinarily prompts α (½{110}) and F (½{111}) superlattice spots (Muller et al., 2012). Thus, a translational evenness mark is a significant and vital property, however once more, additional consideration is requested for a fitting investigation.

3.3 Dependence of Phase Transition Behaviour on Hydrostatic Pressure

Distinctions change in volume among antiferroelectric and ferroelectric stages comparative with that of volume of paraelectric stage incite results on strength of these stages as component of outer tension applied an increase in hydrostatic tensions leads to higher value of densities (and consequently more modest model unit cell volumes). This pressure expands solidness of the perovskite antiferroelectric stages to higher values of temperature, though the stability of ferroelectric stage is stifled to bring down progress temperatures. Consequently, a paraelectric-antiferroelectric transition has $dT_c/dp > 0$, while Paraelectric–Ferroelectric transition (Chattopadhyay et al., 1998) is $dT_c/dp < 0$. If soft mode theory. The hydrostatic pressure dependence on the dielectric anomaly for $PbZrO_3$ ceramics is shown experimentally by Samara in Fig. 2.

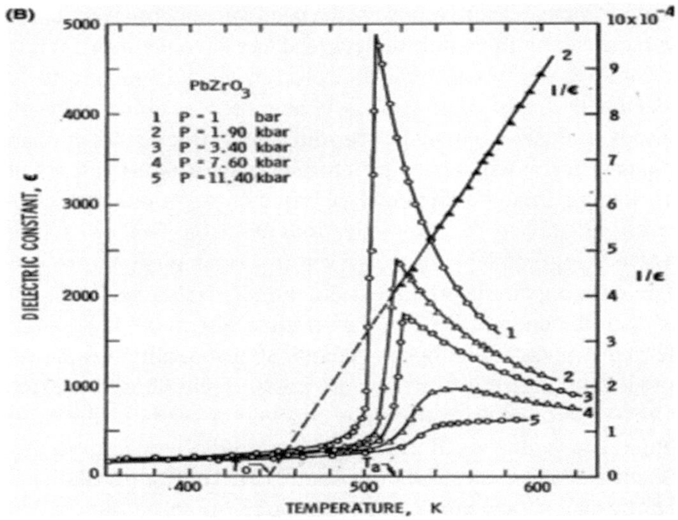

Fig. 2 Dependence of behavior of phase transitions of AFE $PbZrO_3$ on isobaric hydrostatic pressures (1 bar to 11.4 Kbar) and temperature and Curie–Weiss plot of dielectric permittivity in PE regime

3.4 Soft Modes, Structural Phase Transition, and Flexoelectricity

For a common equivalent AFE $PbZrO_3$ (Woldman & Landis, 2019) flighty middle stage with cycloidal request of polarization was gotten during the powerful design examination utilizing the negative-round deviation imaging method. The development of unconstrained polarization (PS) in the middle of the road stage is come about because of the breaking of spatial reversal balance by the antiferro distortive-ferro distortive change and various leveled advancement of the oxygen octahedra. In instance, a curtate cycloid can be used to approximate the tails of the polarization vectors (P_{sx} and P_{sy}). By focusing on the cross-sectional elements of single crystal $PbZrO_3$ and using the X-beam dissipating technique and Brillouin light dispersion, Zubko et al. (2013) resubjected the characteristics of antiferroelectricity (Kalantari et al., 2011). In $PbZrO_3$, the AFE state is a missing disproportionate stage, and the progression from the paraelectric to the AFE stage is attributed to mode mellowing via flexoelectric coupling. The flexoelectric influence of $PbZrO_3$ and $AgNbO_3$ at room temperature, however, is too little, according to later trial estimates of the flexoelectric coefficient and flexo-coupling.

4 Recent Advances in AFE

4.1 Antiferroelectric for Explosive Energy Conversion

Minimal expense and eco-accommodating energy transformation assumes a significant part in tending to the difficulties of limited regular fills. Ferroelectric materials stand out in this space attributable to their remarkable reaction to outside improvements. For instance, the pressure-driven depolarization conduct in poled FE materials empowers the quick change of mechanical energy to electrical energy, which has been broadly read up in this field for a long time. The field-driven FE-AFE stage progress of AFE materials (Chowdary & Molli, 2018) is by all accounts the most direct methodology due to the antipolar idea of the AFE state. This implies that the bound electric charges of the field-initiated metastable FE stage can be totally delivered inside a brief period (microseconds), producing a sharp voltage/current heart beat. In this manner, AFE materials have been read up since 1960s because of their various applications as touchy or beat power energy transformation gadgets in sustainable power, clinical ventures, safeguards, and mining regions.

4.2 The AFE for Solid State Cooling

For the last many years, there has been expanding interest in the improvement of elective cooling advancements with higher effectiveness in view of an eco-accommodating system to tackle our intensity issue and to diminish the utilization of ozone harming substances that are experienced and vigorously utilized in current refrigeration. Among these other options, electrocaloric materials go through a huge adiabatic temperature change ΔT close to the stage progress when an electric field is consistently applied and removed. The ΔT can be determined utilizing a circuitous strategy in view of the thermodynamic Maxwell relations.

4.3 Thin-Film AFE-Embedded Power Capacitors

Despite the fact that during this season of composing there is no commercialization of AFE dainty movies, there has been broad examination of flimsy film AFE capacitors. The energy densities of AFE are incredibly high, answered to be between 20 and 80 J cm^{-3}, as the breakdown strength is commonly controlled by warm breakdown. Notwithstanding, in the logical writing, tragically the checking on process does not continuously consider charge-release cycling and now and again, huge quick energy thickness capacity values are not essentially sensible for pragmatic application. Notwithstanding, most sensible energy densities of the order of 30 J cm^{-3} are as yet intriguing and high (He et al., 2005). These antiferroelectric are suitable as power plane decoupling capacitors. This was distinguished application and was taken into notice for high recurrence multi-chip-module bundles (Vlazan et al., 2019) with installed dainty film capacitors with square hysteresis circles, for example, This can be acquired with creations like (Sn, Nb) doped PZT. These antiferroelectric (Hao et al., 2014; Lagerwall et al., 2004; Ciuchi et al., 2016) can actually supply an aeronautical capacitance 32 μF cm^{-2}.

4.4 AFE Actuator Applications

Huge change in volume related to antiferroelectric/ferroelectric stage exchanging generates interest in utilization of these antiferroelectric materials in high strain actuators. As an illustration, Fig. 3 shows strain vs electric field hysteresis curve for volume, longitudinal and cross over exchanges in PLZT, which has tetragonal balance that transforms to rhombohedral ferroelectric. In both longitudinal and cross-over cases, the succession of stage progression and space reorientation stages indicates the decoupling of the strain and polarization. As mentioned above, the Electric field invests in completing the antiferroelectric domains. With precise in situ neutron diffraction analysis, Lu et al. provided additional insights into these cycles. They verified that field exchanging of antiferroelectric - ferroelectric stage progress isn't completely reversible and initiated strain fields are curved and drive favoured directions of space states (Weberszpil & Chen, 2017).

Some lead based and lead-free compositions of antiferroelectric materials are listed in Table 1.

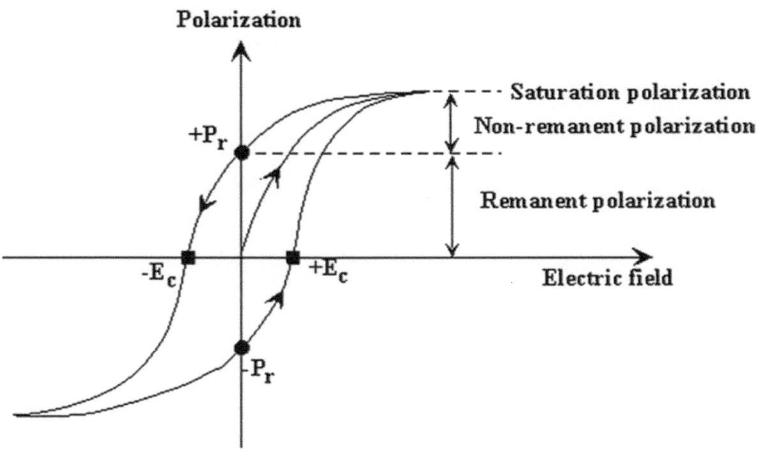

Fig. 3 Strain vs electric field hysteresis curve for volume, longitudinal and cross over exchanges in PLZT

Table 1 Application based antiferroelectric materials

Applications	*Materials*	*Compositions*	*References*
Energy storage, Tunable high frequency devices and dielectric energy storage	Lead based materials	$(Pb_{0.965}La_{0.035})(Zr_{0.9}Ti_{0.1})_{0.99125}O_3$	Jo et al., 2016
		$(Pb_{0.88}La_{0.08})(Zr_{0.91}Ti_{0.09})O_3$	Li et al., 2017
		$(Pb_{0.97}La_{0.02})(Zr_{0.95}Ti_{0.05})O_3$	Lui et al., 2017
		$Pb_{0.97}La_{0.02}(Zr_{0.58}Sn_{0.35}T_{0.07})O_3$	Zhang. et al., 2015
		$Pb_{0.93}Ba_{0.1}La_{0.02}(Zr_{0.68}Sn_{0.24}T_{0.08})O_3$	Zhang. et al., 2015
		$(Pb_{0.858}Ba_{0.1}La_{0.02}Y_{0.008}(Zr_{0.65}Sn_{0.3}Ti_{0.05})$	Tian et al., 2016
		O_3-$(Pb_{0.97}La_{0.02})(Zr_{0.9}Sn_{0.05}Ti_{0.05})O_3$	Zhao et al., 2017
	Lead free materials (Ag and Na based materials)	$AgNbO_3$	Zhao et al., 2018
		$Ag(Nb_{0.85}Ta_{0.15})O_3$	Tian et al., 2017
		$AgNbO_3$-0.1 wt%MnO_2	Yang et al., 2017a, b
		$AgNbO_3$-0.1 wt%WO_3	Yu et al., 2017
		$(Ag_{0.97}Bi_{0.01})NbO_3$	Gao et al., 2011
		$0.9Bi_{0.48}La_{0.02}Na_{0.48}Li_{0.02}Ti_{0.98}Zr_{0.02}O_3$-$0.1Na_{0.73}Bi_{0.09}NbO_3$	Yang et al., 2017a, b
		$0.94(0.75Na_{0.5}Bi_{0.5}TiO_3$-$0.25Bi_{0.5}K_{0.5}TiO_3)$-$0.06BiAlO_3$	Zhou et al., 2018
		$0.89Na_{0.5}Bi_{0.5}TiO_3$-$0.06BaTiO_3$-$0.05(K_{0.5}Na_{0.5})NbO_3$	Qi et al., 2019
		$0.92(0.65BaTiO_3$-$0.35Bi_{0.5}Na_{0.5}TiO_3)$-$xNa_{0.73}Bi_{0.09}NbO_3$	Gao et al., 2018
		$Ag_{1\text{-}3x}Bi_xNbO_3$	Tian et al., 2016
		$Ag_{1\text{-}3x}La_xNbO_3$	Qi et al., 2019
		$Ag_{1\text{-}3x}Sm_xNbO_3$	Luo et al., 2019
		$NaNbO_3$	Tian et al., 2017
		$Na_{0.7}Bi_{0.1}NbO_3$	Gao et al., 2019

5 Conclusion

As a dielectric materials subgroup that was found long back, there is still a lot to be gained from AFE based materials. Antiferroelectric are still being explored as compared to their Ferroelectric partners. Additionally, as far as compositions and precious stone designs, there are not many AFE materials comparative with FE materials. The fact that this survey helps makes it trusted the future investigation of new AFE materials. Thinking and foreseeing what is to come is dependably a dangerous undertaking, however, some vision on patterns that AFE exploration could give over the course of the following 10 years or so is discussed. Antiferroelectric have restricted range of applications as compared to ferroelectrics. Be that as it may, the new presentation of the mass-produced DC-connect capacitor will lessen forthcoming vulnerabilities in business choices related to these materials. With numerous of antiferroelectric with great performance being connected with toxic perovskites, there are still worries over poisonousness. Thus, the revelation of sans lead choices with superior execution would be of extraordinary interest to the business to stay away from future issues under Restriction of Hazardous Substances (RoHS) regulation.

References

Čakare, L., Malič, B., Kosec, M. and Sternberg, A. (2000). Antiferroelectric $PbZrO_3$ films prepared by sol-gel processing. *Ferroelectrics*, **241(1):** 107–114.

Chaiyo, N., Muanghlua, R., Niemcharoen, S., Boonchom, B. and Vittayakorn, N. (2011). Solution combustion synthesis and characterization of lead-free piezoelectric sodium niobate (NaNbO3) powders. *Journal of Alloys and Compounds*, **509(5):** 2445–2449.

Chattopadhyay, S., Ayyub, P., Palkar, V.R., Multani, M.S., Pai, S.P., Purandare, S.C. and Pinto, R. (1998). Dielectric properties of oriented thin films of $PbZrO_3$ on Si produced by pulsed laser ablation. *Journal of applied physics*, **83(12):** 7808–7812.

Chowdary, B. and Molli, M. (2018, August). Doping and temperature dependent thermoelectric properties of $NaNbO_3$. In AIP Conference Proceedings, **2005(1)**, AIP Publishing.

Ciuchi, I.V., Mitoseriu, L. and Galassi, C. (2016). Antiferroelectric to ferroelectric crossover and energy storage properties of $(Pb_{1-x}La_{ax})(Zr_{0.90}Ti_{0.10})_{1-x/4}O_3$ $(0.02 \leq x \leq 0.04)$ ceramics. *Journal of the American Ceramic Society*, **99(7):** 2382–2387.

Fulanović, L., Zhang, M.H., Fu, Y., Koruza, J. and Rödel, J. (2021). $NaNbO_3$-based antiferroelectric multilayer ceramic capacitors for energy storage applications. *Journal of the European Ceramic Society*, **41(11):** 5519–5525.

Gao, F., Dong, X., Mao, C., Liu, W., Zhang, H., Yang, L. and Wang, G. (2011). Energy-storage properties of 0.89 $Bi_{0.5}Na_{0.5}TiO_3$–0.06 $BaTiO_3$–0.05 $K_{0.5}Na_{0.5}NbO_3$ lead-free anti-ferroelectric ceramics. *Journal of the American Ceramic Society*, **94(12):** 4382–4386.

Gao, J., Zhang, Y., Zhao, L., Lee, K.Y., Liu, Q., Studer, A. and Li, J.F. (2019). Enhanced antiferroelectric phase stability in La-doped $AgNbO_3$: perspectives from the microstructure to energy storage properties. *Journal of Materials Chemistry A*, **7(5):** 2225–2232.

Gao, J., Zhao, L., Liu, Q., Wang, X., Zhang, S. and Li, J.F. (2018). Antiferroelectric-ferroelectric phase transition in lead-free $AgNbO_3$ ceramics for energy storage applications. *Journal of the American Ceramic Society*, **101(12):** 5443–5450.

Hao, X., Zhai, J., Kong, L.B. and Xu, Z. (2014). A comprehensive review on the progress of lead zirconate-based antiferroelectric materials. *Progress in Materials Science*, **63:** 1–57.

He, H., and Tan, X. (2005). Electric-field-induced transformation of incommensurate modulations in antiferroelectric $Pb_{0.99}Nb_{0.02}[(Zr_{1-x}Sn_x)_{1-y}Ti_y]_{0.98}O_3$. *Physical Review B*, **72(2):** 024102.

Jiang, Y., Zhu, C., Zhao, P., Bi, K., Liu, J., Guo, L. and Li, L. (2021). Excellent energy storage performance of $NaNbO_3$-based antiferroelectric ceramics with ultrafast charge/discharge rate. *Journal of the European Ceramic Society*, **41(13):** 6465–6473.

Jo, H.R. and Lynch, C.S. (2016). A high energy density relaxor antiferroelectric pulsed capacitor dielectric. *Journal of Applied Physics*, **119(2)**.

Johnson, R.R. (1990). Multichip modules: Next-generation packages. *IEEE Spectrum*, **27(3):** 34–36.

Kalantari, K., Sterianou, I., Karimi, S., Ferrarelli, M.C., Miao, S., Sinclair, D.C. and Reaney, I.M. (2011). Ti-doping to reduce conductivity in $Bi_{0.85}Nd_{0.15}FeO_3$ ceramics. *Advanced Functional Materials*, **21(19):** 3737–3743.

Lagerwall, S.T. (2004). Ferroelectric and antiferroelectric liquid crystals. *Ferroelectrics*, **301(1):** 15–45.

Li, B., Liu, Q.X., Tang, X.G., Zhang, T.F., Jiang, Y.P., Li, W.H. and Luo, J. (2017). Antiferroelectric to relaxor ferroelectric phase transition in PbO modified $(Pb_{0.97}La_{0.02})(Zr_{0.95}Ti_{0.05})O_3$ ceramics with a large energy-density for dielectric energy storage. *RSC Advances*, **7(68):** 43327–43333.

Liu, Y.Y. and Li, J.Y. (2009). Energetic analysis of ferroelectric domain patterns by equivalent inclusion method. *Journal of Materials Science*, **44:** 5214–5224.

Liu, Z., Bai, Y., Chen, X., Dong, X., Nie, H., Cao, F. and Wang, G. (2017). Linear composition-dependent phase transition behavior and energy storage performance of tetragonal PLZST antiferroelectric ceramics. *Journal of Alloys and Compounds*, **691:** 721–725.

Luo, N., Han, K., Cabral, M.J., Liao, X., Zhang, S., Liao, C. and Wei, Y. (2020). Constructing phase boundary in $AgNbO_3$ antiferroelectrics: pathway simultaneously achieving high energy density and efficiency. *Nature communications*, **11(1):** 4824.

Luo, N., Han, K., Zhuo, F., Xu, C., Zhang, G., Liu, L. and Wei, Y. (2019). Aliovalent A-site engineered $AgNbO_3$ lead-free antiferroelectric ceramics toward superior energy storage density. *Journal of Materials Chemistry A*, **7(23):** 14118–14128.

Muller, J., Boscke, T.S., Schroder, U., Mueller, S., Brauhaus, D., Bottger, U. and Mikolajick, T. (2012). Ferroelectricity in simple binary ZrO_2 and HfO_2. *Nano letters*, **12(8):** 4318–4323.

Müller, U. (2013). Symmetry relationships between crystal structures: applications of crystallographic group theory in crystal chemistry (Vol. 18). OUP Oxford.

Pan, W., Zhang, Q., Bhalla, A. and Cross, L.E. (1989). Field-forced antiferroelectric-to-ferroelectric switching in modified lead zirconate titanate stannate ceramics. *Journal of the American Ceramic Society*, **72(4):** 571–578.

Salje, E. (1990). Phase transitions in ferroelastic and co-elastic crystals. *Ferroelectrics*, **104(1):** 111-120.

Qi, H. and Zuo, R. (2019). Linear-like lead-free relaxor antiferroelectric $(Bi_{0.5}Na_{0.5})TiO_3$–$NaNbO_3$ with giant energy-storage density/efficiency and super stability against temperature and frequency. *Journal of Materials Chemistry A*, **7(8):** 3971–3978.

Qi, H., Zuo, R., Xie, A., Tian, A., Fu, J., Zhang, Y. and Zhang, S. (2019). Ultrahigh energy-storage density in $NaNbO_3$-based lead-free relaxor antiferroelectric ceramics with nanoscale domains. *Advanced Functional Materials*, **29(35):** 1903877.

Tian, Y., Jin, L., Zhang, H., Xu, Z., Wei, X., Politova, E.D. and Yan, H. (2016). High energy density in silver niobate ceramics. *Journal of Materials Chemistry A*, **4(44):** 17279–17287.

Tian, Y., Jin, L., Zhang, H., Xu, Z., Wei, X., Viola, G. and Yan, H. (2017). Phase transitions in bismuth-modified silver niobate ceramics for high power energy storage. *Journal of Materials Chemistry A*, **5(33):** 17525–17531.

Valant, M., Suvorov, D., Hoffmann, C. and Sommariva, H. (2001). Ag (Nb, Ta) O3-based ceramics with suppressed temperature dependence of permittivity. *Journal of the European Ceramic Society*, **21(15):** 2647–2651.

Vlazan, P., Rus, S.F., Poienar, M. and Sfirloaga, P. (2019). The effect of dopant concentrations on the structural, morphological and optical properties of $NaNbO_3$ semiconductor materials. *Materials Science in Semiconductor Processing*, **102:** 104602.

Weberszpil, J. and Chen, W. (2017). Generalized Maxwell Relations in Thermodynamics with Metric Derivatives. *Entropy*, **19(8):** 407.

Woldman, A.Y. and Landis, C.M. (2019). Thermo-electro-mechanical phase-field modeling of paraelectric to ferroelectric transitions. *International Journal of Solids and Structures*, **178:** 19–35.

Yang, H., Yan, F., Lin, Y., Wang, T. and Wang, F. (2017). High energy storage density over a broad temperature range in sodium bismuth titanate-based lead-free ceramics. *Scientific reports*, **7(1):** 8726.

Yang, H., Yan, F., Lin, Y., Wang, T., Wang, F., Wang, Y. and Wei, H. (2017). Lead-free $BaTiO_3$-$Bi_{0.5}Na_{0.5}TiO_3$-$Na_{0.73}Bi_{0.09}NbO_3$ relaxor ferroelectric ceramics for high energy storage. *Journal of the European Ceramic Society*, **37(10):** 3303–3311.

Yu, Z., Liu, Y., Shen, M., Qian, H., Li, F. and Lyu, Y. (2017). Enhanced energy storage properties of $BiAlO_3$ modified $Bi_{0.5}Na_{0.5}TiO_3$–$Bi_{0.5}K_{0.5}TiO_3$ lead-free antiferroelectric ceramics. *Ceramics International*, **43(10):** 7653–7659.

Zhang, G., Zhu, D., Zhang, X., Zhang, L., Yi, J., Xie, B. and Jiang, S. (2015). High-energy storage performance of $(Pb_{0.87}Ba_{0.1}La_{0.02})(Zr_{0.68}Sn_{0.24}Ti_{0.08})O_3$ antiferroelectric ceramics fabricated by the hot-press sintering method. *Journal of the American Ceramic Society*, **98(4):** 1175–1181.

Zhang, L., Jiang, S., Fan, B. and Zhang, G. (2015). Enhanced energy storage performance in $(Pb_{0.858}Ba_{0.1}La_{0.02}Y_{0.008})(Zr_{0.65}Sn_{0.3}Ti_{0.05})O_3$–$(Pb_{0.97}La_{0.02})(Zr_{0.9}Sn_{0.05}Ti_{0.05})O_3$ antiferroelectric composite ceramics by Spark Plasma Sintering. *Journal of Alloys and Compounds*, **622:** 162–165.

Zhao, L., Gao, J., Liu, Q., Zhang, S. and Li, J. F. (2018). Silver niobate lead-free antiferroelectric ceramics: enhancing energy storage density by B-site doping. *ACS applied materials & interfaces*, **10(1):** 819–826.

Zhao, L., Liu, Q., Gao, J., Zhang, S. and Li, J. F. (2017). Lead-free antiferroelectric silver niobate tantalate with high energy storage performance. *Advanced Materials*, **29(31):** 1701824.

Zheludev IS. Domain structure of ferroelectrics and antiferroelectric. Physics of crystalline dielectrics. Boston: Springer.

Zhou, M., Liang, R., Zhou, Z. and Dong, X. (2018). Superior energy storage properties and excellent stability of novel $NaNbO_3$-based lead-free ceramics with A-site vacancy obtained via a Bi_2O_3 substitution strategy. *Journal of Materials Chemistry A*, **6(37):** 17896–17904.

Zubko, P., Catalan, G. and Tagantsev, A.K. (2013). Flexoelectric effect in solids. *Annual Review of Materials Research*, **43:** 387–421.

Chapter 15
Performance Improvement in p-Si/n-CdS/ALD-ZnO Heterojunction Solar Cell by Introducing $MoSe_2$ Ingenious Hole Transport Layer

Atish Kumar Sharma, Ankita Srivastava, Prakash Kumar Jha, Manish Kumar, Nitesh K. Chourasia, and Ritesh Kumar Chourasia

1 Introduction

In the renewable energy sector, the research community as well as the industrialists are mostly inclined towards silicon solar photovoltaic technology because of its eco-friendly character as well as abundance, low cost, and non-toxic nature. The first ever p-type silicon solar cells were created more than five decades ago which was based on the diffusion of the p–n junction. The power conversion efficiency (PCE) of the cell was as reported by the researchers around 4.5% (Green, 2009). Subsequently, the performance of the silicon solar cell has been on an upward curve and it has now reached upto 26.7% (Andreani et al., 2019) just 5.3% less than its theoretical thermodynamic limit of about 32% (Bhattacharya & John, 2019). One of the studied solar cell types, heterojunction solar cells, is selected due to the solution being processable and probably having lowest possible the manufacturing

A. K. Sharma · P. K. Jha
University Department of Physics, Lalit Narayan Mithila University, Darbhanga 846004, India

A. K. Sharma · A. Srivastava · P. K. Jha · R. K. Chourasia (✉)
Post Graduate Department of Physics, Samastipur College (A Constituent College of L.N.M.U.-Darbhanga), Samastipur 848134, India
e-mail: riteshphyzics@gmail.com; dr.riteshchaurasia@gmail.com

M. Kumar
Experimental Research Laboratory, Department of Physics, ARSD College, University of Delhi, New Delhi 110021, India

N. K. Chourasia (✉)
Department of Physics, Faculty of Science, Shree Guru Gobind Singh Tricentenary University, Gurugram 122505, Haryana, India
e-mail: niteshphyzics@gmail.com

R. K. Kotnala et al. (eds.), *Advanced Functional Materials for Sustainable Environments*, https://doi.org/10.1007/978-3-031-62620-3_15

costs (Wang et al., 2012). The base or absorber material of a heterojunction solar cell should have an energy gap between 1.1 eV and 1.9 eV to achieve the ultimate PCE (Liu et al., 2020). These cells provide high PCE with stability and can sustain up to a relatively longer duration (Blakers et al., 2013). The highest PCE of the silicon solar cell is still to a lesser extent because of different types of parasitic loss mechanisms present in the device. These loss mechanisms may be non-radiative recombination due to trap centers present in the mid-band region and shallow band region of the silicon (Srivastava et al., 2021). It may also have absorption loss of near-infrared radiation due to its indirect nature of electronic band gap (Fell et al., 2022). To overcome these issues the p-Si absorber layer is being passivated at both the surfaces front and rear by using electron-selective and hole-selective materials, respectively. In recent years, researchers fabricated a front surface passivation layer or electron transport layer (ETL) on the p-type crystalline wafer. For this purpose, they concerned the transparent oxides or sulfides of various metals such as cerium oxide (CeO_2) (Ali and Ali, 2022), zinc sulfide (ZnS) (Qiu et al., 2017), cadmium sulfide (CdS) (Srivastava et al., 2021), magnesium oxide (MgO) (Wan et al., 2017), silicon oxide (SiO_2) (Singh et al., 2019), etc. These materials have a higher energy band gap (approx. 2 to 3 eV) relative to the absorber layer providing superior absorption of the blue photons. Furthermore, the researchers also deposited an additional antireflection layer devising even higher bandgap (approx. >3 eV) materials like aluminum doped zinc oxide (AZO) (Hosen et al., 2022), Fullerene doped tin oxide (FTO) (Hosen et al., 2022), titanium oxide (TiO_2) (Sabbah et al., 2022), atomic layer deposited zinc oxide (ALD-ZnO) (Sharma et al., 2023a, Sharma et al., 2023c), etc. Likewise, electron transport layer for the enhanced transport of holes along with minimizing the loss of infrared radiation, the researchers used relatively lower band gap materials as the HTL. These layers furnish the passivation at the rear surface of the absorber layer. It also imparts protection to the absorber from moisture. Some of the materials used recently to fabricate the HTL are Cu_2O (Hosen et al., 2022), Spiro-OMeTAD (organic material HTL) (Noh et al., 2013), Sb_2Se_3 (Deng et al., 2022), $MoSe_2$, MoS_2 (NGOUPO et al., 2022), etc. The affordability and ease of fabrication of n-CdS/p-Si heterojunction solar cells implementing p-type c-Si substrate appear interesting. Further to have a remarkable amount of conductivity with transparency an atomic layer deposition method is employed for the deposition of zinc oxide. This method provides homogeneous and uniform surface texturing that helps in maintaining good conductivity. To get better management of carriers at the back of the absorber layer, the $MoSe_2$ material must have the right electrical characteristics such as the electronic band gap, hole mobility, and electron affinity. As reported in literature for the $MoSe_2$ the indirect band gap is 1.1 eV and the electron affinity is 4.0 eV (Tongay et al., 2012; Zhang et al., 2019). Therefore, in our investigation, we considered $MoSe_2$ to have a shallow doping density of 10^{20} cm^{-3} for HTL with a thickness of 0.1 μm.

After reviewing several works of literature, ultimately in our present study, we have designed a model of the proposed solar cell device using p-Si for the absorber layer followed by adding the thin layer of n-type cadmium sulfide (n-CdS) as an electron transport layer. Further additional window layer of ALD-ZnO is chosen. For the hole transport layer, $MoSe_2$ material with a band gap smaller than the absorber layer reduces the absorption loss of infrared photons. Further simulations are performed using SCAPS-1D simulation software. The theoretical discussions are explained in the next Sect. 2. The details of the device modeling parameters used in the simulation tool are prescribed in Sect. 3. After the simulation and optimization of the adopted device configuration, the outcomes are discussed in Sect. 4 with the help of theories stated in Sect. 2. Ultimately, we concluded our investigations in Sect. 5.

2 Theoretical Modeling

As the solar radiation strikes at the forward-facing of the device i.e., at the window layer ALD-ZnO, the shorter wavelength photons are absorbed by the window layer and longer wavelength photons travel into the device and get absorbed in the p-Si absorber layer. The absorption of radiation generates electron–hole pairs following the criteria that the wavelength of the photons is shorter than the threshold wavelength (λ_c) for the absorption in the absorber layer according to the relation in Eq. 1 (Singh & Ravindra, 2012; Jha et al., 2023).

$$\lambda_c = 1240\left[\frac{1}{E_g(eV)}\right](nm) \tag{1}$$

where, E_g is the band gap of the p-Si absorber layer in eV. In our study, the threshold wavelength is found to be in the region of near-infrared. Therefore, the generation of carriers will take place in the whole visible region. Now the photo-generated carriers will drift owing to the presence of the built-in potential at the p–n junction formed by the p-Si and the n-CdS. The electrons will drift towards the front electrode through the ALD-ZnO (n + layer) and the holes will drift towards the back metal electrode. This gives rise to a photogenerated current in the device and the fundamental performance parameters are obtained such as short circuit current density (J_{sc}) (Jha et al., 2023), open circuit voltage (V_{oc}), fill factor (FF), and power conversion efficiency (PCE) (Sharma et al., 2022) as expressed in Eqs. (2) to (5), respectively.

$$J_{sc} = q\left[\int_{E_g}^{\infty}\left(\frac{dN_{\mathrm{ph}}}{dh\nu}\right)dh\nu\right] \tag{2}$$

$$V_{oc} = \frac{2kT}{q}\ln\left(\frac{\tau}{n_i}G\right) \tag{3}$$

$$FF = \frac{P_{mpp}}{V_{oc}J_{sc}} \tag{4}$$

$$PCE = \frac{V_{oc}J_{sc}FF}{P_{in}} \tag{5}$$

where, q is the electronic charge, N_{ph} is photogenerated carriers, v the frequency of irradiance and h is the Plank's constant. In Eq. (3), k is Boltzmann's constant, T is working temperature, τ is carrier lifetime, n_i is intrinsic carrier concentration and G is the rate of generation of photogenerated carriers. Further in Eqs. (4) and (5), P_{mpp} is the electrical power produced by the cell when it is at its highest point of power, and P_{in} is the input power from the irradiation. The band offsets at the interfaces in the device hinder the carriers as the barriers for electrons in the conduction band and the holes in the valance band. At the interface of the absorber and the HTL, the band offsets such as the valance band offset (VBO) and conduction band offset (CBO) are as expressed below in Eqs. (6) and (7) (Sharma et al., 2023b).

$$VBO_{HTL/Absorber} = |(\chi + E_g)_{HTL} - (\chi + E_g)_{Absorber}| \quad (6)$$

$$CBO_{HTL/Absorber} = |\chi_{HTL} - \chi_{Absorber}| \quad (7)$$

where, χ_{HTL} is the electron affinity of the HTL and $\chi_{Absorber}$ is the electron affinity of absorber layer material.

Since the p-Si absorber layer has a work function larger than that of the back metal electrode this increases the probability of parasitic recombination at the back electrode (Za'abar et al., 2021). This recombination arises because of a lower conduction band offset (CBO) at the interface of the metal back electrode and p-Si absorber layer. To reduce the interfacial recombination at the back electrode we have inserted an ingenious $MoSe_2$ HTL layer at the interface. Because the inserted HTL layer offers a higher conduction band than the absorber layer, there is reduced recombination loss at the interface between the metal back electrode and the absorber layer. Moreover, this HTL also creates an easy pathway for the hole to move in the direction of the rear electrode which enhances the current in the proposed solar cell (Sharma et al. 2023b).

3 Device Simulation Methodology

The proposed device has been modeled and simulated using the SCAPS-1D simulation tool. This tool allows us to model solar cell devices by accepting the numerical values of different parameters for any materials to be used in the modeling of the device. Then it generates several points (meshing) in the device where the provided action is to be performed i.e., the calculations are done by solving the semiconductor equations as mentioned in the earlier Sect. 2 (Burgelman et al., 2021). The required parametric specifications for the modeling of various layers of the proposed heterojunction photovoltaic device are gathered from the kinds of literature cited elsewhere or assumed in their respective viable range. The specifications used in modeling numerous layers of the projected solar cell device are mentioned in Table 1. The top/front layer is of transparent conducting oxide (TCO) material ALD-ZnO (atomic layer deposited zinc oxide) with the highest energy bandgap value of 3.3 eV (relative to the other materials used in the device) (Jha et al., 2023). Then 15 nm thin layer of CdS is added which acts as the buffer layer whose bandgap is 2.4 eV and the electron mobility is 50 cm^2V^{-1} s^{-1} (Sharma et al., 2022; Jha et al., 2022). Now the third layer from the top is the c-Si of 300 μm this serves the purpose of the absorber layer. The electron affinities of the buffer and the absorber is the same (~4.1 eV) (Jha et al. 2022) so the conduction band has a uniform level near the interface of the buffer and the absorber layer can be seen in Fig. 1. This allows the photogenerated carriers (electrons) in the conduction band to pass through the interface smoothly. Apart from these layers, we insert an ingenious $MoSe_2$ HTL having a bandgap of 1.1 eV, the lowest in the device layers (Tongay et al., 2012). The HTL layer possesses hole mobility of 100 cm^2V^{-1} s^{-1} and an electron affinity of 4.0 eV (Za'abar et al., 2021; Zhang et al., 2019). For the electrical connections, the metal electrodes with suitable work functions are stacked at the rear layer and the front layer. The device schematic layout of the proposed heterojunction solar photovoltaic device is illustrated in Fig. 2.

4 Results and Discussion

The impact of the various efficient properties of the HTL film was observed for the working parameters of the adopted heterojunction photovoltaic device. The obtained results are illustrated in the following subsections along with a thorough discussion based on the theoretical modeling as discussed in section 2. The analysis based on the thickness, shallow acceptor density, and electron affinity of the HTL layer of the proposed heterojunction photovoltaic device are covered in subsections 4.1–4.3, respectively. In subsection 4.4, it is analyzed how the standard and proposed heterojunction solar cell configurations compare based on their respective I-V characteristics.

4.1 Impact of the Thickness of the Ingenious $MoSe_2$ HTL on the Proposed Solar Cell Configuration

At first, the ingenious $MoSe_2$ HTL layer is introduced in the standard device. In this section, to seek information regarding the device performance the proposed device is simulated concerning the thickness of the HTL, and the results are shown in Fig. 3. It can be seen (Fig. 3b) that the V_{oc} shows a mostly independent trend with the HTL thickness it remains at 0.674 V whereas J_{sc} increases slightly with the HTL thickness. The increased HTL thickness makes the volume of the layer increase. Thus, the slight gain in the short circuit current (J_{sc}), (Fig. 3c) can be attributed to the increased carriers in the layer because of the larger no of carriers in the larger volume available. Alternatively, the HTL layer thickness also adds to the series resistance of

Table 1 Specifications for the modeling of the proposed heterojunction photovoltaic device

Parameters/layers*		*$MoSe_2$ (HTL)* (Za'abar et al., 2021; Tongay et al., 2012; Zhang et al., 2019)	*p-Si (absorber)* (Jha et al., 2022)	*n-CdS (buffer)* (Jha et al., 2022)	*p-Si/n-CdS Interface* (Sharma et al., 2022; Jha et al., 2022)	*ALD-ZnO (window)* (Jha et al., 2023)
t (μm)		0.100	300	0.015		0.012
E_g (eV)		1.1	1.12	2.4		3.3
χ (eV)		4.0	4.1	4.1		4.45
ϵ_r		7.9	11.9	10		9
N_C (cm^{-3})		2.8×10^{18}	2.8×10^{19}	2.8×10^{19}		2.2×10^{18}
N_V (cm^{-3})		1.8×10^{19}	1.04×10^{19}	2.8×10^{19}		1.8×10^{19}
v_{thn} (cms^{-1})		1×10^7	1×10^7	1×10^7		1×10^7
v_{thp} (cms^{-1})		1×10^7	1×10^7	1×10^7		1×10^7
μ_n ($cm^2V^{-1}\ s^{-1}$)		25	1099	50		100
μ_p ($cm^2V^{-1}\ s^{-1}$)		100	427	5		25
ND (cm^{-3})		0	0	1×10^{14}		1×10^{20}
NA (cm^{-3})		1×10^{20}	1×10^{16}	0		0
N_{t1}	State (eV)	$E_v + 0.6$	$E_c - 0.79$	$E_v + 0.6$	$E_v + 0.6$	$E_v + 0.6$
	N_t (cm^{-3} or cm^{-2})	1×10^{14}	1×10^{11}	1×10^{15}	1×10^{10}	1×10^{15}
N_{t2}	State (eV)		$E_c - 0.57$			
	N_t (cm^{-3} or cm^{-2})		1×10^{12}			

t = thickness, E_g = bandgap, χ = electron affinity, ϵ_r = relative dielectric permittivity, N_C = conduction band effective density of state, N_V = valence band effective density of state, v_{thn} = thermal velocity of electron, v_{thp} = thermal velocity of holes, μ_n = electron mobility, μ_p = hole mobility, ND = shallow donor density, NA = shallow acceptor density, N_{t1} = defect 1, N_{t2} = defect 2 and N_t = defect density.

the device. Hence the FF also bears a little negative slope for the increased thickness of the HTL (Bag et al., 2020). Further in Fig. 3a, the PCE is improved by minutely from 20.94% to 21.18% with increasing thickness of HTL. This improvement in the PCE of the proposed heterojunction solar cell device may be accounted for the reduced non-radiative recombination at the interface of HTL and the absorber layer (Bag et al., 2020). Therefore, the optimized thickness of HTL for the proposed configuration is selected as 0.1 μm (100 nm) for further simulation in this study.

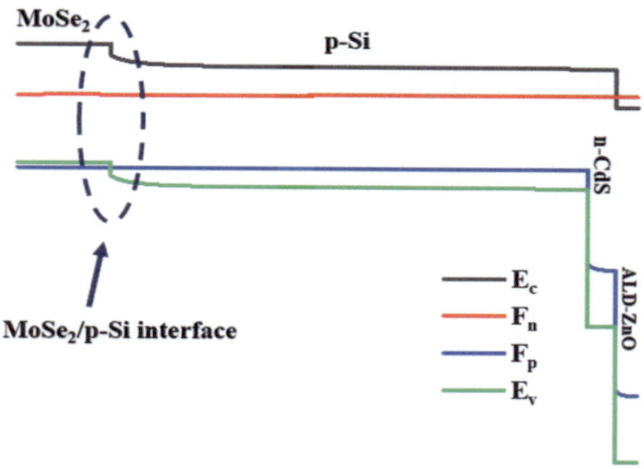

Fig. 1 Band alignment of proposed heterojunction photovoltaic device

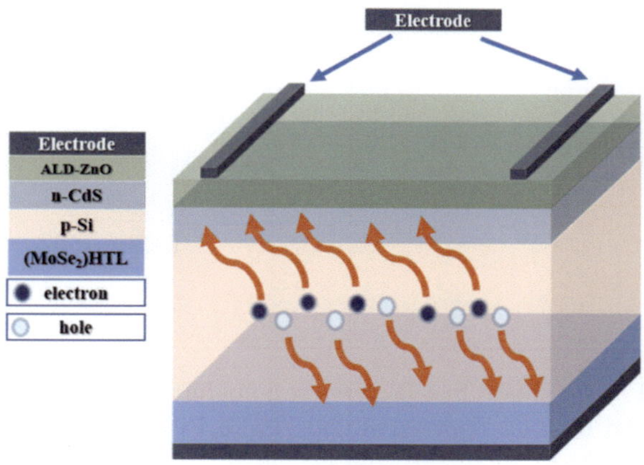

Fig. 2 Device model of the adopted heterojunction photovoltaic device

4.2 Impact of Shallow Acceptor Doping Level on the Performance of Adopted Cell Configuration

Since the doping level of any material controls various effective parameters of the materials. Therefore, we have simulated the proposed device as a function of shallow acceptor density of the HTL from 10^{15} cm^{-3} to 10^{20} cm^{-3} for the performance of the proposed solar cell. The V_{oc} gets improved from around 0.636 V to 0.672 V it shows a very slow rate of increment in the range of 10^{15} to 10^{16} cm^{-3} then it has large slop up to 10^{18} cm^{-3} after this level of doping the V_{oc} shows nearly saturation behavior till 10^{20} cm^{-3} as depicted in Fig. 4b (Bag et al. 2020). The larger doping level increases the carriers in the HTL material to be generated when it is illuminated. Therefore, the improved generation rate also increases the V_{oc} can be verified from Eqs. (2) and (3). On the other hand, the J_{sc} also shows the identical trend of slops as V_{oc}, it increased from 37.79 mA/cm^2 to 40.20 mA/cm^2 as shown in Fig. 4c. In

Fig. 3 Effect of the thickness of HTL from 0.01 to 0.1 μm on the cell performance parameters (3a) PCE, (3b) V_{oc}, (3c) J_{sc}, and (3d) FF of the adopted heterojunction photovoltaic device

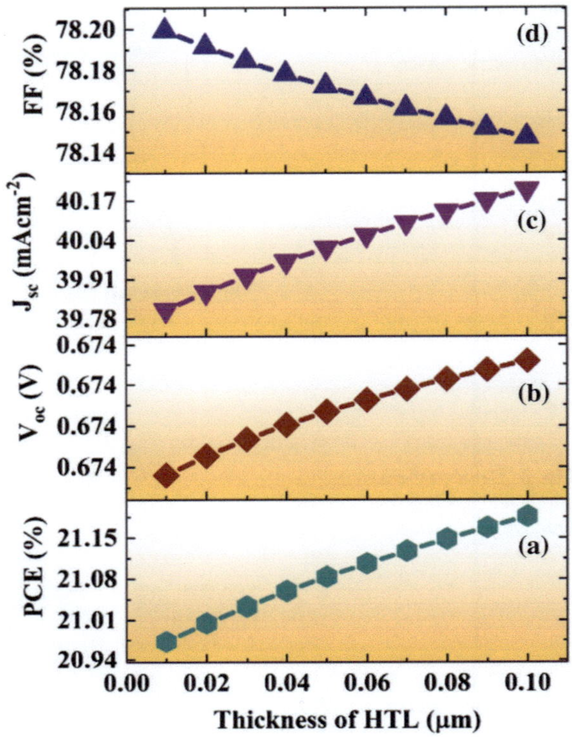

Fig. 4d shows that the FF has a negligible slope with FF = 77.95% till 10^{19} cm^{-3} and then it improved slightly to 78.15% up to 10^{20} cm^{-3}. As the conductivity of the device increased with the larger doping level consecutively the series resistance gets suppressed a little. Thus, the FF shows a slight increase due to lowered series resistance (Jha et al., 2023; Sharma et al., 2023b). Thus, the PCE of the device gets enhanced from 18.76% to 21.18% for the doping density of HTL from 10^{15} cm^{-3} to 10^{20} cm^{-3}. Therefore, the best-performing doping level of the ingenious $MoSe_2$ HTL layer is selected to be 10^{20} cm^{-3} for further simulation.

4.3 *Impact of Electron Affinity of HTL on FF and PCE of the Proposed Cell Configuration*

In heterojunction solar cells, the layers of the device like the absorber layer, electron transport layer or buffer layer, window layer, hole transport layer, etc. have different energy band gaps. These differences create band offsets (valance band offset (VBO) and conduction band offset (CBO) in their respective bands) at the interfaces which act as the potential barriers for the photogenerated carriers to be collected at respective

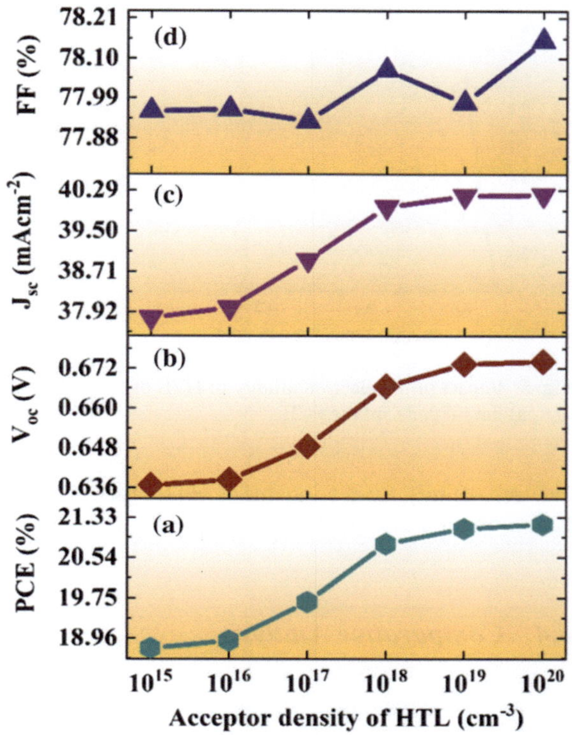

Fig. 4 Impact of doping level (shallow acceptor density) of HTL on (4a) PCE, (4b) V_{oc}, (4c) J_{sc}, and (4d) FF of the proposed heterojunction solar cell

electrodes. Therefore, to reduce the potential barriers due to band offsets the HTL is introduced in the proposed device. Furthermore, for analyzing the offsets the device is simulated in regards to the electron affinity of the HTL (χ_{HTL}) from 3.5 eV to 4.5 eV. In Fig. 5, the FF and PCE are plotted concerning the electron affinity of the HTL layer. The FF increases rapidly from 71.24% to 78.15% for the electron affinity up to 3.7 eV. After 3.7 eV the FF becomes almost independent of the higher electron affinities of HTL. The lower FF at $\chi_{HTL} = 3.5$ eV may be due to the formation of a new recombination pathway (form conduction band of the absorber to valance band of the HTL) at HTL/absorber interface for the photogenerated minority carriers. However, the bands of the HTL layer move downward due to the increased electron affinity of the layer. Further, it makes a barrier in the recombination path which was created due to the lower electron affinity. Ultimately this reduces the back recombination of the minority carriers (electrons). Also, the valance band creates a potential difference between the valance bands of the HTL and absorber layer which support the majority carriers (hole) to be collected at the back electrode as it can be assisted by Eqs. (6) and (7). This is why the FF increases rapidly and then saturates (Jha et al., 2022; Sharma et al., 2023b). Similarly, the PCE also has the minimum at 3.5 eV of electron affinity and increases rapidly finally attaining saturation after 3.7 eV. Thus, the optimum

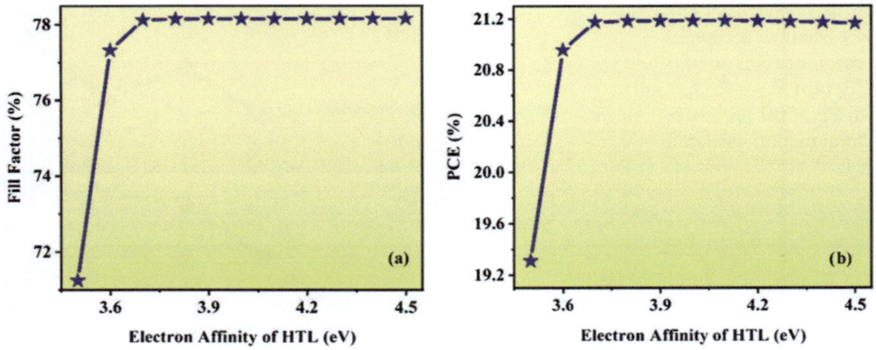

Fig. 5 Impact of the electron affinity of HTL of the proposed heterojunction photovoltaic device on (5a) the FF and (5b) the PCE

electron affinity for the ingenious HTL layer is selected at 4.0 eV for the proposed heterojunction solar cell.

4.4 Comparative Analysis

Hence, we analyze the performances after the introduction of the HTL layer in the standard solar cell configuration. From Fig. 6, it is evident that J_{sc} and Voc both get improved for the proposed configuration. Researchers reported the silicon solar cell device without the HTL layer and the observations were as $V_{oc} = 0.636$ V, $J_{sc} = 37.72$ mA/cm^2, FF = 75.82% and PCE = 18.19% (Sharma et al., 2022). Gao et al., (2018) reported the performance of MoO_3/p-Si/n-CdS/AZO configuration of the device with MoO_3 as the HTL layer. Their findings were as $V_{oc} = 0.543$ V, $J_{sc} = 28.75$ mA/cm^2, FF = 68.13% and PCE = 10.64% (Gao et al. 2018). Moreover, the performance parameters for some of the similar solar cells are given in Table 2.

The I-V characteristic of the proposed solar cell and the best-optimized performance $V_{oc} = 0.674$ V, $J_{sc} = 40.20$ mA/cm^2, FF = 78.15% and PCE = 21.18% also shown in Fig. 7.

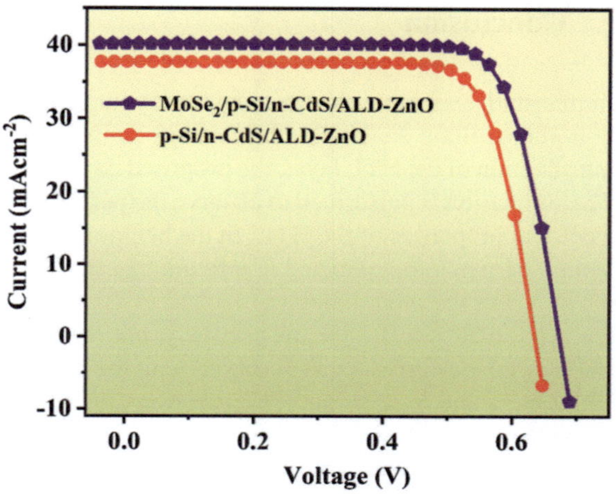

Fig. 6 Comparison of I-V curves for standard p-Si/n-CdS/ALD-ZnO solar cell configuration and proposed $MoSe_2$/p-Si/n-CdS/ALD-ZnO heterojunction solar cells

Table 2 Performance of different configurations of heterojunction silicon solar cells

Configurations	V_{oc} *(V)*	J_{sc} *(mA/cm^2)*	*FF (%)*	*η (%)*	*References*
MoO_3/p-Si/n-CdS/AZO	0.5430	28.75	68.13	10.64	(Gao et al., 2018)
p-Si/n-CdS/ITO	0.6360	37.72	75.82	18.19	(Sharma et al., 2022)
$MoSe_2$/p-Si/n-CdS/ALD-ZnO	0.6740	40.20	78.15	21.18	Our study

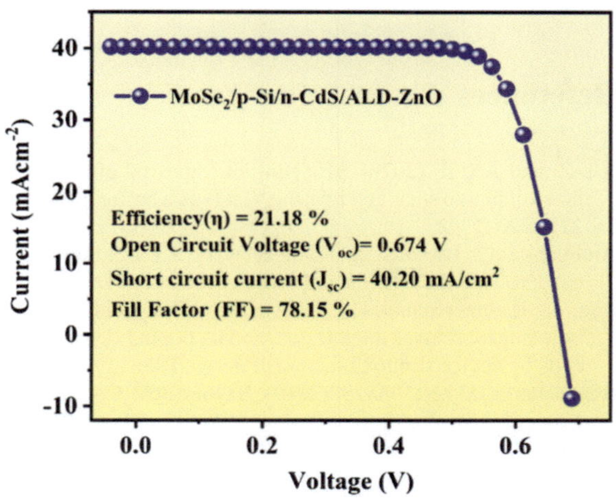

Fig. 7 The I-V curve for adopted $MoSe_2$/p-Si/n-CdS/ALD-ZnO cell configuration after optimization and the obtained performance parameters

5 Conclusion

In this study, we have chosen a device layout of p-Si/n-CdS/ALD-ZnO heterojunction photovoltaic device that introduced an ingenious $MoSe_2$ HTL layer. After the introduction of the HTL layer, the proposed device is simulated utilizing SCAPS-1D. The simulation is performed to observe the performance concerning the physical and electrical properties of the HTL. In the beginning, the effect of the thickness of HTL from 0.01 μm to 0.1 μm and it is found that the V_{oc} remains almost the same about 0.674 V. On the other hand, the J_{sc} and PCE grew a little from 39.80 to 40.20 mA/cm^2 and 20.97 to 21.18%, respectively, but the FF shows a trivial drop by 0.06%. Further, the performance is observed with the doping level of the HTL from 10^{15} to 10^{20} cm^{-3}, and V_{oc}, J_{sc} and PCE are noted to be increased from 0.636 to 0.674 V, 37.79 to 40.20 mA/cm^2, and 18.75 to 21.18%, respectively, whereas the FF rise trivially. Finally, the influence of the electron affinity from 3.5 to 4.5 eV of the HTL on the FF and the PCE is examined and it is seen that both the FF and PCE boosted quickly from 71.24 to 78.15% and 19.30 to 21.16%, respectively having the proper band alignment in the adopted device configuration. Conclusively, it is revealed that the optimum properties of the $MoSe_2$ HTL introduced in the proposed $MoSe_2$/p-Si/n-CdS/ALD-ZnO heterojunction photovoltaic device should have a thickness of 0.1 μm, shallow acceptor density of 10^{20} cm^{-3} and the electron affinity of 4.0 eV.

Acknowledgements The authors Atish Kumar Sharma and Ankita Srivastava contributed equally to this research work.

References

Ali, Z. and Ali, K. (2022). Modeling of improved efficiency and spectral response of a Si-based heterojunction solar cell by using CeO_2 as a buffer layer. *Journal of Computational Electronics*, **21:** 1320–1328.

Andreani, L.C., Bozzola, A., Kowalczewski, P., Liscidini, M. and Redorici, L. (2019). Silicon solar cells: toward the efficiency limits. *Advances in Physics: X*, **4:** 1548305.

Bag, A., Radhakrishnan, R., Nekovei, R. and Jeyakumar, R. (2020). Effect of the absorber layer, hole transport layer thicknesses, and its doping density on the performance of perovskite solar cells by device simulation. *Solar Energy*, **196:** 177–182.

Bhattacharya, S. and John, S. (2019). Beyond 30% Conversion Efficiency in Silicon Solar Cells: A Numerical Demonstration. *Scientific Reports*, **9:** 12482.

Blakers, A., Zin, N., McIntosh, K.R. and Fong, K. (2013). High-Efficiency Silicon Solar Cells. *Energy Procedia*, **33:** 1–10.

Burgelman, M., Decock, K., Niemegeers, A., Verschraegen, J. and Degrave, S. (2021). SCAPS manual: Single shot calculation. *Department of Electronics and Information Systems (ELIS) University of Gent*, pp. 55–60.

Deng, H., Chen, S., Ishaq, M., Cheng, Y., Sun, Q., Lin, X., Zheng, Q., Zhang, C. and Cheng, S. (2022). Efficient All-Inorganic Sb_2S_3 Solar Cells with Matched Energy Levels Using Sb_2Se_3 as Hole Transport Layers. *Solar RRL*, **6:** 2101017.

Fell, A., Greulich, J., Feldmann, F., Messmer, C., Schön, J., Bivour, M., Schubert, M.C. and Glunz, S.W. (2022). Modeling parasitic absorption in silicon solar cells with a near-surface absorption parameter. *Solar Energy Materials and Solar Cells*, **236:** 111534.

Gao, B., Zhao, Y., Cai, L., Liu, P., Liang, Z. and Shen, H. (2018). Fabrication of cadmium sulfide/ p type silicon heterojunction solar cells under 300 °C with more than 10% efficiency. *Solar Energy*, **173:** 635–639.

Green, M.A. (2009). The path to 25% silicon solar cell efficiency: History of silicon cell evolution. *Progress in Photovoltaics*, **17:** 183–189.

Hosen, A., Rashel, S. and Ahmed, A. (2022). Performance analysis of SnS solar cell with a hole transport layer based on experimentally extracted device parameters. *Journal of Alloys and Compounds*, **909:** 164823.

Jha, P.K., Chourasia, N.K., Sharma, A.K. and Chourasia, R.K. (2022). Optimization of electrical properties for performance analysis of p-Si/n-CdS/ITO heterojunction photovoltaic cell. *Materials Today: Proceedings*, **67:** 620–624.

Jha, P.K., Chourasia, N.K., Srivastava, A., Sharma, A.K., Kumar, R., Sharma, S., Kumar, M. and Chourasia, R.K. (2023). Study of Eco-Friendly Organic-Inorganic Heterostructure $CH_3NH_3SnI_3$ Perovskite Solar Cell via SCAPS Simulation. *Journal of Electronic Materials*, **52:** 4321–4329.

Liu, Y., Li, Y., Wu, Y., Yang, G., Mazzarella, L., Moya, P.P., Tamboli, A.C., Weber, K., Boccard, M., Isabella, O., Yang, X. and Sun, B. (2020). High-Efficiency Silicon Heterojunction Solar Cells: Materials, Devices, and Applications. *Materials Science and Engineering: R: Reports*, **142:** 100579.

NGOUPO, A.T. and NDJAKA, JM. B. (2022). Performance enhancement of Sb_2Se_3-based solar cell with hybrid buffer layer and $MoSe_2$ as a hole transport material using simulator device. *Discov Mechanical Engineering*, **1:** 5.

Noh, J.H., Jeon, N.J., Choi, Y.C., Nazeeruddin, Md.K., Grätzel, M. and Seok, S.I. (2013). Nanostructured $TiO_2/CH_3NH_3PbI_3$ heterojunction solar cells employing spiro-OMeTAD/Co-complex as hole-transporting material. *Journal of Materials Chemistry A*, **1:** 11842–11847.

Qiu, K., Qiu, D., Cai, L., Li, S., Wu, W., Liang, Z. and Shen, H. (2017). Preparation of ZnS thin films and ZnS/p-Si heterojunction solar cells. *Materials Letters*, **198:** 23–26.

Sabbah, H., Arayro, J. and Mezher, R. (2022). Numerical Simulation and Optimization of Highly Stable and Efficient Lead-Free Perovskite $FA_{1-x}Cs_xSnI_3$-Based Solar Cells Using SCAPS. *Materials*, **15:** 4761.

Sharma, A.K., Chourasia, N.K. and Chourasia, R.K. (2022). Optical, temperature, and bulk analysis theoretically in p-Si/n-CdS heterojunction solar cell. *Materials Today: Proceedings*, **67:** 632–636.

Sharma, A.K., Kumar, R., Jha, P.K., Kumar, M., Chourasia, N.K. and Chourasia, R.K. (2023a). Bulk Parameters Effect and Comparative Performance Analysis of p-Si/n-CdS/ALD-ZnO Solar Cell. *Silicon*, 1876–9918.

Sharma, A.K., Chourasia, N.K., Jha, P.K., Kumar, R., Kumar, M. and Chourasia, R.K. (2023b). Characteristic Features and Performance Investigations of a PTB7:PC71BM/PFN:Br Pure Organic Solar Cell Using SCAPS-1D. *Journal of Electronic Materials*, **52:** 4302–4311.

Sharma, A.K., Srivastava, A., Jha, P.K., Kumar, R., Kumar, M., Kulriya, P.K., Chourasia, N.K. and Chourasia, R.K. (2023c). Bulk/Interface Defects Engineering and Comparative Performance Analysis of p-Si/n-CdS/ALD-ZnO Heterojunction Solar Cell. *Energy Technology*, **11:** 2300169.

Singh, P. and Ravindra, N.M. (2012). Temperature dependence of solar cell performance-an analysis. *Solar Energy Materials and Solar Cells*, **101:** 36–45.

Singh, V.K., Nagaraju, J. and Avasthi, S. (2019). Radial junction silicon solar cells with micro-pillar array and planar electrode interface for improved photon management and carrier extraction. *Current Applied Physics*, **19:** 341–346.

Srivastava, S., Singh, S. and Singh, V.K. (2021). Bulk and interface defects analysis of n-CdS/p-Si heterojunction solar cell. *Optical Materials*, **111:** 110687.

Tongay, S., Zhou, J., Ataca, C., Lo, K., Matthews, T.S., Li, J., Grossman, J.C. and Wu, J. (2012). Thermally Driven Crossover from Indirect toward Direct Bandgap in 2D Semiconductors: $MoSe_2$ versus MoS_2. *Nano Letters*, **12:** 5576–5580.

Wan, Y., Samundsett, C., Bullock, J., Hettick, M., Allen, T., Yan, D., Peng, J., Wu, Y., Cui, J., Javey, A. and Cuevas, A. (2017). Conductive and Stable Magnesium Oxide Electron-Selective Contacts for Efficient Silicon Solar Cells. *Advanced Energy Materials*, **7:** 1601863.

Wang, Y., Wei, W., Liu, X. and Gu, Y. (2012). Research progress on polymer heterojunction solar cells. *Solar Energy Materials and Solar Cells*, **98:** 129–145.

Za'abar, F.I., Yusoff, Y., Mohamed, H., Abdullah, S.F., Mahmood Zuhdi, A.W., Amin, N., Chelvanathan, P., Bahrudin, M.S., Rahman, K.S., Samsudin, N.A. and Wan Abdullah, W.S. (2021). A Numerical Investigation on the Combined Effects of $MoSe_2$ Interface Layer and Graded Bandgap Absorber in CIGS Thin Film Solar Cells. *Coatings*, **11:** 930.

Zhang, Q., Zhang, S., Sperling, B.A. and Nguyen, N.V. (2019). Band Offset and Electron Affinity of Monolayer $MoSe_2$ by Internal Photoemission. *Journal of Electronic Materials*, **48:** 6446–6450.

Chapter 16
Fabrication of Silicon Nanowires for Solar Cell and Sensor Applications

Vrishty Kundu, Sathi Das, Omita Nanda, Dalip Singh Mehta, and Kanchan Saxena

1 Introduction

Nowadays, silicon nanowires (Si-NWs) are mostly considered for the development of electronic devices and commonly used in various optoelectronics, hybrid solar cell and sensing applications (Chartier et al., 2008). Silicon is an eco-friendly material, and its structure and surface can be easily engineered using a diverse range of nanofabrication methods (Li et al., 2014). The electrical and optical characteristics of Si-NWs have been found to be advantageous particularly in the fabrication of solar cells (Gayathri et al., 2022). Various methods have been utilized for the synthesis of Si-NWs, such as, metal-mediated vapor–liquid–solid technique, physical or chemical vapor deposition, etc. (Grover et al., 2020). The fabrication of Si-NWs using these techniques requires high vacuum equipment (Matsumoto et al., 2020). The other technique, namely, photolithography is used to produce the regular Si-NWs which requires mask for patterning of Si wafer (Mohajeri et al., 2015; Li et al., 2020). The metal-assisted chemical etching (MACE) approach (Bunthof et al., 2018) eliminates complicated vacuum systems and cumbersome instrumentation. The MACE process (Huang et al., 2009) is an easy and cheaper fabrication method and enables large-area manufacturing (Priolo et al., 2014). In addition, homogenous Si-NWs structure with high-crystallinity and large aspect ratio is achieved using MACE technique (Hong et al., 2013). Cong et al. (2018) studied the optical properties of Si-NWs using MACE technique (Cong et al., 2018). Das et al. (2023a, b) have fabricated Ag coated Si-NWs using MACE method and applied it for surface enhanced Raman spectroscopy (SERS) applications. Ag nanoparticles are used as metal catalysts during the etching process of silicon wafer using H_2O_2 and HF solution (Das et al., 2022a, b). Although

V. Kundu · O. Nanda · K. Saxena (✉)
Amity Institute of Renewable and Alternative Energy, Amity University Sector, 125 Noida, Noida, Uttar Pradesh 201303, India

S. Das · D. S. Mehta
Physics Department, Indian Institute of Technology Delhi Hauz Khas, New Delhi 110016, India

R. K. Kotnala et al. (eds.), *Advanced Functional Materials for Sustainable Environments*, https://doi.org/10.1007/978-3-031-62620-3_16

MACE technique was used by many researchers for the fabrication of Si-NWs, however, the dependence of the metal nanoparticles (NP) size on the formation of NWs is not described (Karaağaç et al., 2022; Agnihotri et al., 2020; Cao et al., 2010 and Das et al., 2023a, b).

The present study provides a brief description of simple two-step MACE fabrication process that produces Si-NWs without the need for lithography, using p and n type Si wafers. The impact of size and inter-particle spacing between Ag nanoparticles (Ag NP) on Si-NWs formation was analyzed. According to the investigation, the etching time affected the size of the Si-NWs which is largely dependent on the size and concentration of Ag NP. The details of MACE method and further discussion on the morphology of fabricated Si-NWs are subsequently provided in different sections. The FDTD simulation was used to determine the absorption efficiency of the fabricated Si-NWs. The results showed an increased absorption cross-section compared to a flat Si wafer.

2 Experimental Details

2.1 Fabrication of Si-NWs Using p-type Si Wafer

A p-type single crystalline <100> oriented Si wafer was cleaned using propanol, deionized water (DI) ultrasonically for removal of dust and other contamination. Next, the wafers were cleaned using piranha solution (comprising of H_2SO_4 and H_2O_2 with molar ratio (3:1)) for 30 min to remove organic and inorganic contamination. The native oxide layer formed on the Si wafers were further removed using HF buffer solution followed by rinsing thoroughly in DI water and consequently dried by nitrogen gun.

The cleaned wafers were utilized for the preparation of Si-NWs by a two-step MACE method, as described in Fig. 1. The wafers were first dipped inside an etchant solution (HF + $AgNO_3$) for electro-less Ag NP deposition. A batch of Si (p-type) wafers were immersed into the solution having 22.6 M HF and 12.5 mM $AgNO_3$ for 40 s and the other batch of wafers were dipped into a solution of 22.6 M HF and 15 mM for 40 s. The wafer color changes to shiny grey due to the deposition of Ag NP. Further after rinsing with DI water the wafers were dried and placed inside another fresh etching solution of 22.6 M HF, 9.8 M H_2O_2 and DI water (2:1:1) for various time intervals (75 min. and 100 min.). Hence the Si-NWs were fabricated, indicating a color change of the wafers to brown. To remove extra Ag^+ ions the wafers washed with DI water and kept inside 70% HNO_3 for 10 min. Finally, the wafers were again dipped inside buffer solution to remove the native oxide layers, washed with DI water and dried using the nitrogen gun.

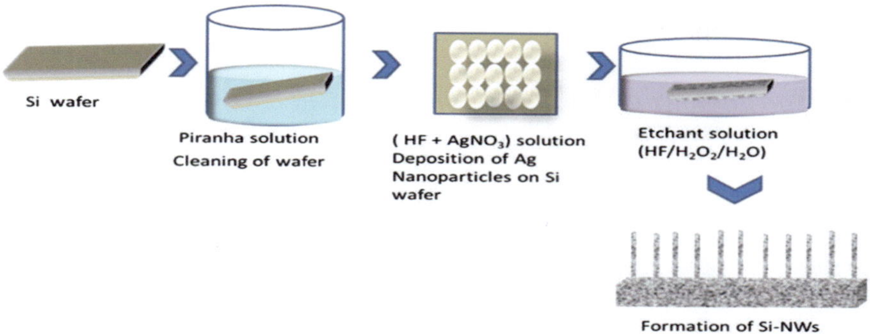

Fig. 1 Fabrication of Si-NWs using MACE method

2.2 *Fabrication of Si-NWs Using n-type Si Wafer*

The cleaning step of n-type Si wafer was similar as discussed in the previous section. The cleaned wafers were placed inside an etched solution of 4.6 M HF and 15 mM $AgNO_3$ for 60 s. Next the wafers were immediately placed inside another etchant solution of 4.6 M HF and 0.4 M H_2O_2 in a volumetric ratio of 1:1 and kept for two different time intervals (75 min and 90 min). Hence, the Si-NWs were formed. Next the wafers were dipped inside 70% HNO_3 solution to further remove the Ag^+ ions and placed inside the buffer HF for native oxide removal. Finally, the wafers were thoroughly washed using DI water and dried with nitrogen gun. The fabrication details of all the samples for p-type (P1 and P2) and n-type (N1 and N2) silicon wafers are mentioned in Table 1.

Further, the analysis of absorption cross-section of fabricated Si-NWs was performed using three-dimension finite differential time domain (3D-FDTD) methods.

Table 1 Fabrication of Si-NWs using different concentrations of $AgNO_3$, HF and H_2O_2 for p-type (P1, P2) and n-type (N1, N2) Si wafer for different etching duration

Sample	Composition for synthesis of Ag nanoparticles		Si-wafer etchant composition and etching time		
	HF (M)	$AgNO_3$ (mM)	HF (M)	H_2O_2 (M)	Time (min.)
P1	22.6	12.5	22.6	9.8	75
P2	22.6	15	22.6	9.8	100
N1	4.6	15	4.6	0.4	75
N2	4.6	15	4.6	0.4	90

3 Results and Discussion

3.1 Morphology of Nanostructures

The size and concentration of Ag nanoparticles affects the morphology of etched Si wafers. Figure 2 ((a) and (b)) shows the SEM images of different sizes of Ag nanoparticle at different $AgNO_3$ concentrations on p-type Si wafer. Figure 3 (a) and (b) shows the SEM images of p-type Si-NWs fabricated for different etching time and different concentration of $AgNO_3$. Figure 3 ((c) and (d)) show the SEM images of n-type Si-NWs for different etching time.

The scanning electron microscopy (SEM) images displayed the morphology of the Ag NP coated on Si wafer (Fig. 2 (a) and (b)) in the etching step 1 that strongly affected the morphology of etched Si wafer. Three things are essential for the two-step MACE method: (i) Ag NP used for metal catalysis process; (ii) HF as a complexing agent; and (iii) H_2O_2 as an oxidant that has larger reduction potential than silicon. In a two-step MACE process, the first step includes the deposition of Ag NP on Si wafer, while the second step involves the formation of Si-NWs. Ag^+ ions induce holes on the Si wafer when immersed in the etchant solution of HF and H_2O_2. During this process Si gets oxidized and SiO_2 is formed which can be dissolved in HF solution giving rise to the formation of Si nanowires. As observed, from Fig. 2 ((a) and (b)), the concentration of HF and $AgNO_3$ controls the size of the deposited Ag NP. For 12.5 mM $AgNO_3$ in 22.6 M HF etchant solution, the deposited Ag NP size was 300 nm, and for 15 mM $AgNO_3$ in 22.6 M HF solution, the size of the Ag NP was 200 nm. Thus, the increase in concentration of $AgNO_3$ results in a decrease in the inter-particle separation and a reduction in particle size, accordingly. Figure 3 ((a) and (b)) show the SEM images of p-type Si-NWs for different concentration of $AgNO_3$ (12.5 mM, 15 mM) for different etching time (75 min, 100 min), respectively. The sequential investigation with etching time resulted in the formation of nanowires having long aspect ratio (from 800 nm to 6 μm), for longer etching time. The increase in concentration of

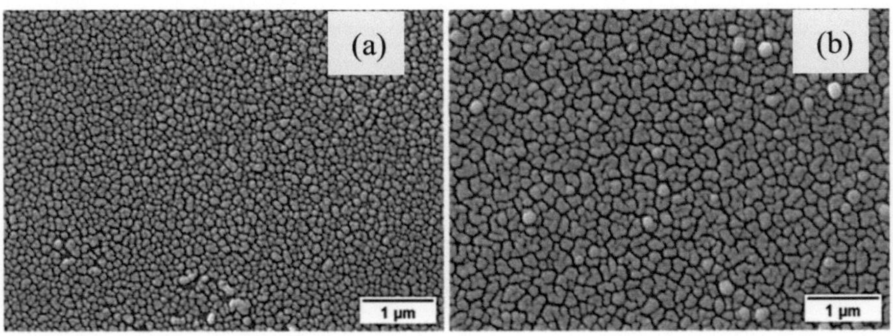

Fig. 2 SEM images of Ag nanoparticles deposited on p-type Si wafers for: (a) 300 nm Ag nanoparticle size and (b) 200 nm Ag nanoparticle size

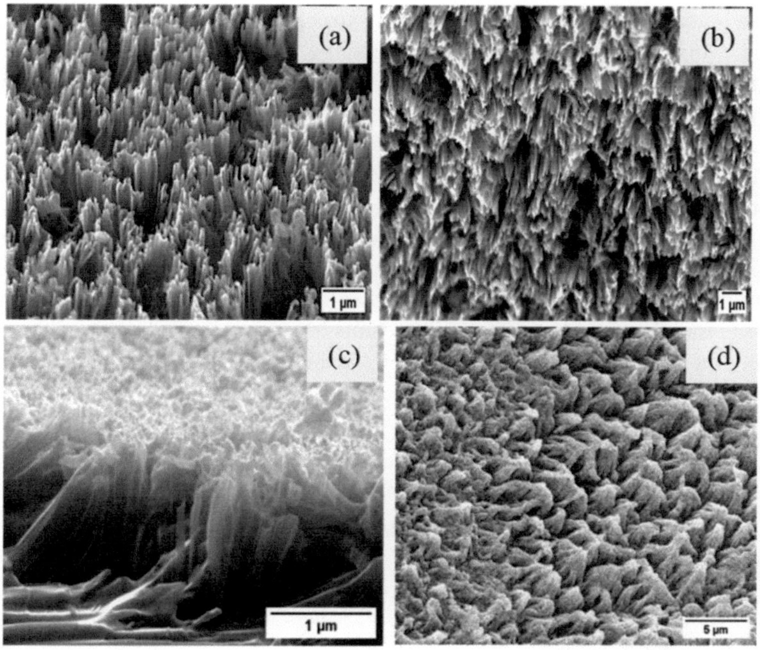

Fig. 3 SEM images of morphology of: (a) p-type Si-NWs fabricated from batch P1, (b) p-type Si-NWs from batch P2, (c) n-type Si-NWs from batch N1 and (d) n-type Si-NWs from batch N2

$AgNO_3$ results densely spaced Si-NWs further implying the coating of tightly spaced Ag NP on Si wafer in the first etching step. Figure 3 ((c) and (d)) represent the SEM images of n type Si-NWs fabricated from 15 mM $AgNO_3$ solution and 0.4 M H_2O_2 for different etching time (75 min, 90 min). As observed from figure, the formation of long nanowires obtained for 75 min etching time, where the length of Si-NWs is 1.1 μm. As the etching time increased, the individual Si-NWs got accumulated and formed bundles at various locations as shown in Fig. 3 (d).

3.2 3D-FDTD Simulation Analysis

For the analysis of absorption cross-section of fabricated Si-NWs using three-dimension finite differential time domain (3D-FDTD) methods, the simulation geometry of Si-NWs was taken considering the morphology captured from SEM images. The length and radius of the nanowire was taken to be 500 nm and 100 nm, respectively, with a gap of 40 nm. The optical parameters were considered from Palik data (Das et al., 2021; Robak et al., 2014; Das et al., 2022a, b). The boundary condition was considered as perfectly matched layer (PML). The simulation mesh was set to be auto-non uniform where the entire structure was covered with an override mesh of 3 nm. The incident light source having plane wave front was considered ranging from 200 to 1200 nm. To restrict the simulation size, only four nanowires were constructed. The simulation geometry is shown in Fig. 4.

To compare the absorption efficiency of Si-NWs, a flat Si wafer geometry was chosen considering all other parameters same.

Figure 5 shows the absorption cross section of Si-NWs and a flat Si wafer, using 3D FDTD simulation. The simulation results showed an increased amount of absorption cross-section in Si-NWs geometry compared to the flat Si wafer. The increase in absorption for Si-NWs is due to the nanowire geometry. The nanowires acting as a cavity can trap the incident electric field results in an increased amount of absorption (Stratakis et al., 2013). Thus, the Si-NWs exhibited enhanced optical properties compared to a flat Si wafer.

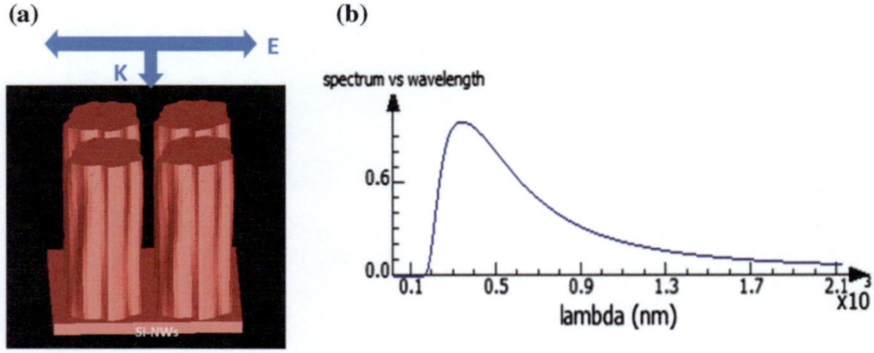

Fig. 4 (a) Simulation set-up and (b) source spectra of total-field scattered-field chosen for the simulation

Fig. 5 Simulated absorption spectra of Si-NWs and a flat Si wafer

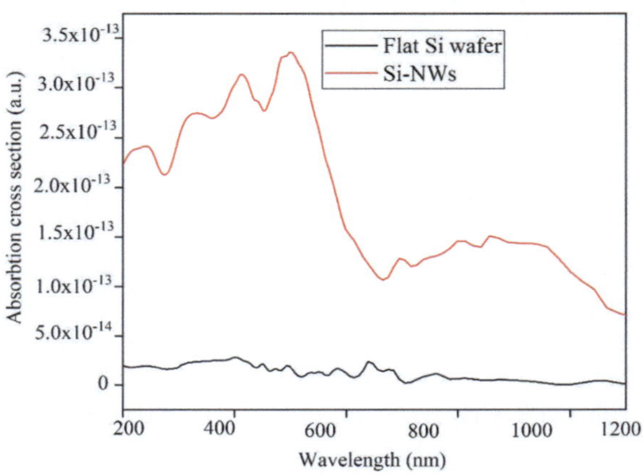

4 Conclusion

Silicon nanowires have been successfully fabricated using MACE method for solar cells and sensor application. Fabrication protocol of Si-NWs from p- and n-type Si wafers was developed. It is well-known that the maximum absorption of solar radiation and large hetero-junction area are the key factors for achieving high efficiency of a hybrid solar cell. The fabricated Si-NWs offer a large surface area and increased light absorption due to its inherent anti-reflection properties as compared to plain Si wafer. Thus, the effective large area offers more contact surface for the hetero-junction solar cell applications. Also, the FDTD simulation revealed an enhanced absorption cross-section of the fabricated Si-NWs compared to a flat Si wafer. In future, the performance of fabricated Si-NWs as inorganic layer for hybrid solar cell applications will be investigated thoroughly. The performance of Si-NWs with coating using plasmonic nano-particles will also be inspected for different sensing applications.

Acknowledgements The authors are grateful to Dr. Ashok K. Chauhan, Founder President, Amity University, Noida, India for his continuous encouragement in this work.

References

Agnihotri, S.K., Samajdar, D.P. and Arefinia, Z. (2020). Design of INP-based truncated nanopyramid solar cells with conformal coating of PEDOT: PSS for improved light harvesting efficiency. *Optical Materials,* **110:** 110475.

Bunthof, L.A.A., Haverkamp, E.bJ., Woude, D.V.D., Bauhuis G.J., Corbeek, W.H.M., Veelenturf , S., Vlieg, E. and Schermer, J.J. (2018). Influence of laterally split spectral illumination on multi-junction CPV Solar Cell Performance. *Solar Energy,* **170:** 86–94.

Chartier, C., Bastide, S. and Lévy-Clément, C. (2008). Metal-assisted chemical etching of silicon in Hf–H_2O2. *Electrochimica Acta,* **53:** 5509–5516.

Cong, L.T., Lam, N.T.N., Giang, N.T., Kien, P.T., Dung, N.D., Ha, N.N. (2018). N-type silicon nanowires prepared by silver metal-assisted chemical etching: Fabrication and Optical Properties. *Materials Science in Semiconductor Processing,* **90:** 198–204.

Cao, L., Fan, P., Vasudev, A.P., White, J.S., Yu, Z., Cai, W., Schuller, J.A., Fan, S. and Brongersma, M.L. (2010). Semiconductor nanowire optical antenna solar absorbers. *Nano Letters,* **10:** 439-445.

Das, S., Saxena, K., Tinguely, J.C., Pal, A., Wickramasinghe, N.L., Khezri, A., Dubey, V., Ahmed, A., Vivekanandan, P., Ahmad, R., Wadduwage, D.N., Ahluwalia, B. S. and Mehta, D.S. (2023). SERS nanowire chip and machine learning enabled instant identification and classification of clinically relevant wild-type and antibiotic resistant bacteria at species and strain level. *bioRxiv.*

Das, S., Saxena, K., Khezri, A., Dubey, V., Ahmad, A., Tinguely, J.C., Mehta, D.S., Ahmad, R. and Ahluwalia, B. (2023). Rapid identification of antimicrobial drug resistance strains of E-coli using SERS nanowire chip. *Plasmonics in Biology and Medicine XX.*

Das, S., Saxena, K., Goswami, L.P., Gayathri, J. and Mehta, D.S. (2022). Mesoporous AG–Tio_2 based nanocage like structure as sensitive and recyclable low-cost SERS substrate for biosensing applications. *Optical Materials,* **125:** 111994.

Das, S., Goswami, L.P., Gayathri, J., Tiwari, S., Saxena, K. and Mehta, D.S. (2021). Fabrication of low cost highly structured silver capped aluminium nanorods as SERS substrate for the detection of biological pathogens. *Nanotechnology,* **32:** 49.

Das, S., Saxena, K. and Mehta, D.S. (2022). A highly sensitive SERS substrate based on a mesoporous AG–Tio_2 thin film for the detection of Dye Molecules. *Materials Advances,* **3:** 5337–5343.

Gayathri, J., Mehta, D.S. and Saxena, K. (2022). Recent advances in hybrid organic–inorganic perovskite solar cells with different halides and their combinations. *Springer Proceedings in Energy,* 21–29.

Grover, R., Gupta, N., Nanda, O. and Saxena, K. (2020). Aligned zinc oxide nanostructures for dye-sensitized solar cells: A Review. *Advances in Solar Power Generation and Energy Harvesting,* 9–16.

Hong, L., Rusli, Wang, X., Zheng, H., Wang, H., Yu Yu, H. (2013). Design guidelines for slanting silicon nanowire arrays for solar cell application. *Journal of Applied Physics,* **114**.

Huang, R., Fan, X., Shen, W. and Zhu, J. (2009). Carbon-coated silicon nanowire array films for high-performance lithium-ion battery anodes. *Applied Physics Letters,* **95**.

Karaağaç, H., Peksu, E., Alhalaili, B. and Islam, M.S. (2022). One-dimensional silicon nano-/microstructures based opto-electronic devices. *Topics in Applied Physics,* 731–766.

Li, S., Ma, W., Zhou, Y., Chen, X., Xiao, Y., Ma, M. Z. W. and Wei, F. (2014). Fabrication of porous silicon nanowires by Mace method in HF/H_2O2/AGNO3 system at room temperature. *Nanoscale Research Letters,* 9.

Li, J., Chen, X., Yi, Z., Yang, H., Tang, Y., Yi, Y., Yao, W., Wang, J. and Yi, Y. (2020). Broadband solar energy absorber based on monolayer molybdenum disulfide using tungsten elliptical arrays. *Materials Today Energy,* **16:** 100390.

Matsumoto, A., Son, H., Eguchi, M., Iwamoto, K., Shimada, Y., Furukawa, K. and Yae, S. (2020). General Corrosion during metal-assisted etching of *n*-type silicon using different metal catalysts of silver, Gold, and platinum. *RSC Advances,* **10:** 253-259.

Mohajeri, A. and Omidvar, A. (2015). Fullerene-based materials for solar cell applications: Design of novel acceptors for efficient Polymer Solar Cells – a DFT study. *Physical Chemistry Chemical Physics,* **17:** 22367-22376.

Priolo, F., Gregorkiewicz, T., Galli, M. and Krauss, T.F. (2014). Silicon Nanostructures for photonics and Photovoltaics. *Nature Nanotechnology,* **9:** 19-32.

Robak, E., Grześkiewicz, B. and Kotkowiak, M. (2014). Absorption enhancement in silicon nanowire-optical nanoantenna system for photovoltaic applications. *Optical Materials,* **37:** 104-109.

Stratakis, E. and Kymakis, E. (2013). Nanoparticle-based Plasmonic Organic Photovoltaic Devices. *Materials Today,* **16:** 133-146.

Chapter 17
Thermoelectric Response of Vacancy Order Double Perovskite K_2PtBr_6 for Energy Harvesting Applications

Mudasir Younis Sofi, Mohd Shahid Khan, Javid Ali, and M. Ajmal Khan

1 Introduction

In the race against global energy constraints, sustainable and renewable energy sources have consistently been emphasized (Ajia et al., 2018; Kojima et al., 2009; Sofi et al., 2022). In addition to solar energy, converting waste heat into electrical energy is a significant renewable energy source. We require good thermoelectric materials to fully utilise this waste heat (Jing et al., 2019; Tan et al., 2016). The advantages of thermoelectric generators are their durability, scalability, compact size and lack of moving parts. To have the maximum thermoelectric figure of merit (zT), the material must have a high Seebeck coefficient (S), a high electrical conductivity (σ) and a low thermal conductivity (k); $zT = \frac{S^2\sigma}{k} T$. However, it is difficult to achieve a high zT in a single system due to tight coupling between these parameters and associated trade-off relationship.

In recent times, metal halide perovskites have shifted the paradigm in the pursuit of high-performance materials. This might be due to their diverse structural and compositional properties, which allow a wide range of functional capabilities (Kojima et al., 2009). Due to their distinctive structural characteristics and lattice dynamics, halide perovskites have recently received an immense interest in thermoelectric energy conversion and solar cell applications. The ultra-low lattice thermal conductivity observed in lead halide perovskites makes them important for thermoelectric device applications (Xie et al., 2020). These appealing materials, however, have long-term instability and lead poisoning problems. These disadvantages have prompted scientists to investigate other perovskites with different composition and structures. In this aspect, vacancy-ordered lead-free halide double perovskites have proven to be a major breakthrough, outperforming lead halide perovskites in structural, thermodynamic and dynamic stabilities. Recently, Mahmood et al. (2022) scrutinized the

M. Y. Sofi · M. S. Khan · J. Ali · M. A. Khan (✉)
Department of Physics, Jamia Millia Islamia, New Delhi 110025, India
e-mail: majkhan@jmi.ac.in

R. K. Kotnala et al. (eds.), *Advanced Functional Materials for Sustainable Environments*, https://doi.org/10.1007/978-3-031-62620-3_17

thermoelectric and optoelectronic characteristics of $K_2Pd\ (Cl/Br)_6$ and concluded that these materials exhibit excellent thermoelectric and optical outputs. Moreover, it was discovered that these materials are dynamically stable.

Motivated by these efforts, this article investigated the thermoelectric and optical response of vacancy-ordered double perovskite K_2PtBr_6. The structural, elastic and optical properties of this material we are previously studied by (Naceur et al., 2022). However, they employed the HSE functional in their calculations, which occasionally overestimated band gap owing to PAW pseudopotentials. In this study, we implemented the GGA + mBJ functional, which is entirely *ab-initio* and performs exceedingly well for the *d/f*-electron systems (Sofi et al., 2022). The adoption of mBJ culminates in a band gap of 1.29 eV, which is in the visible range of the spectrum, prompting us to look into the thermoelectric and optical characteristics of proposed material. As of now, no theoretical studies have been undertaken to study thermoelectric response of the present material. Therefore, we performed calculations within Wien2k algorithm to determine the thermoelectric and optical properties of vacancy-ordered K_2PtBr_6 double halide perovskite.

2 Computational Details

The electronic, thermoelectric and optical characteristics of vacancy-ordered double perovskite K_2PtBr_6 are investigated using the Wien2k code (Schwarz et al., 2002), which utilizes the FP-LAPW method, in which the crystal unit cell is divided into non-overlapping muffin tin spheres and interstitial spaces. The Brillouin zone is integrated using a 1000-*k* point compact mesh. The core and valence states have an energy difference of –6.0 Ry. Convergence is controlled by the cut-off variable R_{MT} $K_{max} = 7$ Ry, where K_{max} is the reciprocal lattice vector and R_{MT} shows the MT sphere radii. The convergence criterion for energy is set at 10^{-4} Ry, and the total charge difference between the two following contacts is calibrated to be less than 10^{-4} e/a.u.^{3}. The BoltzTraP code (Madsen et al., 2006) as integrated with the Wien2k programme has been used to explore the thermoelectric properties. The calculations of transport coefficients were accomplished by employing a denser mesh of 100,000-*k* points.

K_2PtBr_6 crystallises in the face-centered Fm-3 m space group, with K atoms at the 8c Wyckoff positions and (0.25, 0.25, 0.25) coordinates, Pt atoms at the 4a Wyckoff positions and (0, 0, 0) coordinates, and Br atoms at the 24e Wyckoff positions and (x, 0, 0) coordinates, where x is roughly 0.25. While the Pt atoms are located at the corners and face-centered locations of the $[PtBr_6]$ octahedra, each K atom is sandwiched between the $[PtBr_6]$ octahedra and is surrounded by 12 Br atoms. The presence of K atoms in the octahedral cage behaves as heavy rattlers, leading to lattice anharmonicity. The cubic K_2PtBr_6 is shown in Fig. 1(a).

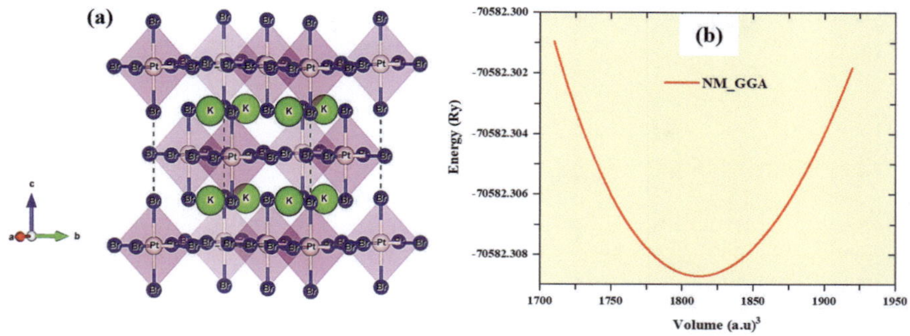

Fig. 1 (a) Crystal structure and (b) Optimized energy-volume plot of vacancy-ordered K_2PtBr_6 double perovskites

3 Results and Discussion

The ground state relaxed structure of the given material has been obtained through structural optimizations within the GGA scheme by performing the least square fit of the crystal energy using the Birch Murnaghan EOS equation, defined elsewhere (Sofi et al., 2022). The optimsed E-V curve is shown in Fig. 1(b). Table 1 gives the retrieved structural parameters such as the relaxed lattice constant, bulk modulus, pressure derivative of bulk modulus, and ground state energy. These are found to be in excellent accord with other comparable studied materials.

Furthermore, the electronic band structure and density of states of the given material have been computed to identify the band gap and nature of electronic states, which must be known in order to determine its thermoelectric and optoelectronic characteristics. The mBJ potential has been adopted to assess the band gap accurately since GGA usually underestimates it for highly correlated systems. The calculated band structure and density of states are shown in Fig. 2(a–b). The proposed material clearly displays semiconducting characteristics, having a small band gap of 1.29 eV in the GGA + mBJ approximation. The calculated band gap values exhibit decent agreement with those reported for other similar compounds in the existing literature (Bhamu et al., 2018; Naceur et al., 2022; Q. Mahmood et al., 2022). The previously reported band gap value for the present material was determined to be 1.53 using the HSE potential (Naceur et al., 2022), which relies on the PAW approach and involves

Table 1 Calculated values of lattice constant (a_0 in Å), bulk modulus (B_0), pressure derivative of bulk modulus (B_0') and ground state energy (E_0 in Ry) of cubic K_2PtBr_6 compound

Compound		B_0 (GPa)	B_0'	E_0
K_2PtBr_6	10.24	35.50	4.87	–70,582.308723
K_2PdBr_6 (Q. Mahmood et al., 2022)	10.57	20.9	–	–

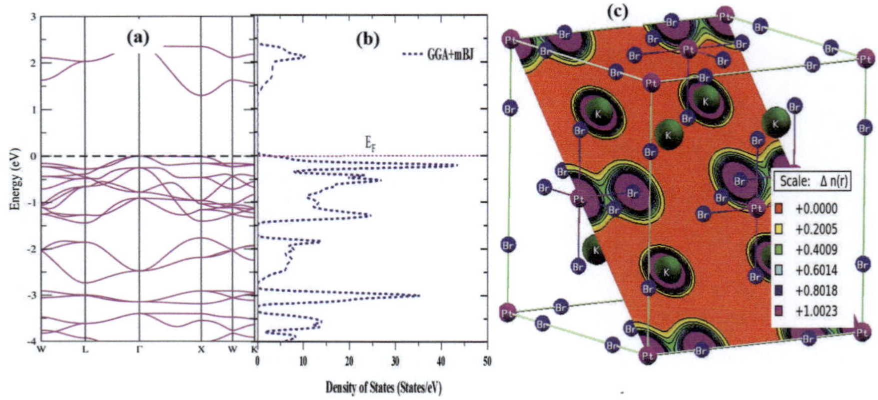

Fig. 2 (a–b) Plot of the band structure and density of states and (c) Charge density graphics of double halide K_2PtBr_6 perovskite

empirical input. In this study, we opted for the purely *ab-initio* mBJ potential, known to perform effectively for systems with *d/f* electrons.

Moreover, it is worth noting that the valence band maximum (VBM) and conduction band minimum (CBM) share identical symmetry points, resulting in a direct band gap value of 1.29 eV. This observation underscores the dispersive nature of electronic states near the Fermi level, suggesting that carriers exhibit low effective mass. This, in turn, implies enhanced mobility, which is conducive to the favorable thermoelectric and optoelectronic performance of the present material.

The charge density plot along 011 plane shown in Fig. 2(c) indicates that polar covalent nature of the bonding (admixture of covalent and ionic bonding) is preserved within the lattice structure of K_2PtBr_6 perovskite. The thermoelectric coefficients, including the Seebeck coefficient (S), electrical conductivity (σ/τ), thermal conductivity (κ/τ) and figure of merit (zT), have been computed using the BoltzTraP method (Madsen et al., 2006) under the assumption of constant relaxation time. The obtained thermoelectric plots against chemical potential ($-2 \leq \mu \geq 3$) at different temperatures are shown in Fig. 3(a–d). The chemical potential refers to the energy needed to either add or remove electrons from a system, taking into account the presence of Coulomb repulsion. This results in a positive or negative chemical potential, respectively. When electrons are added to the system, the electron density increases, leading to an *n*-type behavior. Conversely, removing electrons increases the number of void spaces or hole carriers, leading to a *p*-type behavior. Moreover, the electronic density of states (DOS) analysis of the material reveals that the valence band has a higher number of accessible states compared to the conduction band, indicating a substantial presence of hole carriers and supporting the predominantly *p*-type nature of the studied material.

Seebeck estimates the induced thermoelectric voltage (ΔV) as a function of temperature difference (ΔT) across the material and expresses it as $S = \frac{\Delta V}{\Delta T}$, where the addition or removal of electrons (doping) is depicted in contrast to the repulsive

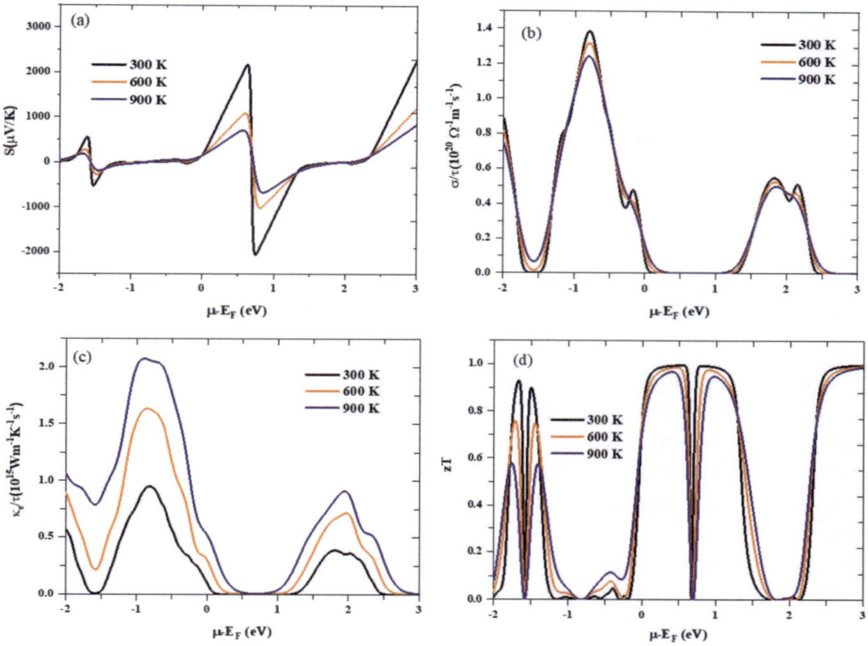

Fig. 3 Transport coefficient plots of: (a) Seebeck, (b) electrical conductivity, (c) thermal conductivity and (d) figure of merit (zT) against chemical potential at different temperatures (300 K, 600 K, 900 K)

forces of electrons previously present in the material. The Seebeck coefficient of the given material has been plotted against chemical potential at different temperatures, (Fig. 3(a)). It is clear that over a selected range of chemical potential, Seebeck exhibits several peaks and troughs in its behavior. At 300 K, the region of *n*-type doping yields the highest value of 2000 $\mu V/K^{-1}$. Table 2 presents the documented values of the Seebeck coefficient, computed alongside other theoretical findings. The results clearly indicate that the material under study exhibits higher Seebeck coefficients compared to previously reported perovskites.

Table 2 Calculated values of 'Seeback coefficient (S)' and figure of merit (zT) for the present study in comparison with other results available in the literature

Parameter	K_2PtBr_6 (Present)	K_2PdBr_6 (Q. Mahmood et al., 2022)	Cs_2PdBr_6 (Bhamu et al., 2018)
Seebeck coefficient (S)	2000	240	300
Figure of merit (zT)	1.00	0.75	0.83

As the temperature escalates, both types of carriers get thermally stimulated (bipolar effect), leading to a drop in Seebeck, as anticipated by the Mott formula (M. Jonson and Mohan 1980). $S = \frac{\pi^2 k_B^2 T}{3e}\left\{\frac{1dn(\varepsilon)}{nd\varepsilon} + \frac{1dn(\varepsilon)}{\mu d\varepsilon}\right\}_{\varepsilon=\mu}$.

The first term of this equation states that steepness in electronic DOS enhances thermopower while second term in Mott formula incorporates the derivatives of carrier density with respect to energy and chemical potential, and their relationship with charge carrier mobility, to describe the contribution of mobility to the Seebeck coefficient.

The present material's electronic DOS reveals that the *n*-type region covers fewer electronic states (carriers) than the *p*-type, resulting in a greater thermopower for the *n*-type region since Seebeck varies inversely with carrier concentration (n), as illustrated by the Mott relation. Nevertheless, Seebeck coefficient, on the other hand, decreases with increasing temperature and chemical potential, which might be related to thermal excitations of both types of carriers (bipolar effect). However, bipolar effect appears to have less effect in materials with band gaps greater than 8 K_BT. Since the proposed material bears a band gap value of 1.29 eV and hence the bipolar effect is not observed below 900 K, indicating that this material might be viable for thermoelectric technology.

The electrical conductivity over relaxation time (σ/τ) is graphically shown in Fig. 3(b). Electrical conductivity exhibits peaks in both doping zones, although the highest value of 1.4×10^{20} ($\Omega^{-1}\ m^{-1}\ s^{-1}$) is retrieved in the *p*-type doping region at 300 K. These peaks coincide with electronic DOS, indicating that the existence and absence of electronic states precisely reflects electrical conductivity's factual behaviour. Electronic states terminate when $\mu = 0$, and hence conductivity drops to zero.

The total thermal conductivity is contributed by electronic and phononic components. Both these components have been determined for the given material. The graphical interpretation of electronic thermal conductivity divided by relaxation time (κ_e/τ is shown in Fig. 3(c). This plot depicts that κ_e/τ displays the similar trend as σ/τ since these two are correlated by Weidman Franz law ($\kappa = \sigma$LT). At 300 K, the obtained value of κ_e/τ is 0.7×10^{15} ($Wm^{-1}\ K^{-1}\ s^{-1}$) in the *p*-type doping region, which is quite low and hence gratifies better thermoelectric performance of the given material. The escalation in temperature excites more carriers and hence boosts thermal conductivity.

The lattice thermal conductivity (κ_l) has been determined using the Slack model (Sofi et al., 2022), as shown in Fig. 4(b). Electronic thermal conductivity is clearly seen to rise with temperature, but lattice thermal and total thermal conductivity show a downward trend with temperature. The electronic thermal conductivity increases as temperature rises because more carriers are thermally stimulated, whereas lattice thermal conductivity decreases due to excessive phonon scattering at higher temperatures driven by increased phonon–phonon and anharmonic interactions.

Finally, the true competency of the given material has been determined by computing its thermoelectric figure of merit (zT) in response to chemical potential,

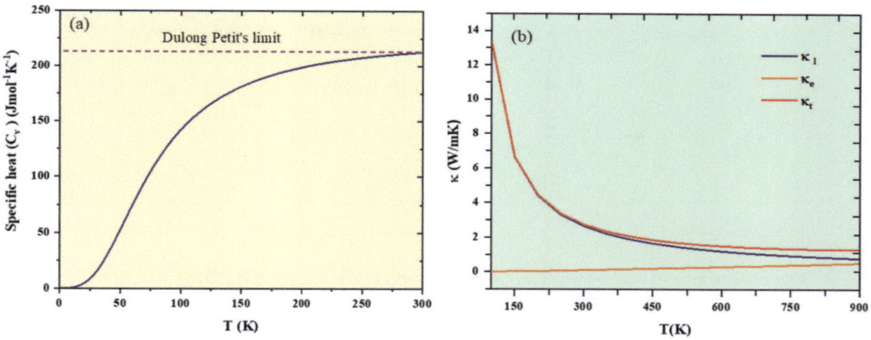

Fig. 4 The computed: (a) Specific heat capacity and (b) thermal conductivity (lattice, electronic and total) against temperature

(Fig. 3(d)). A good zT value necessitates robust Seebeck and electrical conductivity values, in addition to a low thermal conductivity value. All of these characteristics are appropriate for the specified material. The given material's zT value at 300 K is ~1, which seems to be extremely good and hence recommends it for energy harvesting applications in thermoelectrics. The estimated electronic figure of merit (zT) for the present compound surpasses other existing theoretical results (highlighted in Table 2). This remarkable achievement is primarily attributed to exceptionally higher Seebeck coefficients and satisfactory electrical conductivity values.

In addition to thermoelectric analysis, the specific heat capacity (C_v) against the temperature has been designed to disclose the heat retaining capacity of the present material (Fig. 4(a)). The C_v ascends linearly till it reaches a particular constant value (Dulong Petit's limit). The calculated value of C_v for the given material is 220.

Optical Properties

The optical properties of the given material have been studied in terms of dielectric constant, optical absorption and optical conductivity to investigate their visible light harvest. The dielectric function of material consists of both real and imaginary parts and is represented as; $\varepsilon_1(\omega) + i\varepsilon_2(\omega)$, where $\varepsilon_1(\omega)$ is the real part of the dielectric and $i\varepsilon_2(\omega)$ represents the imaginary part of the dielectric constant. Figure 5(a) shows graphical interpretations of both real and imaginary dielectric constant as function of photon energy.

The real part of the dielectric function $\varepsilon_1(\omega)$, climbs from $\varepsilon_1(0)$ to a maximum value of 7.85, before it starts to decrease. Even at high energies, the curve for K_2PtBr_6 stays above zero, indicating that the given compound is a practical semiconductor in the entire energy range. The imaginary component of the dielectric function $\varepsilon_2(\omega)$ versus photon energy (Fig. 5(a)), may be used to compute the amount of light energy absorbed by a material. The two primary peaks of the investigated double perovskites show a wide variety of absorption zones. The peak value of $\varepsilon_2(\omega)$ is obtained at 2.5 eV. Also, we computed the frequency dependent optical conductivity ($\sigma(\omega)$ against photon energy (Fig. 5(b)). The optical conductivity exhibits a range of

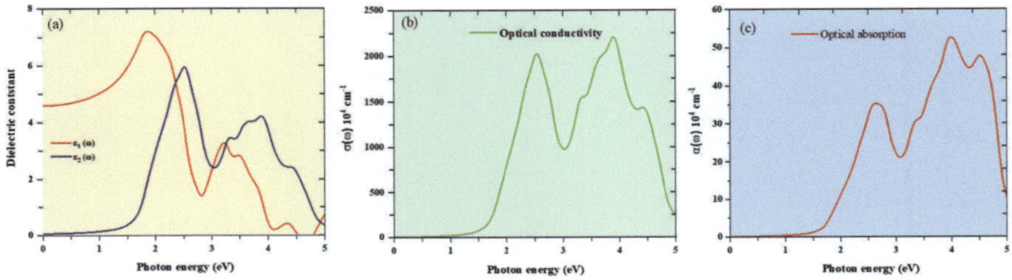

Fig. 5 The computed: (a) $\varepsilon_1(\omega) + \varepsilon_2(\omega)$, (b) $\sigma(\omega)$ and (c) $\alpha(\omega)$ against photon energy

peaks and valleys, increases gradually, and peaks at 4.2 eV. The DOS of the specified material is connected to these peaks, which are caused by interband transitions. The optical bandgap of 1.29 eV corresponds to the electrical bandgap determined by BS, evidencing the accuracy of our computational approach. Lastly, the optical absorption coefficient ($\alpha(\omega)$) of the studied material has been graphically shown in Fig. 5(c), revealing insights into the material's capacity to absorb light. Since band gap and absorption threshold are inversely correlated, materials with wider band gaps exhibit narrower absorption in the visible range. However, as in Fig. 5(c), the studied material has a small band gap, which results in more apparent absorption. The given material exhibits a high absorption coefficient that spans the whole visible spectrum, from infrared to ultraviolet. The absorption spectrum steadily broadens with increasing photon energy, attaining its utmost peak and maximal absorption at 4.2 eV. The visible spectrum resulted from electrons excited from the valence to conduction band.

Table 3 presents a comprehensive comparison of the computed optical parameters with other theoretical results. The obtained values demonstrate a decent agreement with the existing literature, indicating the reliability and consistency of our findings.

Table 3 Intended values of dielectric constant, optical conductivity (Ω^{-1} cm^{-1}) and optical absorption (α (w)10^4 cm^{-1}) in comparison with other published results

Parameter	K_2PtBr_6 (Present)	K_2PdBr_6 (Q. Mahmood et al., 2022)	Cs_2PdBr_6 (Bhamu et al., 2018)
Real part of dielectric $\varepsilon_1(\omega)$	7	8	8
Imaginary part of dielectric $\varepsilon_2(\omega)$	6	7	7
Optical conductivity $\alpha(\omega)$	2250	2100	–
Optical absorption	55	60	95

4 Conclusion

The thermoelectric and optical characteristics of the vacancy-ordered K_2PtBr_6 double perovskite have been meticulously examined utilizing density functional theory (DFT) within the Wien2k software framework. The structural arrangement exhibits Fm-3 m cubic symmetry, showcasing remarkable stability in its non-magnetic phase. Its electronic behavior is that of a semiconductor, displaying a direct band gap of 1.29 eV. Notably, the thermoelectric coefficients exhibit a zT value of 1.00, indicating significant advancements in the realm of thermoelectric device applications. Furthermore, this compound demonstrates exceptional light absorption capabilities within both the visible and infrared spectra, positioning it as a highly promising material for solar cell applications.

Acknowledgements Mudasir Younis Sofi (A PMRF Fellow) would like to gratefully acknowledge the Prime Minister Research Fellowship (PMRF) program (PMRF ID; 3302522) for the financial support.

References

Ajia, I.A., Edwards, P.R., Pak, Y., Belekov, E., Roldan, M.A., Wei, N., Liu, Z., Martin, R. W. and Roqan, I.S. (2018). Generated carrier dynamics in V-pit-enhanced InGaN/GaN light-emitting diode. *ACS Photonics*, **5:** 820–826.

Bhamu, K.C., Soni, A., and Sahariya, J. (2018). Revealing optoelectronic and transport properties of potential perovskites Cs_2PdX_6 (X = Cl, Br): A probe from density functional theory (DFT). *Solar Energy*, **162:** 336–343.

Jing, Z., Wang, H., Feng, X., Xiao, B., Ding, Y., Wu, K. and Cheng, Y. (2019). Superior thermoelectric performance of ordered double transition metal MXenes: $Cr_2TiC_2T_2$ (T = −OH or −F). *J. Phys. Chem. Lett.*, **10:** 5721–5728.

Jonson, M. and Mahan, G.D. (1980). *Phys. Rev. B*, **21:** 4223–4229.

Kojima, A., Teshima, K., Shirai, Y. and Miyasaka, T. (2009). Organometal halide perovskites as visible-light sensitizers for photovoltaic cells. *J. Am. Chem. Soc.*, **131:** 6050–6051.

Madsen, G.K.H. and Singh, D.J. (2006). BoltzTraP. A code for calculating band-structure dependent quantities. *Comput. Phys. Commun.* **175(1):** 67–71.

Mahmood, M. et al. (2022). First principles study of electronic, optical, and thermoelectric properties of $K_2Pd\ (Cl/Br)_6$ for solar cells and renewable energy. *Phys. Scr.* **97:** 035803.

Naceur, Y., Bourbaba, H., Ghebouli, M.A. et al. (2022). *Ab-initio* Study of structural, elastic, electronic and optical properties of hexahalometallate single crystals K_2XBr_6(X = Se, Pt). *Sci Rep* **12:** 8345.

Schwarz, K., Blaha, P. and Madsen, G.K.H. (2022). Electronic structure calculations of solids using the WIEN2k package for material sciences. *Comput. Phys. Commun.* **147:** 71–76.

Sofi, M.Y. and Gupta, D.C. (2022). Scrutinized the inherent spin half-metallicity and thermoelectric response of *f*-electron-based $RbMO_3$ (M = Np, Pu) perovskites: a computational assessment. *Sci Rep* **12:** 19476.

Sofi, M.Y., Khan, M.S., Ali, J. and Ajmal Khan, M. (2023). Novel Thermoelectric Material Ba_2AlNbO_6 for Energy Harvesting Applications. In: Khan, Z.H., Jackson, M., Salah, N.A. (eds) Recent Advances in Nanotechnology. ICNOC 2022. Springer Proceedings in Materials, vol 28. Springer, Singapore. pp 85–91.

Tan, G., Zhao, L.D. and Kanatzidis, M. G. (2016). Rationally designing high-performance bulk thermoelectric materials. *Chem. Rev.*, **116:** 12123–12149.

Xie, H., Hao, S., Bao, J., Slade, T.J., Snyder, G.J., Wolverton, C. and Kanatzidis, M.G. (2020). All-inorganic halide perovskites as potential thermoelectric materials: Dynamic cation off-centering induces ultralow thermal conductivity. *J. Am. Chem. Soc.*, **142:** 9553–9563.

Chapter 18
Growth, Characterization and Pyroelectric Study of Pure and Lanthanum Doped PZT Film

Vandana, Monika Tomar, and Reema Gupta

1 Introduction

As the technology is progressing at a large scale, the power demand is also continuously increasing, resulting in excessive use of non-renewable sources (Bowen et al., 2014). The devices based on power generation are used to supply energy generally reject approximately two-third of energy in the form of waste heat (Sharma et al., 2015; Kim et al., 2013). Basically, the rejected waste heat is the excess amount of low grade heat in temperature range 100–230°C. In order to supply excess energy demand, there can be some way to utilize this waste heat. Thermoelectric and pyroelectric are two devices which can harvest this waste heat in a useful manner. But in case of thermoelectric devices, due to poor efficiency and low figure of merit, there is a limit on the energy conversion in temperature range less than 100°C (Kim et al., 2013; Charlot et al., 2015; Kang et al., 2016). Therefore, pyroelectric devices are one of the promising alternative to convert waste heat into useful voltage. Pyroelectric materials are basically polar dielectrics having non-centrosymmetry which exhibit strong coupling between polarization and change in temperature of the material (Siao et al., 2018; Poplavko et al., 2020). There are two pyroelectric parameters such as pyroelectric coefficient and figure of merit which can evaluate the performance of any

Vandana
Department of Physics and Astrophysics, University of Delhi, Delhi 110007, India

Department of Physics, Punjab Engineering College (Deemed to Be University), Chandigarh 160012, India

M. Tomar
Department of Physics, Miranda House, University of Delhi, Delhi 110007, India

R. Gupta (✉)
Department of Physics, Hindu College, University of Delhi, Delhi 110007, India
e-mail: reema.gupta.25@gmail.com

R. K. Kotnala et al. (eds.), *Advanced Functional Materials for Sustainable Environments*, https://doi.org/10.1007/978-3-031-62620-3_18

pyroelectric device. Table 1 shows the literature survey on various multicomponent materials which are used for pyroelectric energy harvesting applications.

From Table 1, it is noticed that there are various multicomponent materials which are used for pyroelectric energy harvesting purpose. However, amongst various ferroelectric materials, "Lead Zirconium Titanate" exhibits good pyroelectric coefficient of 3.52×10^{-4} Cm^{-2} K^{-1} with high dielectric constant 1800 (Sappati et al., 2013). These properties can further be tailored and improved with doping. In literature, PZT has been doped with various dopant like Fe^{3+}, Sc^{3+}, La^{3+}, Nb^{5+}, Ce^{3+}, Ta^{5+}, K^{+}, Na^{+} in "A" or "B" sites of PZT. It has been shown that with doping, the ferroelectric, dielectric and pyroelectric properties increases. Qiao et al. (2019) studied pyroelectric properties of $Pb_{0.95}La_{0.03}(Zr_{0.86}Ti_{0.14})O_3$ and reported pyroelectric coefficient to be enhanced to 5.2×10^{-4} Cm^{-2} K with La doping (Qiao et al., 2019). Therefore, pure and La doped PZT film have been taken for pyroelectric study. Pyroelectric properties have been analysed using current source meter connected with the heat radiating unit.

Table 1 Brief Literature survey on the multicomponent materials used for thermal harvesting applications.

Multicomponent material	Multifunctional properties	Reference
$Ba_{0.75}Sr_{0.25}TiO_3$ (BST)	$\epsilon_r = 2000$ $p = 0.82 \times 10^{-4}$ Cm^{-2} K^{-1} $P_r = 7$ $\mu C/cm^2$	Stancu et al., 2019
$PbZr_{0.68}Fe_{0.14}Nb_{0.14}Ti_{0.04}O_3$	$P_r = 15$ $\mu C/cm^2$ $p = 1.69 \times 10^{-4}$ Cm^{-2} K^{-1}	Stancu et al., 2019
Manganese doped $(Pb_{0.93}La_{0.07})(Zr_{0.82}Ti_{0.18})O_3$	$p = 2 \times 10^{-3}$ Cm^{-2} K^{-1} $P_{max} = 31$ $\mu C/cm^2$ $P_r = 2.2$ $\mu C/cm^2$, $\epsilon_r = 2{,}128$	Kumar et al., 2019a, b
Lead magnesium niobate-lead titanate (PMN-PT)	$p = 2.8 \times 10^{-5}$ Cm^{-2} K^{-1}	Lam et al., 2005
$PbZr_{0.5}Ti_{0.5}O_3$	$\epsilon_r = 1800$ $P_s = 80$ $\mu C/cm^2$ $p = 3.52 \times 10^{-4}$ Cm^{-2} K^{-1}	Sappati et al., 2013
$Sr_xBa_{1-x}Nb_2O_6$ $(x = 0.50)$	$p = 2 \times 10^{-4}$ Cm^{-2} K^{-1}	Zhang et al., 2013
Gadolinium Doped $Sr_{0.53}Ba_{0.47}Nb_2O_6$	$P_r = 8.8$ $\mu C/cm^2$ $p = 2.85 \times 10^{-4}$ Cm^{-2} K^{-1}	Said et al., 2017
$Pb_{0.95}La_{0.03}(Zr_{0.86}Ti_{0.14})O_3$	5.2×10^{-4} Cm^{-2} K	Qiao et al., 2019
$Pb_{0.94}La_{0.06}(Zr_{0.5}Ti_{0.43})O_3$	3.99×10^{-4} Cm^{-2} K^{-1}	Bajpai et al., 2019
$(Pb_{0.93}La_{0.07})(Zr_{0.82}Ti_{0.18})O_3$	2×10^{-3} Cm^{-2} K^{-1}	Kumar et al., 2019a, b

P_r = remnant polarization, P_s = saturation polarization, p = pyroelectric coefficient, ϵ_r = dielectric constant.

2 Experimental Section

Pure and La doped PZT films were prepared using chemical solution deposition technique. Mole ratio of Zr/Ti = 40/60 was chosen for preparation of PZT sol and 2% La doping for PLZT sol. First, PZT and PLZT sols were prepared and their corresponding powders. Afterwards, by mixing appropriate quantity of powders into corresponding sols, the composite slurries of PZT and PLZT were prepared. Spin coating technique was used for preparation of PZT and PLZT film. Chemicals used in the experiments were lead acetate trihydrate (Sigma Aldrich), zirconium acetylacetonate (Sigma Aldrich), titanium isopropoxide (Sigma Aldrich) and lanthanum acetate dehydrate.

3 Results and Discussion

3.1 Structural Properties of Pure and La Doped PZT Film

X-Ray diffraction pattern (Make: Rigaku, Model: Ultima IV) of pure and La doped PZT film is shown in Fig. 1. Both the films have been found to be polycrystalline in nature without any secondary pyrochlore phase indicating perovskite structure. All the peaks are according to JCPDS card no: 33–0784 indicating the formation of tetragonal structure with P4mm space group (Jeyaseelan et al., 2020). It is observed from Fig. 1 that characteristic diffraction peak corresponding to (200) and (111) diffraction planes of nickel substrate are observed at $2\theta = 51.73°$ and 44.48°, respectively, for pure and doped PZT film (Vandana et al., 2022). The crystallographic planes (100), (110), (111), (201), (211) at $2\theta = 22.01°$, 31.36°, 38.67, 50.00° and 55.40°, respectively, corresponding to PZT film are shown in Fig. 1. The crystallite size is also calculated for pure and doped PZT film along the dominant (110) plane and found to be slightly increased from 25 nm to 27 nm with La doping.

3.2 Atomic Force Microscopic Images

The surface topologies of pure and La doped PZT films grown at optimized annealing temperature of 650°C were analysed using Atomic Force Microscopy (AFM) (Make: Park, Model: NX10) and are shown in Fig. 2. It was observed from AFM images that pure and doped PZT film exhibit low root mean square roughness within the range of 8 nm which may be beneficial for good dielectric and ferroelectric properties. Both the films show distinct grain boundaries with dense microstructure. The obtained topographical images for CSD grown La doped PZT films annealed at 650°C are in agreement with the results reported by Zhang et al. (2011). Hence, pure and doped PZT film are expected to perform better in terms of pyroelectric properties.

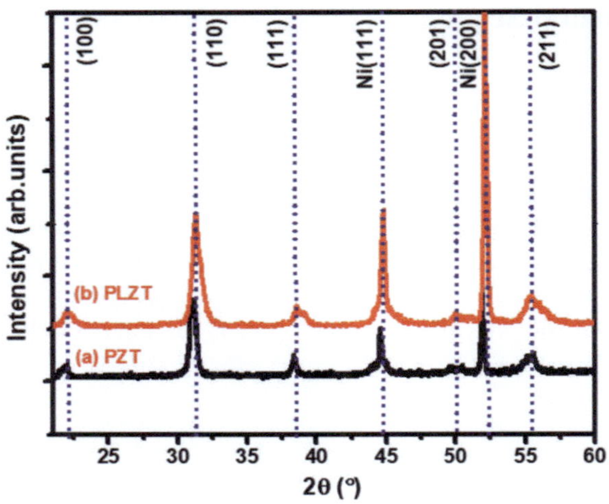

Fig. 1 The XRD pattern pure and La doped PZT film

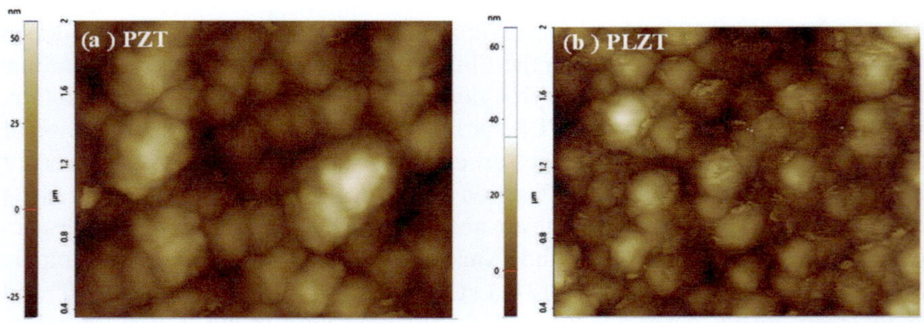

Fig. 2 Surface topographical images of: (a) pure and (b) La doped PZT thick film annealed at fixed annealing temperature of 650°C

3.3 Pyroelectric Measurement of Pure and La Doped PZT Film

In order to study the pyroelectric effect in pure and doped PZT film, thermal fluctuations are provided periodically to the top electrode of films and pyroelectric current with time was measured through Keithley 4200 semiconductor characterization unit connected with the films. In order to investigate the influence of lanthanum doping on pyroelectric properties of PZT films, 2% lanthanum content has been chosen for doping in PZT films. The pyroelectric current variation with time for both films is shown in Figure 3.

The behaviour of pyroelectric current with time is similar for pure and doped PZT film annealed at 650°C. However, it was observed that with La doping, the

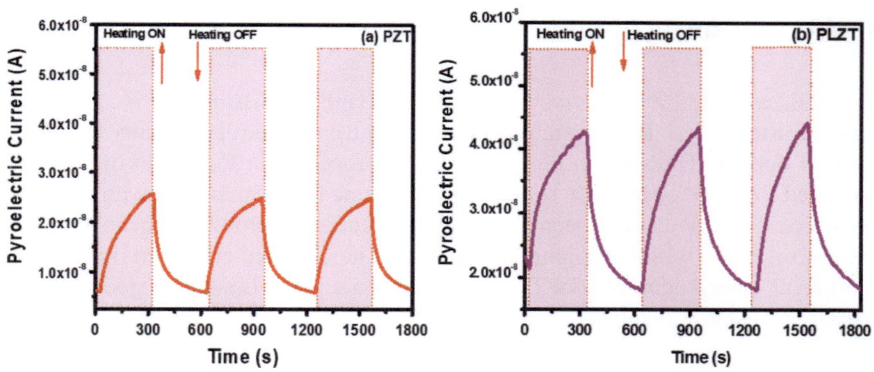

Fig. 3 Pyroelectric current variation with time for: (a) PZT and (b) PLZT (2% La) thick film

pyroelectric current (I_p) is observed to increase (Fig. 3). The change in pyroelectric current was observed to be 1.93×10^{-8} A and 2.45×10^{-8} A for PZT and La doped PZT films, respectively. The pyroelectric coefficient and figure of merit are also calculated from pyroelectric current versus time for pure and La doped PZT films and are given in Table 1. It is found that pyroelectric coefficient (p) is increased from 230 mC m^{-2} K^{-1} to 240 mC m^{-2} K^{-1} with La doping and figure of merit is also enhanced with La doping. The increase in pyroelectric coefficient with lanthanum doping is due to enhanced polarization in PLZT thick films with lanthanum doping resulting in large dipole–dipole interaction. With lanthanum doping in PZT, the pinning centres are decreased resulting in high value of pyroelectric coefficient. However, voltage responsivity and energy harvesting figure of merit are decreased with La doping due to increase in dielectric constant with La doping as discussed in previous work (Zhang et al., 2011). These results suggest that PLZT is a very promising pyroelectric material to be employed in the pyroelectric converters (Table 2).

Table 2 Pyroelectric parameters determined for the PLZT films annealed at 650°C

Lanthanum Content (%)	Pyro electric Coefficient (mCm^{-2} K^{-1})	Current Responsivity (F_i) $\times 10^{-10}$ mV^{-1}	Voltage Responsivity (F_v) m^{2}C^{-1}	Energy Harvesting Figure of Merit (F_e) $\times 10^{-8}$ m^{3}J^{-1}
PZT	230	920	32	296
PLZT	240	960	0.1	1.7

4 Conclusion

Pure and La doped PZT films were successfully synthesized using chemical solution deposition technique. Polycrystalline PZT film with improved crystallinity was found with La doping. The smooth morphology and proper grain formation of pure and La doped PZT thick films was revealed from atomic force microscopic images. The pyroelectric coefficient has been observed to be enhanced from 230 mCm^{-2} K^{-1} to 240 mCm^{-2} K^{-1} with La doping in PZT film. The obtained enhanced pyroelectric coefficient values achieved for PLZT film indicate these films as a good thermal energy harvester in real time applications.

Acknowledgements Authors are thankful to University of Delhi and DST, Government of India (CRG/2019/001709) for technical and financial Support.

References

Bajpai, K.K., Sreenivas, K., Gupta, A.K. and Shukla, A.K. (2019). Cr-doped lead lanthanum zirconate titanate (PLZT) ceramics for pyroelectric and energy harvesting device applications, *Ceram. Int.,* **45:** 14111–14120. https://doi.org/10.1016/j.ceramint.2019.04.111.

Bowen, C.R., Kim, H.A., Weaver, P.M. and Dunn, S. (2014). Piezoelectric and ferroelectric materials and structures for energy harvesting applications. *Energy Environ. Sci.,* **7:** 25–44, 2. https://doi.org/10.1039/C3EE42454E.

Charlot, B., Coudouel, D., Very, F., Combette, P. and Giani, A. (2015). Droplet generation for thermal transient stimulation of pyroelectric PZT element. *Sensors Actuators, A Phys.,* **225:** 103–110. https://doi.org/10.1016/j.sna.2015.02.011.

Jeyaseelan, A. and Dutta, S. (2020) Improvement in piezoelectric properties of PLZT thin film with large cation doping at A-site. *Journal of Alloys and Compounds*, **826:** 153956. https://doi.org/10.1016/j.jallcom.2020.153956.

Kang, M. and Yeatman, E.M. (2016). Thermal Energy Harvesting Using Pyroelectric and Piezoelectric Effect, *J. Phys. Conf. Ser.* **773:** 1–4. https://doi.org/10.1088/1742-6596/773/1/012073.

Kim, S.B., Park, H., Kim, S.H., Wikle, H.C., Park, J.H. and Kim, D.J. (2013). Comparison of MEMS PZT cantilevers based on d31 and d 33 modes for vibration energy harvesting. *J. Microelectromechanical Syst.,* **22:** 26–33. https://doi.org/10.1109/JMEMS.2012.2213069.

Kumar, A., Yoon, J.Y., Thakre, A., Peddigari, M., Jeong, D., Kong, Y. and Ryu J. (2019). Dielectric, ferroelectric, energy storage, and pyroelectric properties of Mn-doped $(Pb_{0.93}La_{0.07})(Zr_{0.82}Ti_{0.18})O_3$ anti-ferroelectric ceramics. *J. Korean Ceram. Soc.*, **56:** 412–420. https://doi.org/10.4191/kcers.2019.56.4.10.

Lam K.H. and Chan, H.L.W. (2005). Piezoelectric and pyroelectric properties of 65PMN-35PT/P(VDF-TrFE) 0–3 composites. Compos. *Sci. Technol.* **65:** 1107–1111, 2005, https://doi.org/10.1016/j.compscitech.2004.11.006.

Poplavko, Y. and Yakymenko, Y. (2020). Pyroelectricity. Funct. Dielectr. Electron., 131–159,. https://doi.org/10.1016/b978-0-12-818835-4.00004-3.

Qiao, P., Zhang, Y., Chen, X., Zhou, M., Wang, G. and Dong, X. (2019). Enhancing pyroelectric properties in $(Pb_{1-1.5x}\ La_x)(Zr_{0.86}Ti_{0.14})\ O_3$ ceramics through composition modulated phase transition. *Ceram. Int.* **45:** 7114–7119, 2019, https://doi.org/10.1016/j.ceramint.2018.12.216.

Said, M., Velayutham, T.S. and Abd Majid, W.H. (2017). Dielectric, pyroelectric, and ferroelectric properties of gadolinium doped $Sr_{0.53}Ba_{0.47}Nb_2O_6$ ceramic. *Ceram. Int.*, **43:** 9783–9789. https://doi.org/10.1016/j.ceramint.2017.04.156.

Sappati K.K. and Bhadra, S. (2018). Piezoelectric Polymer and Paper Substrates: A Review. *Sensors*, **18:** 1–30. https://doi.org/10.3390/s18113605.

Sharma, M., Chauhan, A., Vaish, R. and Chauhan, V.S. (2015). Pyroelectric materials for solar energy harvesting: A comparative study' *Smart Mater. Struct.,* **24:** 1–10. https://doi.org/10.1088/0964-1726/24/10/105013.

Siao, A.S., McKinley, I.M., Chao, C.K., Hsiao, C.C. and Pilon, L. (2018). Pyroelectric waste heat energy harvesting using the Olsen cycle on Pb(Zr, Ti)O_3-Pb(Ni, Nb)O_3 ceramics. *J. Appl. Phys.,* **124:** 1–12. https://doi.org/10.1063/1.5037112.

Stancu, V., Amarande, L., Botea, M., Iuga, A., Leonat, L.N., Tomulescu, A.G., Cioangher, M., Balescu, L.M., Pintilie, L. (2019). Comparison between dielectric and pyroelectric properties of PZFNT and BST type ceramics, *Process. Appl. Ceram.*, **13:** 269–276, https://doi.org/10.2298/PAC1903269S.

Vandana, Gupta, R., Tandon, R.P. and Tomar, M. (2022). Electro caloric and energy storage properties of sol gel derived lanthanum doped PZT thick films. Material Science in Semiconductor Processing, 150. https://doi.org/10.1016/j.mssp.2022.106970.

Zhang, J., Wang, G., Gao, F., Mao, C., Cao, F. and Dong, X. (2013). Influence of Sr/Ba ratio on the dielectric, ferroelectric and pyroelectric properties of strontium barium niobate ceramics. *Ceram. Int.*, **39:** 1971–1976: https://doi.org/10.1016/j.ceramint.2012.08.048.

Zhang, S.Q., Wang, L.D., Li, W.L., Li, N. and Fei, W.D. (2011). Effects of lanthanum doping on the preferred orientation, phase structure and electrical properties of sol–gel derived Pb1–3x/2Lax ($Zr_{0.6}Ti_{0.4}$) O_3 thin films. *Journal of Alloys and Compounds*, **509:** 2976–2980. https://doi.org/10.1016/j.jallcom.2010.11.177

Chapter 19
Comparative Performance Analysis of P-Si/n-CdS/ITO Heterojunction Solar Cell by Inserting Cu_2O as a Back Surface Layer Using SCAPS-1D

Prakash Kumar Jha, Ankita Srivastava, Narendra Bihari, Atish Kumar Sharma, Manish Kumar, Nitesh K. Chourasia, and Ritesh Kumar Chourasia

1 Introduction

The per capita energy need of the world is about 78 MMBtu. Due to limited reserves of coal and oil and the ecological imbalance related to hydroelectricity and nuclear power, we will need a viable alternative source of energy to accomplish the perpetual need for energy. It must be ubiquitous, free of cost, and nature-friendly. Due to this property in ensuing years, this must be the cynosure of the human civilization to produce electricity. Solar energy is an example of renewable energy. The solar cell of low cost, more stability, and higher efficiency is the best option for the energy fulfillment of the forthcoming generation. The solar cell has been developed through various steps from the first to the fourth generation. However, the growth of solar cells is impeded by efficiency and cost (Barman et al., 2021). The present market mainly comprise the first two generations. Nowadays, both generations have become

P. K. Jha · N. Bihari · A. K. Sharma
University Department of Physics, Lalit Narayan Mithila University, Darbharnga 846004, India

P. K. Jha · A. Srivastava · N. Bihari · A. K. Sharma · R. K. Chourasia (✉)
Post Graduate Department of Physics, Samastipur College Samastipur-848134 (A Constituent College of L.N.M.U.- Darbhanga), Darbharnga 846004, India

N. K. Chourasia (✉)
Department of Physics, Faculty of Science, Shree Guru Gobind Singh Tricentenary University, Gurugram, Gurugram, Haryana 122505, India
e-mail: niteshphyzics@gmail.com

M. Kumar
Experimental Research Laboratory, Department of Physics ARSD College, University of Delhi, New Delhi 110021, India

R. K. Kotnala et al. (eds.), *Advanced Functional Materials for Sustainable Environments*, https://doi.org/10.1007/978-3-031-62620-3_19

a hot topic for research interest. The first generation includes monocrystalline and polycrystalline solar cells while the second includes thin film-based technologies. Si-based heterojunction solar cells attract worldwide attention for low-temperature processing. Due to easily processible fabrication techniques and the best electrical conversion efficiency (ECE = 27%), the market is mainly based on silicon solar cells (Jha et al., 2022). These solar cells are limited by the Shockley Queisser limit for silicon solar cells, which is 29.8% ECE (Jha et al., 2022).

A p-Si type semiconductor serves as the absorber layer, the n-CdS type is used as the electron selective layer, and ITO is employed as the transport electrode in our prior work simulated device configuration (Jha et al., 2022), in which we achieved the best efficiency upto 18.25%. As we know there are still some challenges at the back contact in the standard cell configuration (p-Si/n-CdS/ITO heterojunction solar cells). In quest of more efficiency, we need to modify our standard device configuration by adding an inorganic hole transport layer Cu_2O as a back contact layer, to provide a smooth and superior path for carrier transportation. To efficiently remove holes from a Si-based solar cell and prevent electron recombination, a hole-transporting material must be used. We utilized spiro-OMeTAD as a hole-transporting layer which is organic and enhances the conductivity and hole mobility (Jha et al., 2023). The frequently used HTM has low charge mobility and thereby low short-circuit current density. It is hygroscopic, has propensity to crystallize, and is susceptible to both heat and moisture. A stable and low-cost HTM should be used in its place. We prefer inorganic HTL because it meets our requirements for a stable HTM, which includes better charge mobility and short-circuit current density (Sharma et al., 2023a, 2022; Eli et al., 2019; Wang et al., 2016). Considering above concern, we fabricated a new photovoltaic cell with p-type Cu_2O as a hole transport layer. Films of Cu_2O were formed with oxygen and argon at a deposition pressure of 8×10^{-3} bar utilizing reactive RF magnetron sputtering at 150W with a $3^{//}$ Cu target (Li et al., 2014). Despite its remarkable gain in PCE, people have shown inorganic solar cells such as crystalline silicon (c-Si 25.72% and 25.6%) and gallium arsenide (28.8%) with more efficiencies than this (Sanchez et al., 2016; Masuko et al., 2014), yet we are considering Cu_2O as HTM because we have obtained a little research work on such best configuration. To the best of our knowledge, the present research work is the first to be theoretically simulated, studied, and presented using SCAPS-1D simulation software.

2 Device Simulation and Analysis

Nowadays, before the experimental step, simulation has become essential for analyzing and optimizing the electrical parameters of all types of well-made solar cells. This saves lot of time and energy. In this study, we analyze the CdS-based solar cells by making use of capacitance simulation structures (SCAPS-1D). The Elis group at the University of Gent has created a free, open-source solar cell device simulator called SCAPS-1D. The SCAPS-1D simulation can solve the basic equations of semiconductor physics and electrodynamics (Chen, 2015; Sharma, 2023c):

$$\frac{d^2}{dx^2}\psi(x) = \frac{e}{\varepsilon_0\varepsilon_r}\rho(x) - n(x) + N_D - N_A + \rho_p - \rho_n \quad (1)$$

$$\frac{dj_n}{dx} = G - R \text{ and } \frac{dj_p}{dx} = G - R \quad (2)$$

$$J_n = D_n\frac{dn}{dx} + \mu_n n\frac{d\varphi}{dx} \text{ and } J_p = D_p + \mu_p p\frac{d\varphi}{dx} \quad (3)$$

where $\psi(x)$: electrostatic potential, e: electrical charge, ε_r: relative permittivity, ε_0: vacuum permittivity, p: hole concentration, n: electron concentrations, N_D: donor type, N_A: acceptor type, ρ_p: hole distribution, ρ_n: electron distribution, G: generation rate and R is the recombination rate (Chen 2015; Sharma, 2023c). All of the simulations are carried out at a working temperature of 300 K and are performed under AM 1.5 G light source.

Figure 1 presents a modified standard solar cell configuration (p-Si/n-CdS/ITO) by inserting a p-type inorganic cuprous oxide layer. The p-type Si absorber layer has an energy band gap of 1.12 eV and n-type CdS is extensively used as an emitter or window layer for solar cells having an energy band gap of 2.42 eV. Compared to the a-Si: H thin film, it is more transparent at shorter wavelengths (energy band gap of about 2.7 eV) (Ravindra et al., 2017). The n-CdS emitter layer has a 2.4 eV band gap while the band gap of ITO is 3.6 eV. the ITO layer promotes absorption and makes high optical transmission for visible light. It also works as an effective front contact for the modified solar configuration. A layer of p^+-type Cu_2O having a band gap of 2.17 eV is added as a back surface passivation layer. When light hits the absorber layer, it causes electron–hole pairs to be created in the layer. The amount of photogenerated charge carriers transient through the heterojunction is determined by the band alignment (Fig. 2). All the parameters related to the device architecture are given in Table 1.

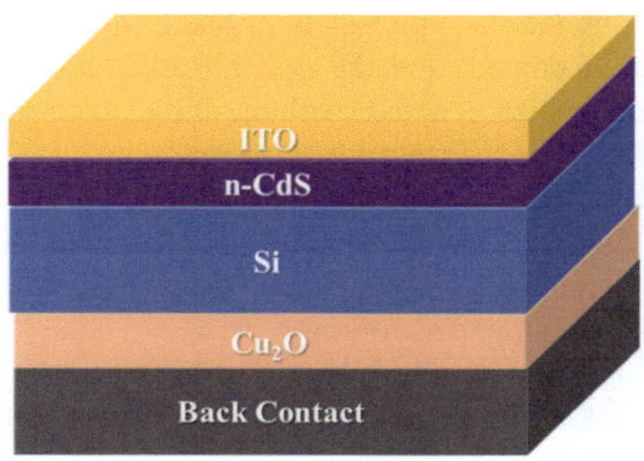

Fig. 1 Newly modified configuration of p-Si/n-CdS/ITO heterojunction solar cell using a back surface layer (Cu_2O)

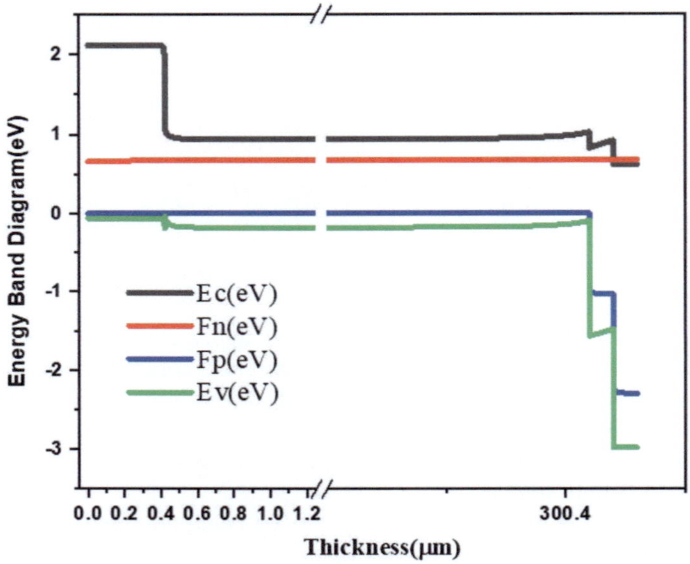

Fig. 2 Energy band diagram of modified cell configuration (as shown in Fig. 1)

3 Theoretical Modeling of Newly Configured Device

In a semiconductor, the p–n junction is a junction diode that has an in-built electric field formed due to immobile charge carriers in the depletion region at the junction. This in-built electric field provides directional flow to charge carriers and the developed voltage across the junction (Sharma et al., 2023b):

$$V = \frac{KT}{q}\ln\frac{N_a N_d}{n_i^2} \tag{4}$$

A similar concept is applied when an additional p+ layer is added amid the back contact and the p-Si absorber layer. There is a formation of a p+ –p junction which generates voltage at the junction. This generated voltage creates a barrier for an electric current at the rear contact. This reduces recombination and enhances efficiency.

An understanding of the electrical characteristics of a solar cell can be gained by examining its I-V curve. The main electrical parameters V_{oc}, J_{sc}, FF, and η which are explained as follows:

Open Circuit Voltage: The highest voltage a solar cell is capable of supplying to a load. This is obtained when the load resistance is infinite (Jha et al., 2023; Sharma et al., 2023a).

Table 1 Main physical and electrical parameters used in simulation (Sharma et al., 2022)

Parameters layer	Cu_2O (Jha et al., 2023) Back surface layer	p-Si (absorber layer)	n-CdS (emitter layer)	p-Si/n-(CdS interface)	ITO (window layer)
Thickness (micrometer)	0.420	300	0.015		0.020
Energy bandgap (eV)	2.17	1.12	2.4		3.6
Electron affinity(eV)	3.2	4.1	4.1		4.6
Relative dielectric permittivity constant	7.1	11.9	10		8.9
Conduction band effective density of states (cm^{-3})	2×10^{17}	2.8×10^{19}	2.8×10^{19}		1×10^{19}
Valence band effective density of states (cm^{-3})	1.1×10^{19}	1.04×10^{19}	2.8×10^{19}		1×10^{19}
Thermal velocity of electrons (cms^{-1})	200	1×10^7	1×10^7		1×10^7
Thermal velocity of holes (cms^{-1})	80	1×10^7	1×10^7		1×10^7
Electron mobility ($cm^2V^{-1}\ s^{-1}$)	0	1099	50		50
Hole mobility ($cm^2V^{-1}\ s^{-1}$)		427	5		50
Donor density (cm^{-3})		0	1×10^{14}		1×10^{20}
Acceptor density (cm^{-3})	1×10^{18}	1×10^{16}	0		0
Position (eV) (Defect 1)	$E_V + 0.1$	E_C–0.79	$E_V + 0.6$	$E_V + 0.6$	$E_V + 0.6$
Total density N_t (cm^{-3}) (Defect 1)	1×10^{14}	1×10^{11}	1×10^{15}	1×10^{10}	1×10^{15}

(continued)

Table 1 (continued)

Parameters layer	Cu_2O (Jha et al., 2023) Back surface layer	p-Si (absorber layer)	n-CdS (emitter layer)	p-Si/n-(CdS interface)	ITO (window layer)
Position (eV) (Defect 2)		E_C–0.57			
Total density N_t (cm^{-3}) (Defect 2)		1×10^{12}			

$$V_{oc} = \frac{\eta kT}{q} \ln\left(\frac{I_L}{I_o} + 1\right) \tag{5}$$

where η is the dimensionless constant, k Boltzmann constant, T: the temperature, q: the charge, I_L: the light-generated current or photovoltaic current, and I_O the saturation current. This relation also provides better insights into understanding the stimulus of operating temperature on the open circuit voltage parameters.

The maximum output current density that a solar cell is capable of providing in conditions where the cell has zero resistance is referred to as the short-circuit current density.

$$J_{sc} = J_O(e^{qV/\eta KT} - 1) \tag{6}$$

where, J_O is the saturation current density.

The Fill Factor of the cell (Jha et al. 2023) is given by

$$FF = \frac{V_{mp} J_{mp}}{I_{oc} V_{oc}} \tag{7}$$

The conversion efficiency (η) (Sharma et al., 2022) is given by

$$\eta = \frac{V_{oc} J_{sc} FF}{p_{in}} \tag{8}$$

The value of shunt and series resistance must be high and low, respectively. The current density over the cells in terms of parallel resistance and shunt resistance can be calculated (Jha et al., 2022) as

$$J = J_{ph} - J_0 e^{\frac{e(V+JR)}{\eta kT}} \tag{9}$$

and again

$$J = J_{ph} - J_0 e^{\frac{eV}{\eta kT}} - \frac{V}{R_{sh}} \tag{10}$$

Electron affinity refers to the difference in energy that exists between the edge of the conduction band and the vacuum level. We are able to comprehend the meaning of the terms "conduction band offset" and "valence band offset"

$$E_c = \chi_{si} - \chi_{CdS} \tag{11}$$

where χ_{si} is the electron affinity of the p-absorber silicon material and χ_{CdS} the electron affinity of the buffer/emitter layer of CdS material (Jha et al., 2022).

4 Results and Discussion

4.1 *Influence of Operating Temperature on Modified Solar Cell Configuration*

The temperature at which the solar cell is operated has a considerable impact on the amount of power that can be extracted from it. We are using a practical temperature range from 290 to 370 K for this investigation as shown in Fig. 3. It was discovered that environmental temperature has a significant impact on solar cells. When the temperature rise, the reverse saturation current also rises, which results in a decrease in the open circuit voltage (Eq. 1). At higher temperatures, more carriers are generated due to the photoelectric effect which leads to a minor increase in short-circuit current density. The fill factor reduces drastically because of additional factors such as the phonon-scattering effect, which is dominant at higher temperatures and provides resistance for free passage of the charge carriers (Burgelman et al., 2000). The overall efficiency of the device degrades with increasing fill factor, rising short-circuit current density, and falling open-circuit voltage (using Eqs. (3)–(6)).

4.2 *Influence of Electron Affinity of ETL and HTL*

In this study, we varied the electron affinity of the Cu_2O and CdS layers, respectively. After simulation, we obtained that the performance parameters are not severely affected by varying electron affinity of the HTL layer (Fig. 4). The open circuit voltage (0.673 V) and the short circuit current density (39.5 mA/cm^2) remain constant while PCE and fill factor have minor variations till 2.8 eV and further remains constant. Moreover, the fill factor and PCE of CdS are greatly affected by varying electron affinity (3.8–4.4 eV) of the ETL layer. The short circuit current density remains constant. There is minor variation in open circuit voltage till 4.0 eV and further remains constant. The curve behavior shown by PCE and fill Factor is similar. By varying the electron affinity of ETL, the PCE value of the cell increased linearly to 4.1 eV. Since the production and recombination rates have remained same, we

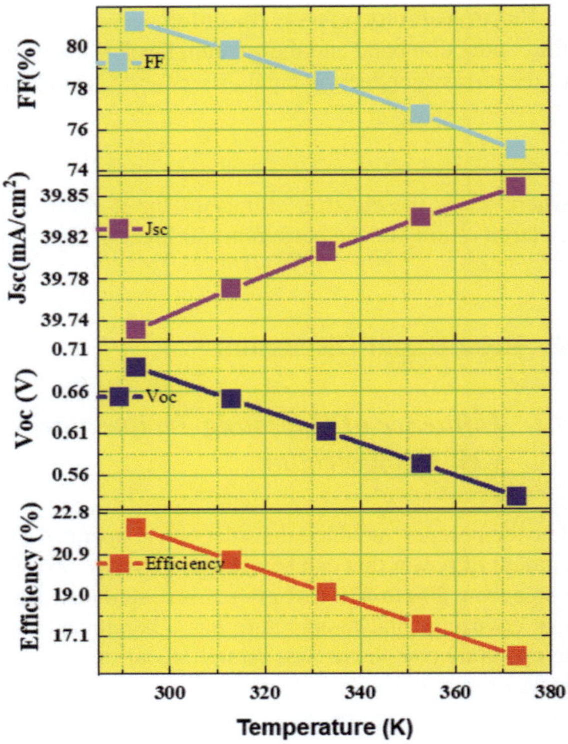

Fig. 3 Temperature vs photovoltaic parameters plot for modified Cu_2O/p-Si/n-CdS/ITO solar cell

concluded that the photogenerated charge carriers have increased linearly. Carrier recombination can be minimized and PSC performance improved through the careful selection of ETMs and HTMs with the right electron affinity.

4.3 Effect of Electron Affinity of Absorber Layer on Electrical Parameters

Figure 5 demonstrates the variation of electrical parameters on changing the value of electron affinity of the p-Si absorber layer. The value of efficiency and the fill factor is constant up to 4.4 eV and declines steeply to 8.8% at 4.6 eV while there is minor variation in the short circuit current density and open circuit voltage in the range of 3.8 to 4.6 eV. Due to the enhanced recombination mechanism at HTL/Cu_2O interface the value of efficiency reduced significantly in the range of 4.4 to 4.6 eV. A positive value of the valance band offset causes a barrier to the following of the photogenerated holes from the absorber to HTL. There are no such barriers in case

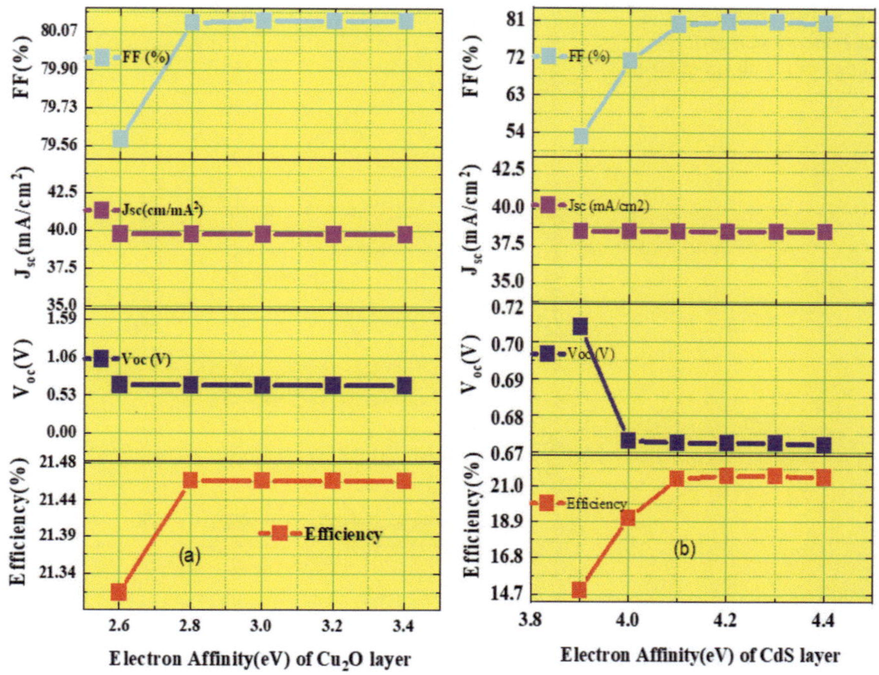

Fig. 4 Performance parameters vs electron affinity plot for (4a) Cu_2O and (4b) CdS layer

of negative VBO and photogenerated holes from the absorber layer can easily flow towards HTL.

4.4 Influence of Shunt and Series Resistance on Device Operation

The solar cell has inherent resistances that significantly affect the fill factor and eventually the PCE of the device. The series resistance arises due to the back and front contact of the device. Also, the cell materials of the device add to the series resistance of the device. To get better performance the devices should have the least possible practical series resistance. Further, shunt resistance is caused by leakage currents, which can never be eliminated in a practical device and hence cannot be prevented. (using Eqs. (8) & (9)). The impact of a variation of shunt resistance is shown in Fig. 6(a) from 1E9 to 1E49 ohm.cm^2. The value of series resistance in Fig. 6(b) is varied from 0 ohm.cm^2 to 10 ohm.cm^2 to better understand how resistances will affect the behavior of the devices. In Fig. 6(a) and 6(b), we observe that there is no improvement in photovoltaic parameters with increasing values of

Fig. 5 Performance parameters vs electron affinity of the absorber layer

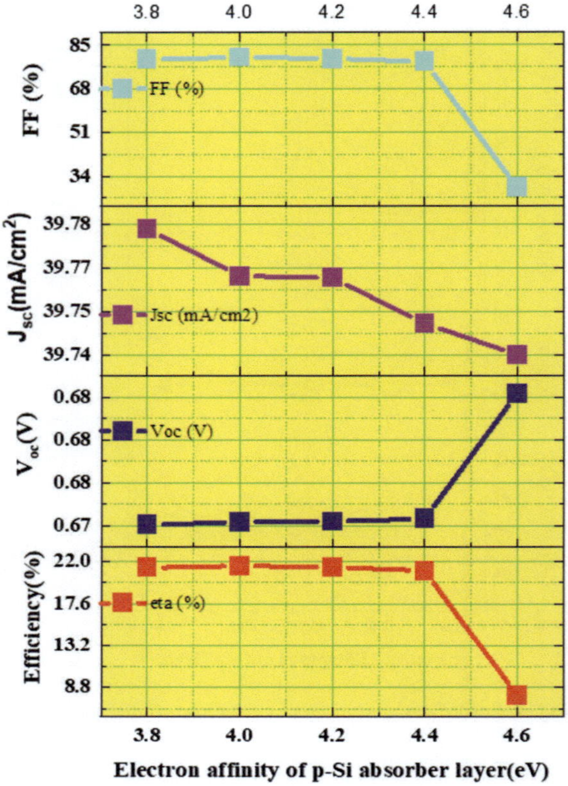

shunt resistance, while the performance parameters decrease with series resistance. In addition, the open circuit voltage remains constant. However, zero resistance is the ideal case.

4.5 *Optimization of Conventional Solar Cell After Inserting Cu_2O Layer*

The induced electric field in p$^+$/p can improve $V_{oc}(V)$ in a cell by creating an in-built voltage at the rear and reduced recombination at the junction and in the bulk**.** The modified cell has shown better and higher efficiency than the standard cell. The p-type inorganic Cu_2O layer has been proven favorable material for cell performance. The overall performance is compared in Table 2 also I-V curve for both the cells presented in Fig. 7.

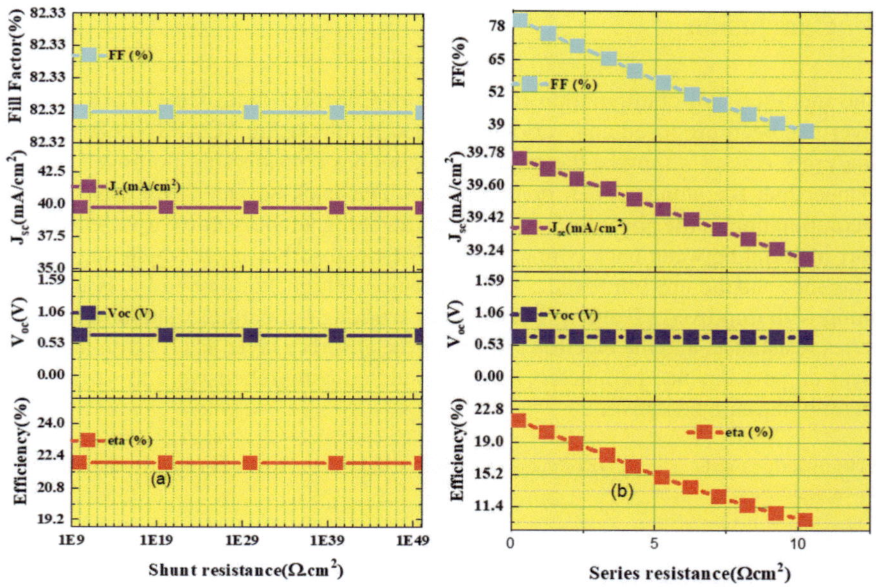

Fig. 6 Variation of performance parameters vs shunt and series resistance

Table 2 Comparative Performance Analysis of p-Si/n-CdS/ITO Heterojunction Solar

Cell configuration	η(%)	FF (%)	V_{oc} (V)	J_{sc} (mA/cm^2)	References
p-Si/n-CdS/ITO	18.20	75.82	0.636	37.72	(Sharma et al., 2022)
Cu_2O/p-Si/n-CdS/ITO	21.58	80.69	0.672	39.73	This work

Fig. 7 The I-V characteristic of simulated device with and without Cu_2O layer

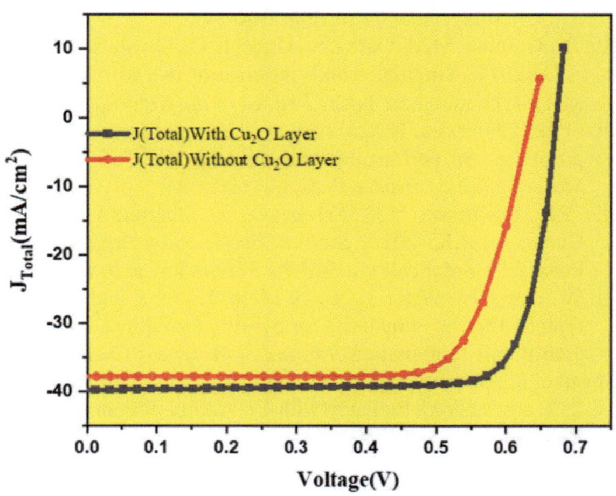

5 Conclusion

The modified cell configuration (Cu_2O/p-Si/n-CdS/ITO) with Cu_2O back surface layer has been proposed and various obtained results using SCAPS-1D have been analyzed. Analysis showed the modified design has considerably improved performance than the standard cell configuration. The back surface layer has reduced recombination in the absorber layer. On sandwiching an inorganic Cu_2O hole transport material between back contact and p-Si type absorber layer which augmented efficiency and thereby improved cell performance. A higher improved V_{oc} is achieved along with improved FF and η. The effect of working temperature, shunt resistance, series resistance, and the electron affinity of the absorber layer, HTL layer, and ETL layer on performance parameters has been studied comprehensively to sort out the recombination mechanism.

Acknowledgements The authors Prakash Kumar Jha and Ankita Srivastava contributed equally to this research work.

References

Barman, B. and Kalita, P.K. (2021). Influence of back surface field layer on enhancing the efficiency of CIGS Solar cell. *Solar Energy*, **216:** 329–337.

Burgelman, M., Nollet, P. and Degrave, S. (2000). Modelling polycrystalline semiconductor solar cells. *Thin Solid Films*, **361:** 527–532.

Chen, W., Wu, Y., Yue, Y., Liu, J., Zhang, W., Yang, X., Chen, H., Bi, E., Ashraful, I., Grätzel, M. and Han, L. (2015). Efficient and stable large-area perovskite solar cells with inorganic charge extraction layers. *Science*, **350:** 944–948.

Eli, D., Onimisi, M.Y., Garba, S., Ugbe, R.U., Owolabi, J.A., Ige, O.O., Ibeh, G.J. and Muhammed, A.O. (2019). Simulation and optimization of lead-based perovskite solar cells with cuprous oxide as a p-type inorganic layer. *Journal of the Nigerian Society of Physical Sciences*, **1:** 72–81.

Jha, P.K., Chourasia, N.K., Sharma, A.K. and Chourasia, R.K. (2022). Optimization of electrical properties for performing analysis of p-Si/n-CdS/ITO heterojunction photovoltaic solar cell. *Material Today: Proceedings*, **67:** 620–624.

Jha, P.K., Chourasia, N.K., Srivastava, A., Sharma, A.K., Kumar, R., Sharma, S., Kumar, M. and Chourasia, R.K. (2023). Study of Eco-friendly Organic-Inorganic Heterostructure $CH_3NH_3SnI_3$ Perovskite Solar cell via SCAPS Simulation. *Journal of Electronic Materials*, **52:** 4321–4329.

Li, W., Dong, H., Wang, L., Li, N., Guo, X., Li, J. and Qiu, Y. (2014). Montmorillonite as bifunctional buffer layer material for hybrid perovskite solar cells with protection from corrosion and retarding recombination. *Journal of Materials Chemistry A*, **2:** 13587–13592.

Masuko, K., Shigematsu, M., Hashiguchi, T. and Fujishima, D. (2014). Achievement of more than 25% conversion efficiency with Crystalline Silicon Heterojunction Solar cell. *IEEE Journal of Photovoltaics*, **4:** 1433–1435.

Ravindra, P., Mukherjee, R. and Avasthi, S. (2017). Hole-Selective Electron-Blocking Copper Oxide Contact for Silicon Solar Cells. *IEEE Journal of Photovoltaics*, **7:** 1278–1283.

Sanchez, R.S. and Marza, E.M. (2016). Light-induced effects on Spiro-OMeTAD films and hybrid lead halide perovskite solar cells. *Solar Energy Materials Solar Cells*, **158:** 189–194.

Sharma, A.K., Chourasia, N.K. and Chourasia, R.K. (2022). Optical, Temperature, and Bulk analysis theoretically in p-Si/n-CdS/ITO heterojunction photovoltaic solar cell. *Material Today: proceedings*, **67:** 620–624.

Sharma, A.K., Chourasia, N.K., Jha, P.K., Chourasia, Srivastava, A., Kumar, R., Sharma, S., Kumar, M. and Chourasia, R.K. (2023a). Characteristics Features and Performance Investigations of a PTB7:PC71BM/PFN:Br Pure organic solar cells using SCAPS-1D. *Journal of Electronic Materials*, **52:** 4302–4311.

Sharma, A.K., Srivastava, A., Jha, P.K., Kumar, R., Kumar, M., Kulriya, P.K., Chourasia, N.K. and Chourasia, R.K. (2023b). Bulk/Interface Defects Engineering and Comparative Performance Analysis of p-Si/n-CdS/ALD-ZnO Heterojunction Solar Cell. *Energy Technology*, **11:** 2300169.

Sharma, A.K., Kumar, R., Jha, P.K., Kumar, M., Chourasia, N.K. and Chourasia, R.K. (2023c). Bulk Parameters Effect and Comparative Performance Analysis of p-Si/n-CdS/ALD-ZnO Solar Cell. *Silicon*, 1876–9918.

Wang, D., Wright, M., Elumalai, N.K. and Uddin, A. (2016). Stability of perovskite solar cells. *Solar Energy Materials and Solar Cells*, **147:** 255–275.

Part V
Carbon Based Structures for Energy and Environment Applications

Chapter 20
Advancing Energy Storage: The Role of Carbon-Based Materials in Energizing Tomorrow

Yogita Maithani, Vinay Maithani, Subhendu Shekhar Khali, and Neha Choudhury

1 Introduction

Energy, the lifeline of the modern world and permeating every aspect of our daily lives plays an important role in driving progress and convenience, from powering our homes and businesses to propelling vehicles and machinery. Just as all living organisms require energy to thrive, we too rely on energy to sustain our activities and nourish our bodies. Plants utilize the sun's radiant energy to grow (Ruban, 2015), while animals and humans, in-turn, derive their energy by consuming these plants. Food becomes the fuel that powers our bodies, enabling us to perform tasks with muscle power and generating heat energy in the process. However, as civilization advances and our reliance on energy grows with digitalization, mobility changes, and global connections, the price we pay for our energy consumption becomes evident (Smil, 2017). The over-reliance on fossil fuels and finite resources has resulted in adverse environmental impacts, highlighting the urgency for a more sustainable energy paradigm (Jefferson, 2006). Achieving sustainability is not just a moral imperative; it is essential for economic growth and will continue to be a paramount consideration for our future. Amidst the challenges, the quest for cleaner energy sources drives the development of sustainable technologies that can shape the future of energy and the environment (Dell & Rand, 2001). In this context,

Y. Maithani (✉) · S. S. Khali
Department of Physics, Indian Institute of Technology Madras, Chennai, Tamil Nadu 600036, India
e-mail: yomaithani1@gmail.com

V. Maithani
Department of Metallurgical and Materials Engineering, Indian Institute of Technology Kharagpur, Kharagpur, West Bengal 721302, India

N. Choudhury
Department of Chemistry, Yonsei University, Seoul 03722, Republic of Korea

R. K. Kotnala et al. (eds.), *Advanced Functional Materials for Sustainable Environments*, https://doi.org/10.1007/978-3-031-62620-3_20

stationary energy storage assumes a critical role in transitioning toward a sustainable energy system, particularly with a higher penetration of renewable energy sources (Dell & Rand, 2001; Nižetić et al., 2023). With a growing population and changing lifestyles, our energy demand will double from 14 to 28 TW by 2050 (Isah, 2018; Larcher & Tarascon, 2015). This chapter explores the multifaceted world of energy storage devices and delves into their paramount importance in shaping the future of sustainable energy.

A systematic overview of energy storage devices, encompassing their evolution and the recent literature about this field, is presented herein. This chapter specifically emphasizes the recent advancements in carbon-based materials, for example, graphene, carbon nanotubes, carbon-based polymers, and carbon-based hybrid materials, which play pivotal roles in energy storage technologies. The discussion encompasses technical capabilities, challenges, and future prospects. Section 2 provides an overview of energy storage devices, while Sect. 3 introduces the materials commonly used in their fabrication. An in-depth analysis of the diverse forms of carbon-based materials for energy storage technologies is given in Sect. 4. Section 5 highlights the future scope, address critical challenges followed by concluding summary (Sect. 6). Through this exploration, we aim to equip readers with a deeper understanding of energy storage devices, their role in sustainability, and their potential to reshape our energy systems for a brighter, more sustainable future.

2 Energy Storage Devices (ESDs)

ESDs are the systems that capture and store energy, ready to release it when needed (Dell & Rand, 2001; Liu et al., 2018; Zhao et al., 2015). These devices come in different types, like batteries, capacitors, fuel cells, pumped hydro storage, etc., each designed for specific applications. By improving energy management, ESDs increase the reliability of our electricity grid and assist renewable energy solutions to create a greener and more sustainable future. To store electrical energy effectively, we must transform it into various types of energy like mechanical, chemical, electrochemical, thermal, and electromagnetic forms (Zhao et al., 2015). Figure 1 depicts the classification of crucial energy storage systems based on the different energy storage methods in these diverse forms. Here, we focus on the use of carbon-based materials for battery and supercapacitor applications.

A battery is an energy storage device that translates chemical energy into electrical energy. The electrolyte facilitates the chemical reaction between the anode and cathode, and electrical energy is stored while charging and released during discharging. The availability of batteries with higher charge storage capacity, longer life cycles, and lower costs are essential criteria for a good battery (Noorden, 2014; Ye et al., 2018).

Supercapacitors have a significant ability to store electric charge. They consist of two main components: electrodes and an electrolyte. Based on their energy storage

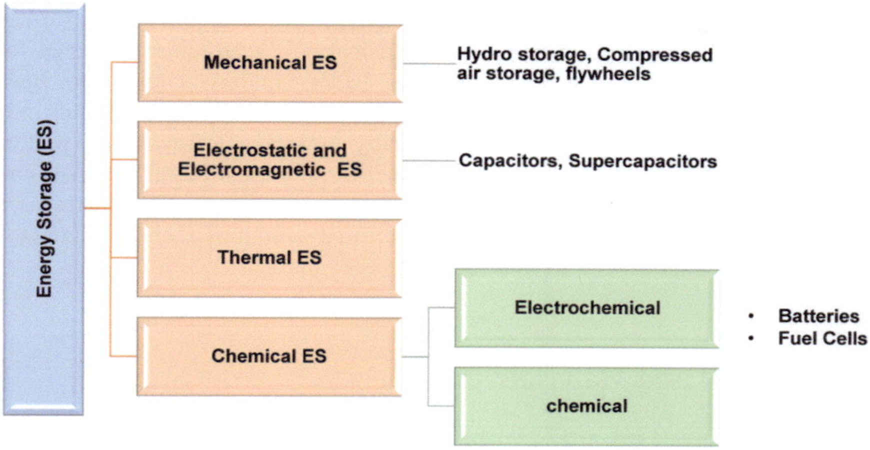

Fig. 1 Classification of energy storage systems

techniques, supercapacitors are classified into two types: electric double-layer capacitors (EDLCs), relying on electrostatic interactions, and pseudo-capacitors, involving chemical reactions between the electrodes and electrolyte ions. These versatile supercapacitors hold significant promise for advancing sustainable energy solutions (Down et al., 2018; Meng et al., 2017; Wu et al., 2023; Zhao et al., 2023).

3 Materials for ESDs

Energy storage materials are fundamental components that drive the efficiency and performance of ESDs. These materials possess unique properties that allow them to store and release energy efficiently, making them vital for various energy storage technologies. Advanced materials like graphene, carbon nanotubes, metal oxides, and different polymers have been extensively studied for their exceptional energy storage capabilities (Iqbal et al., 2019; Liu et al., 2018; Tawalbeh et al., 2023; Zhao et al., 2023). Additionally, research has focused on exploring hybrid materials and composites, combining the strengths of different materials to enhance energy storage performances (Huang et al., 2021; Moorthi et al., 2023; Wang et al., 2021, 2023). Furthermore, nanoscale engineering of these materials has shown promise in increasing their surface area and improving charge transfer kinetics, ultimately leading to enhanced energy storage capacities (Iqbal et al., 2019; Wu et al., 2023). The ongoing advancements in energy storage materials continue to shape the landscape of energy storage technologies, driving us closer to a more sustainable and energy-efficient future.

Carbon-based materials play a pivotal role in advancing ESDs due to their exceptional properties and versatility. These materials have garnered significant attention and research interest for their impressive electrical conductivity, excellent chemical stability, and large surface area (Zhao et al., 2023). Graphene has a high surface area and exceptional electron mobility, which makes it a prominent material for applications of energy storage (Saha & Dutta, 2022). Similarly, carbon nanotubes possess outstanding mechanical strength and unique tubular structure, enabling efficient electron transport and ion diffusion, thereby enhancing energy storage performance (Zhang & Li, 2009). The tunable properties of carbon-based materials and their abundant and sustainable nature position them as frontiers in scientific research for developing efficient and sustainable energy storage solutions, thereby driving the transformation of global energy landscape.

4 Carbon-Based Materials

Carbon-based materials are diverse group of materials composed predominantly of carbon atoms, each exhibiting unique properties and structures (Fig. 2) (Al Sheheri et al., 2019). One-dimensional (1D) carbon materials are characterized by their elongated shape, forming structures like carbon nanotubes and carbon nanofibers. These materials possess exceptional mechanical strength and electrical conductivity, making them ideal for various applications, including reinforcement in composite materials and nanoelectronics (Ganesh et al., 2006; Zhang & Li, 2009). On the other hand, two-dimensional (2D) carbon materials are exemplified by graphene, which is composed of carbon atoms of a single layer organized in a hexagonal pattern. It boasts outstanding electrical conductivity, remarkable thermal conductivity, and an incredibly large surface area, making it important for electronics, sensors, and energy storage devices. Lastly, three-dimensional (3D) carbon materials, for example, activated carbon and carbon aerogels, possess a porous and interconnected structure, affording them a vast surface area and high porosity (Isah, 2018; Wang et al., 2023). These 3D carbon materials are extensively utilized in energy storage applications, gas adsorption, and catalyst supports due to their exceptional capacity for storing gases and liquids. The diverse properties of these carbon-based materials make them an exciting area of scientific research, with potential applications in various industries and technologies.

4.1 Graphene for Energy Storage

Graphene is a super thin material that looks like a flat sheet and is made of carbon atoms with sp^2 hybridized looking like a honeycomb (hexagonal pattern) (Yang et al., 2018). Graphene has become an attractive material for developing energy storage devices because of its enormous surface area, superior electrical conductivity, and mechanical strength (Saha & Dutta, 2022).

Graphene-based metal-organic framework (MOF) hybrids such as MOF-808@S/GEC (Baumann et al., 2020), RGO/$ZnCo_2O_4$-ZnO-C/Ni (Li & Yin, 2015), Fe-MOF/

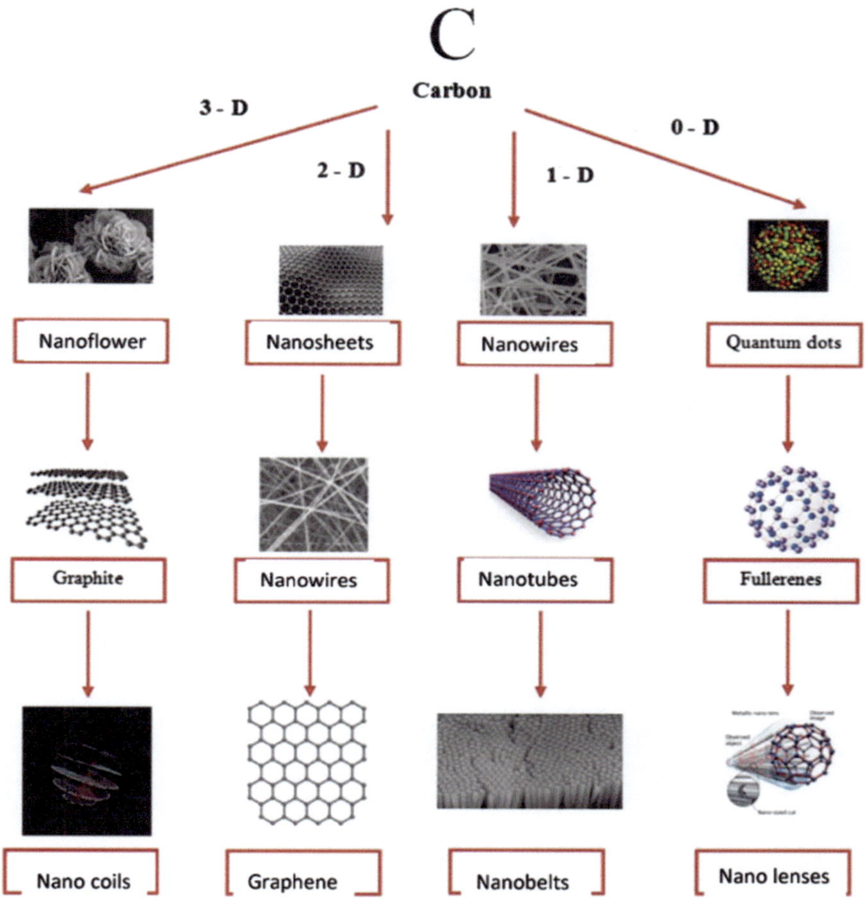

Fig. 2 Different carbon-based materials (Al Sheheri et al., 2019)

RGO (Jin et al., 2016) have been recognized as a significant breakthrough for energy storage applications due to their inimitable qualities. Ashoori et al. (2023) reported a Zn-air battery, which was based on the MOF-RGO cathode, exhibitions large specific energy of over 903 Wh $kgZn^{-1}$ (~290 mWh cm^{-2}), an exceptional power density of 148 $mWcm^{-2}$, and an open-circuit voltage of 1.430 V (Fig. 3a). These hybrids combine the extraordinary ability of graphene to conduct electricity and provide a lot of surface area, alongwith the flexibility and useful features of MOFs. This makes them perfect for storing energy. Down et al. (2018) showed the GO-based supercapacitor having a capacitive performance of 423 Fg^{-1} (Fig. 3b, c). Several research studies have focused on combining graphene with various cathode materials to boost the performance of lithium-ion batteries (Jin et al., 2016). Deng et al. (2016) developed anode materials using nanocomposites SnO_2 and graphene for lithium-ion

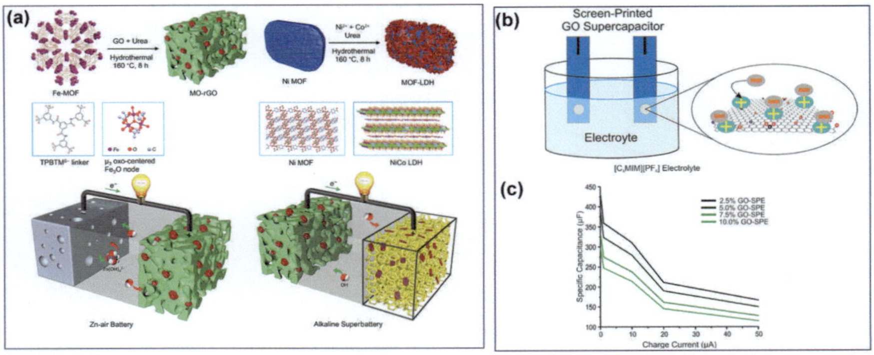

Fig. 3 (a) MOF-RGO framework (Ashoori et al., 2023), (b) GO-based supercapacitor and (c) capacitive performance of the 2.5% GO in the tested electrolytes (Down et al., 2018). Copyright (2023) American Chemical Society

batteries (LIBs). With this, graphene has been used as a conductive additive in the conductive slurry of a LIB (Jin et al., 2016). The 3D graphene networks, or graphene foams (GF), have been studied for their use in LIBs, fuel cells, and supercapacitors (Banciu et al., 2022). Xu et al. (2014) used a mix of graphene and ZrO_2 to create a Li-ion battery. This battery stayed very stable even after being charged and discharged 200 times, and it also worked well when charged or discharged quickly (Xu et al., 2014). Different types of graphene and carbon nanotube (CNT)-based materials for energy shortage devices are summarized in Table 1.

4.2 Carbon Nanotubes (CNTs) for Energy Storage

The CNTs, cylindrical nanostructures composed of carbon atoms, having exceptional properties and wide-ranging applications, attracted significant interest in energy storage devices (Ganesh et al., 2006). They possess high aspect ratios, large surface areas, and excellent electrical conductivity (Zhang & Li, 2009). CNTs are used as electrode materials in supercapacitors and fuel cells, as well as an additive material in active anode components (in batteries) to enhance electronic conductivity (Komatsubara et al., 2022; Simotwo et al., 2016). When CNTs are introduced into other materials, their open structure and greater chirality improve their properties and performances. Xu et al. (2022) created a mixture of spherical graphene and carbon nanotubes for lithium-ion batteries. The composite exhibits a charge and discharge capacity of 1658 mAh g^{-1} and 4409 mAh g^{-1} under a current density of 0.2 A g^{-1}, respectively (Fig. 4a, b). In a study conducted by Simotwo et al. (2016), they utilized electrospinning techniques to create nanofibers of polyaniline (PANI)-based CNT for use in supercapacitor (SC) electrodes. The PANI/CNT hybrid demonstrated the highest performance among the tested electrodes, exhibiting a specific capacity of

Table 1 Different types of graphene and carbon nanotube (CNT)-based materials for energy shortage devices

Reference	*Material*	*Application*	*Performance*
Ashoori et al., 2023	MOF-rGO	Zn-air battery based	High specific energy of over 903 Wh kg Zn^{-1} (~290 mWh cm^{-2})
Lin et al., 2008	Graphene as a conductive additive	Li-ion batteries	Improved conductivity despite low battery capacity
Luo et al., 2013	Bicontinuous mesoporous nanostructure Fe_3O_4 on 3D graphene foam	Li-ion batteries	Capacity of 785 mAh g^{-1} at a 1C rate of upto 500 cycles with no decay
He et al., 2016	3D Li_2S/graphene hierarchical architecture (3DLG)	Li-sulphur batteries	Discharge capacity of 894.7 mAh g^{-1} at 0.1C, Capacity retention of 87.7% after 300 cycles at 0.2C
Simotwo et al., 2016	d-MnO_2/holey reduced graphene oxide	Supercapacitor	Specific capacitance of 245 F g^{-1}
Xu et al., 2020	Sulfur-incorporated carbon nanotubes	Li-sulphur battery cathodes	Discharge capacity of 669 mAh g^{-1}
Huang et al., 2020	3D mesoporous carbon nanosheets (N-doped)	Na-ion batteries K-ion batteries	For SIBs, it showed a capacity of 239 mAh g^{-1} after 10,000 cycles, and for PIBs, the capacity reached 321 mAh g^{-1} after 5,000 cycles at the same 5 A g^{-1} rate
Li et al., 2014	Stacked multiwall carbon nanotubes (MWCNTs)	Li-ion batteries	Discharge capacity of 310 mAh g^{-1} at 0.5 C rate for 300 cycles
Xu et al., 2022	Graphene, carbon nanotubes, silicon particles	Li-ion batteries	Discharge capacity of 4409 mAh g^{-1} under a current density of 0.2 Ag^{-1}
Komatsubara et al., 2022	CNT sheets	Supercapacitor	Highly durable
Simotwo et al., 2016	Polyaniline-based carbon nanotubes	Supercapacitor Supercapacitor	Capacity of 385 F g^{-1}

385 F g^{-1}. Komatsubara et al. (2022) developed a flexible supercapacitor using CNT sheets as electrodes and showed its high durability with >99.9% capacitance retention after 50,000 bending cycles (Fig. 4c, d). Some of the research using CNTs for energy shortage devices is summarized in Table 1.

4.3 Carbon-Based Polymers

Carbon-based polymers have emerged as promising materials for energy storage devices due to their unique properties and versatile applications (Bandyopadhyay

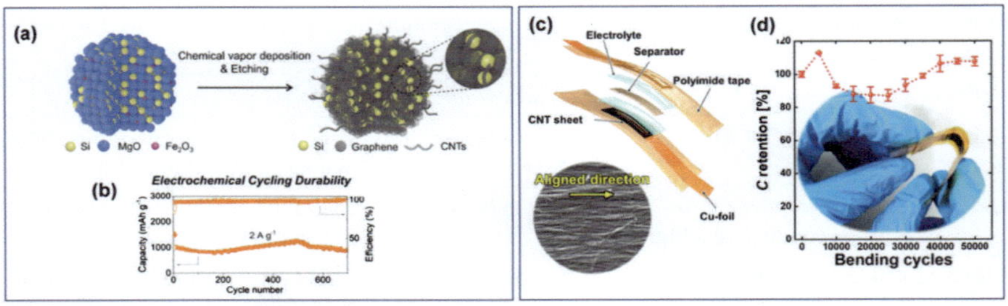

Fig. 4 (a) Forming a spherical graphene and carbon nanotube composite containing the Si particles, MgO and Fe_2O_3 (b) cycling capacity of the fabricated composite. Reprinted with permission from Xu et al., 2022. Copyright (2023) American Chemical Society, (c) fabrication of flexible CNT-SC (d) retention characteristics of CNT-SCs (Komatsubara et al., 2022) copyright (2023) American Chemical Society

et al., 2023; Chang et al., 2023; Wu et al., 2023; Yang et al., 2023). Conductive polymers like polythiophene (PTh), polypyrrole (PPy), PANI, and poly 3,4-ethylene dioxythiophene (PEDOT), etc. are considered potential pseudocapacitive materials because of their numerous active sites and high specific capacitance (Meng et al., 2017). Bandyopadhyay et al. (2023) fabricated an OPF/PEO-based composite film for a solid electrolyte for lithium battery applications, which had a discharge capacity of 130 mAh g^{-1} at 0.2C after 200 cycles (Fig. 5a, b). Chang et al. reported CNTs-based covalently coupled short-chain aniline trimers (AT)/CNT composite (Fig. 5c) for highly stable supercapacitors, which can retain 80% of its original capacitance after 20,000 charge/discharge cycles. It exhibits a specific capacitance of 284 F g^{-1} when tested at a current density of 0.2 mA cm^2 (Fig. 5d) (Chang et al., 2023). Two-dimensional organic polymers have emerged as versatile materials for energy storage devices, especially in supercapacitors. (Wakchaure et al., 2020) discuss the use of hex aminobenzene-derived two-dimensional polymer supercapacitors with a high specific capacitance of 805 Fg^{-1} at 0.5 Ag^{-1} current density in galvanostatic charge−discharge. (Tan et al., 2021) made significant studies in the field of recyclable energy storage devices by developing a solution-processable redox-active conjugated polymer. They reported the polymer electrodes to demonstrate stable functionality throughout 1500 charge–discharge cycles, even when subjected to a mass loading greater than 1 mg/cm^2.

The exceptional properties and tunability of carbon-based polymers offer promising opportunities for enhancing the efficiency and performance of energy storage technologies, paving the way for a sustainable and energy-efficient future. Some of the work using carbon-based polymers for energy shortage devices is summarized in Table 2.

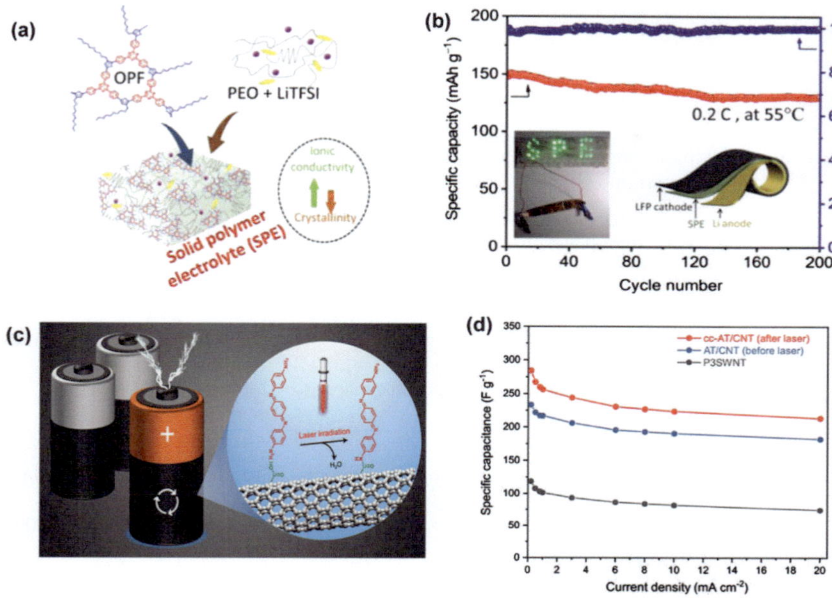

Fig. 5 (a) Synthesis of organic polymer framework (OPE), (b) cycling performance of fabricated solid electrolytes (Inset: Digital image LED powered by a fabricated battery and schematic of battery) (Bandyopadhyay et al., 2023). Copyright (2023) American Chemical Society, (c) AT/CNT based supercapacitor and (d) rate performance study of fabricated AT/CNT composites (Chang et al., 2023). Copyright (2023) American Chemical Society

4.4 Carbon-Based Other Materials

Apart from graphene, carbon nanotubes, and carbon-based polymers, various other carbon materials have been extensively explored for their potential in energy storage devices. This literature survey briefly overviews some of these carbon materials and their applications in different energy storage technologies.

Activated carbon, a widely used carbon material with a highly porous structure, has been extensively investigated for supercapacitors due to its excellent specific surface area and capacitance (Isah, 2018; Wang et al., 2023). Wang et al. (2023) have described the synthesis mechanism based on hydrothermal and thermal activation treatment to fabricate activated carbon fiber with polyaniline thiourea for the application of wearable supercapacitors. The fabricated electrode exhibited a capacitance of 563.8 Fg^{-1} when tested at 1 Ag^{-1}, along with a 89.1% cycling stability over 2000 cycles. Huang et al. (2021) reported heteroatom doping in porous carbons for its use in Zn-ion capacitors. The fabricated material shows an energy density of 131.9 Whk g^{-1} and a capacity of 168.4 mAh g^{-1} at 0.5 Ag^{-1}. Guo et al. (2021) discovered that a double-transition-metal composite (Zn@CoNiP/rGO) demonstrates impressive performance in energy storage. They observed a capacitance of 356.6 Fg^{-1} at

Table 2 Summary of conducting polymer and other carbon-based materials for ESD

Reference	*Material*	*Application*	*Performance*
Bandyopadhyay et al., 2023	OPF/PEO-based composite	Li-ion batteries	Discharge capacity: 130 mAh g^{-1} at 0.2C
Chang et al., 2023	Short-chain aniline trimers and CNTs	Supercapacitor	Specific capacitance: 284 Fg^{-1} at 0.2 mA cm^2 (75% capacitance retention)
Wakchaure et al., 2020	Hexaaminobenzene and pyromelliticdianhydride	Supercapacitor	Specific capacitance: 805 Fg^{-1} at 0.5 Ag^{-1}
Wang et al., 2023	Polyaniline (PANI)-thiourea (TU) -activated carbon fiber	Supercapacitor	Specific capacitance: 563.8 Fg^{-1} at 1 Ag^{-1}
Huang et al., 2021	Porous carbons	Zn-ion capacitors	Capacity: 168.4 $mAhg^{-1}$, energy density: 131.9 $Whkg^{-1}$ at 0.5 Ag^{-1}
Moorthi et al., 2023	(NCD/BMO) composites	Supercapacitor	Specific capacity: 155.8 mA hg^{-1} at 1 Ag^{-1}, capacity retention of ~98% upto 2000 cycles
Wang et al., 2021	Core-shell Co_2VO_4/carbon composite	Na-ion batteries	Current density: 5 Ag^{-1}, capacity: 135.1 mAh g^{-1} >98% capacity retention after 2000 cycles
Huang et al., 2020	Mesoporous 3D carbor nanosheets	Na-ion Batteries K-ion batteries	For SIBs, it holds a charge of 239 $mAhg^{-1}$ after 10,000 cycles at a rate of 5 Ag^{-1}. As for PIBs, they had a capacity of 321 mAh g^{-1} after 5,000 cycles at the same 5 Ag^{-1} rate

a charge/discharge rate of 0.5 Ag^{-1}, maintaining around 92.2% of its capacitance even after undergoing 10,000 charge/discharge cycles at a higher rate of 7.5 Ag^{-1}. Ren et al. (2019) described a material obtained from silkworm feces, which they turned into porous carbon. They mixed this carbon with RGO and CNTs to make a composite having a discharge capacity of 1354 mAh g^{-1} at 0.1 C -rate for the application as a cathode in lithium-sulfur batteries. A study by Wang C. et al. (2021) designed a core-shell Co_2VO_4/carbon composite anode for its application in sodium-ion batteries. The researchers reported a current density of 5 Ag^{-1} and a capacity of 135.1 mAh g^{-1}. Even after undergoing 2000 usage cycles, the composite they created retained over 98% of its initial capacity (Fig. 6a, b). Moorthi et al. (2023) developed carbon dots with nitrogen doping (NCDs) and integrated these with bismuth molybdenum oxide (Bi_2MoO_6; BMO) (NCD/BMO) composites for supercapacitors. They reported the specific capacity of fabricated composite is 155.8 mA h g^{-1} for a current density of 1 Ag^{-1} (Fig. 6c, d). Some of the research using other carbon-based materials for energy storage devices is summarized in Table 2.

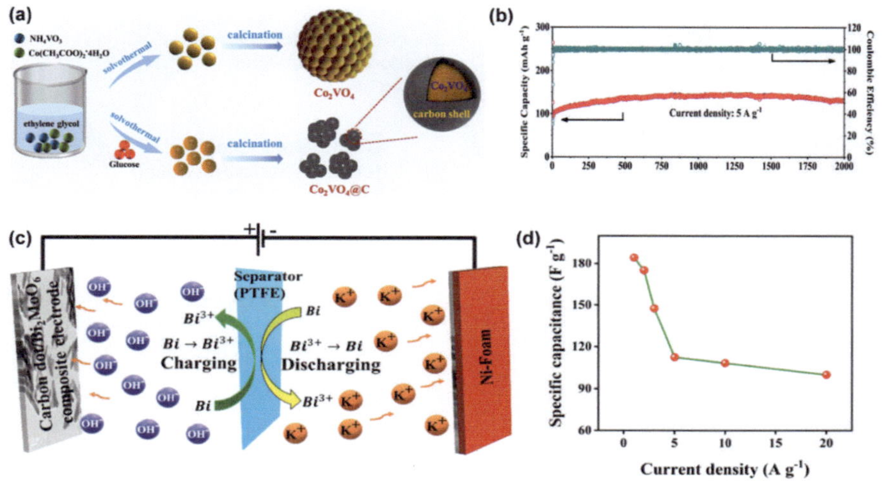

Fig. 6 (a) Co_2VO_4 and Co_2VO_4@C composite synthesis, (b) cycling performance of Co_2VO_4@C at a current density of 5 Ag^{-1} (Wang et al., 2021). Copyright (2023) American Chemical Society, (c) NCD/BMO composite-based supercapacitor and (d) specific capacity vs. current density (Moorthi et al., 2023). Copyright (2023) American Chemical Society

5 Challenges and Future Aspects

As research in energy storage applications progresses, carbon-based materials remain a top priority. There are several challenges, such as the inherent structural changes during charge and discharge cycles, which can lead to degradation and capacity loss over time (Iqbal et al., 2019; Jeerapan & Ma, 2019). Another challenge involves improving the cycling stability and lifespan of carbon-based energy storage devices (Olabi et al., 2023). Furthermore, the cost-effectiveness of large-scale production and integrating carbon-based materials into commercial energy storage devices presents a challenge. Some of the challenges are shown in Fig. 7. In terms of future aspects, the synergistic integration of carbon-based materials with emerging technologies shows tremendous promise. Hybrid structures combining carbon materials with nanoscale additives, metal nanoparticles, or even 2D materials could lead to enhanced electrochemical performance. Fine-tuning the properties and surface functionalities of carbon materials can significantly impact their charge storage capabilities and overall energy storage performance. Exploring environmentally friendly synthesis routes and recycling strategies will align with the broader objective of achieving sustainable and greener energy storage solutions. The future of carbon-based materials in energy storage lies in their integration with emerging technologies, nanoscale engineering advancements, and sustainable synthesis and recycling strategies. By overcoming these challenges and capitalizing on future opportunities, carbon-based materials can play an increasingly transformative role in shaping the landscape of energy storage for a more sustainable and energy-efficient future.

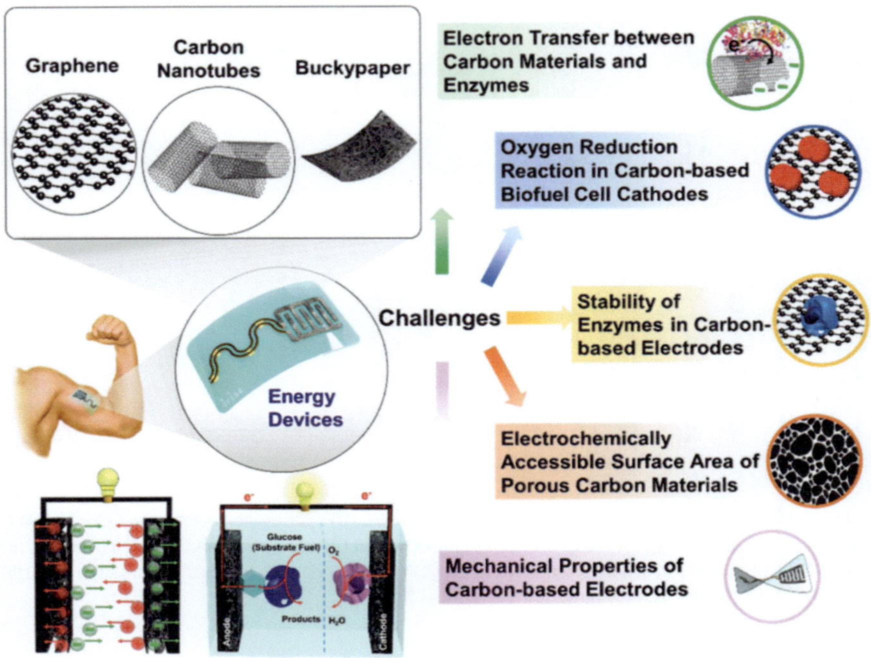

Fig. 7 Carbon nanomaterials for energy storage devices and examples of challenges in device development (Jeerapan & Ma, 2019)

6 Conclusion

This article provides a systematic overview of energy storage devices and the potential of carbon-based materials in revolutionizing energy storage technologies and overviews the advancements, challenges, and future prospects of carbon-based materials in energy storage. As we strive for a sustainable and greener energy future, carbon-based materials will undoubtedly have great potential in influencing the development of upcoming energy storage devices, ushering in an era of clean and efficient energy storage technologies.

References

Al Sheheri, S.Z., Al-Amshany, Z.M., Al Sulami, Q.A., Tashkandi, N.Y., Hussein, M.A. and El-Shishtawy, R.M. (2019). The preparation of carbon nanofillers and their role on the performance of variable polymer nanocomposites. *Designed Monomers and Polymers*, **22(1):** 8–53. https://doi.org/10.1080/15685551.2019.1565664.

Ashoori, A., Noori, A., Rahmanifar, M.S., Morsali, A., Hassani, N., Neek-Amal, M., Ghasempour, H., Xia, X., Zhang, Y., El-Kady, M.F., Kaner, R.B. and Mousavi, M.F. (2023). Tailoring

Metal–Organic Frameworks and Derived Materials for High-Performance Zinc–Air and Alkaline Batteries. *ACS Applied Materials & Interfaces*, **15(25):** 30220–30239. https://doi.org/10.1021/acsami.3c04454.

Banciu, C.A., Nastase, F., Istrate, A.I. and Veca, L.M. (2022). 3D Graphene Foam by Chemical Vapor Deposition: Synthesis, Properties, and Energy-Related Applications. *Molecules*, **27(11)**. https://doi.org/10.3390/molecules27113634.

Bandyopadhyay, S., Gupta, N., Joshi, A., Gupta, A., Srivastava, R. K., Kuila, B.K. and Nandan, B. (2023). Solid Polymer Electrolyte Based on an Ionically Conducting Unique Organic Polymer Framework for All-Solid-State Lithium Batteries. *ACS Applied Energy Materials*, **6(8):** 4390–4403. https://doi.org/10.1021/acsaem.3c00436.

Baumann, A.E., Downing, J.R., Burns, D.A., Hersam, M.C. and Sara Thoi, V. (2020). Graphene-metal-organic framework composite sulfur electrodes for Li-S batteries with high volumetric capacity. *ACS Applied Materials and Interfaces*, **12(33):** 37173–37181. https://doi.org/10.1021/acsami.0c09622.

Chang, X., Lin, C.W., Huang, A., El-Kady, M.F. and Kaner, R.B. (2023). Molecular Engineering of Hierarchical Conducting Polymer Composites for Highly Stable Supercapacitors. *Nano Letters*, **23(8):** 3317–3325. https://doi.org/10.1021/acs.nanolett.3c00284.

Dell, R.M. and Rand, D.A.J. (2001). Energy storage - A key technology for global energy sustainability. *Journal of Power Sources*, **100(1-2):** 2–17. https://doi.org/10.1016/S0378-7753(01)00894-1.

Deng, Y., Fang, C. and Chen, G. (2016). The developments of SnO2/graphene nanocomposites as anode materials for high performance lithium ion batteries: A review. *Journal of Power Sources*, **304:** 81–101. https://doi.org/10.1016/j.jpowsour.2015.11.017.

Down, M.P., Rowley-Neale, S.J., Smith, G.C. and Banks, C.E. (2018). Fabrication of Graphene Oxide Supercapacitor Devices. *ACS Applied Energy Materials*, **1(2):** 707–714. https://doi.org/10.1021/acsaem.7b00164.

Ganesh, E.N., Kumar, V.V. and Huzefa, E.M. (2006). Carbon nano tubes - Overview, simulation of single and multilayer CNTs with it's synthesis and energy storage applications. *NanoSingapore 2006: IEEE Conference on Emerging Technologies - Nanoelectronics - Proceedings*, **2006(Cvd):** 159–168. https://doi.org/10.1109/NANOEL.2006.1609704.

Guo, D., Li, Z., Wang, D., Sun, M. and Wang, H. (2021). Design and Synthesis of Zinc-Activated CoxNi2−xP/Graphene Anode for High-Performance Zinc Ion Storage Device. *ChemSusChem*, **14(10):** 2205–2215. https://doi.org/10.1002/cssc.202100285.

He, J., Chen, Y., Lv, W., Wen, K., Li, P., Qi, F., Wang, Z., Zhang, W., Li, Y., Qin, W. and He, W. (2016). Highly-flexible 3D Li2S/graphene cathode for high-performance lithium sulfur batteries. *Journal of Power Sources*, **327:** 474–480. https://doi.org/10.1016/j.jpowsour.2016.07.088.

Huang, H., Xu, R., Feng, Y., Zeng, S., Jiang, Y., Wang, H., Luo, W. and Yu, Y. (2020). Sodium/Potassium-Ion Batteries: Boosting the Rate Capability and Cycle Life by Combining Morphology, Defect and Structure Engineering. *Advanced Materials*, **32(8):** 1–11. https://doi.org/10.1002/adma.201904320.

Huang, L., Xiang, Y., Luo, M., Zhang, Q., Zhu, H., Shi, K. and Zhu, S. (2021). Hierarchically porous carbon with heteroatom doping for the application of Zn-ion capacitors. *Carbon*, **185:** 1–8. https://doi.org/10.1016/j.carbon.2021.09.019.

Iqbal, S., Khatoon, H., Hussain Pandit, A. and Ahmad, S. (2019). Recent development of carbon based materials for energy storage devices. *Materials Science for Energy Technologies*, **2(3):** 417–428. https://doi.org/10.1016/j.mset.2019.04.006.

Isah, S. (2018). Review article Advanced materials for energy storage devices. *Asian Journal of Nanoscience and Materials*, **1(2):** 90–103.

Jeerapan and Ma. (2019). Challenges and Opportunities of Carbon Nanomaterials for Biofuel Cells and Supercapacitors: Personalized Energy for Futuristic Self-Sustainable Devices. *C—Journal of Carbon Research*, **5(4):** 62. https://doi.org/10.3390/c5040062.

Jefferson, M. (2006). Sustainable energy development: Performance and prospects. *Renewable Energy*, **31(5):** 571–582. https://doi.org/10.1016/j.renene.2005.09.002.

Jin, Y., Zhao, C., Sun, Z., Lin, Y., Chen, L., Wang, D. and Shen, C. (2016). Facile synthesis of Fe-MOF/RGO and its application as a high performance anode in lithium-ion batteries. *RSC Advances*, **6(36):** 30763–30768. https://doi.org/10.1039/c6ra01645f.

Komatsubara, K., Suzuki, H., Inoue, H., Kishibuchi, M., Takahashi, S., Marui, T., Umezawa, S., Nakagawa, T., Nasu, K., Maetani, M., Tanaka, Y., Yamada, M., Nishikawa, T., Yamashita, Y., Hada, M. and Hayashi, Y. (2022). Highly Oriented Carbon Nanotube Supercapacitors. *ACS Applied Nano Materials*, **5(1):** 1521–1532. https://doi.org/10.1021/acsanm.1c04236.

Larcher, D. and Tarascon, J.M. (2015). Towards greener and more sustainable batteries for electrical energy storage. *Nature Chemistry*, **7(1):** 19–29. https://doi.org/10.1038/nchem.2085.

Li, J., Kaur, A.P., Meier, M.S. and Cheng, Y.T. (2014). Stacked-cup-type MWCNTs as highly stable lithium-ion battery anodes. *Journal of Applied Electrochemistry*, **44(1):** 179–187. https://doi.org/10.1007/s10800-013-0625-3.

Li, Z. and Yin, L. (2015). Sandwich-like reduced graphene oxide wrapped MOF-derived $ZnCo_2O_4$-ZnO-C on nickel foam as anodes for high performance lithium ion batteries. *Journal of Materials Chemistry A*, **3(43):** 21569–21577. https://doi.org/10.1039/c5ta05733g.

Lin, B., Wen, Z., Gu, Z. and Huang, S. (2008). Morphology and electrochemical performance of $Li[Ni_{1/3}Co_{1/3}Mn_{1/3}]O_2$ cathode material by a slurry spray drying method. *Journal of Power Sources*, **175(1):** 564–569. https://doi.org/10.1016/j.jpowsour.2007.09.055.

Liu, J., Wang, J., Xu, C., Jiang, H., Li, C., Zhang, L., Lin, J. and Shen, Z.X. (2018). Advanced Energy Storage Devices: Basic Principles, Analytical Methods, and Rational Materials Design. *Advanced Science*, **5(1)**. https://doi.org/10.1002/advs.201700322.

Luo, J., Liu, J., Zeng, Z., Ng, C.F., Ma, L., Zhang, H., Lin, J., Shen, Z. and Fan, H.J. (2013). Three-dimensional graphene foam supported Fe_3O_4 lithium battery anodes with long cycle life and high rate capability. *Nano Letters*, **13(12):** 6136–6143. https://doi.org/10.1021/nl403461n.

Meng, Q., Cai, K., Chen, Y. and Chen, L. (2017). Research progress on conducting polymer based supercapacitor electrode materials. *Nano Energy*, **36**(February), 268–285. https://doi.org/10.1016/j.nanoen.2017.04.040.

Moorthi, K., Yesuraj, J., Kim, K. and Mohan, S. (2023). Electrochemical Performance of Chemically Integrated N-Doped Carbon Dots/Bi_2MoO_6 Nanocomposites for Symmetric Supercapacitors. *Energy and Fuels*, **37(9):** 6841–6853. https://doi.org/10.1021/acs.energyfuels.3c00770.

Nižetić, S., Arıcı, M. and Hoang, A.T. (2023). Smart and Sustainable Technologies in energy transition. *Journal of Cleaner Production*, **389**. https://doi.org/10.1016/j.jclepro.2023.135944.

Noorden, R. Van. (2014). The rechargeable revolution A better battery. *Nature*, **507:** 26–28.

Olabi, A.G., Abbas, Q., Abdelkareem, M.A., Alami, A.H., Mirzaeian, M. and Sayed, E.T. (2023). Carbon-Based Materials for Supercapacitors: Recent Progress, Challenges and Barriers. *Batteries*, **9(1)**. https://doi.org/10.3390/batteries9010019.

Ren, M., Lu, X., Chai, Y., Zhou, X., Ren, J., Zheng, Q. and Lin, D. (2019). A three-dimensional conductive cross-linked all-carbon network hybrid as a sulfur host for high performance lithium-sulfur batteries. *Journal of Colloid and Interface Science*, **552:** 91–100. https://doi.org/10.1016/j.jcis.2019.05.042.

Ruban, A.V. (2015). Evolution under the sun: Optimizing light harvesting in photosynthesis. *Journal of Experimental Botany*, **66(1):** 7–23. https://doi.org/10.1093/jxb/eru400.

Saha, J.K. and Dutta, A. (2022). A Review of Graphene: Material Synthesis from Biomass Sources. In *Waste and Biomass Valorization* (Vol. 13, Issue 3). Springer Netherlands. https://doi.org/10.1007/s12649-021-01577-w.

Simotwo, S.K., Delre, C. and Kalra, V. (2016). Supercapacitor Electrodes Based on High-Purity Electrospun Polyaniline and Polyaniline-Carbon Nanotube Nanofibers. *ACS Applied Materials and Interfaces*, **8(33):** 21261–21269. https://doi.org/10.1021/acsami.6b03463.

Smil, V. (2017). Energy and civilization: A history. In *Energy and Civilization: A History*. Cambridge: MIT Press. https://doi.org/10.1353/tech.2018.0067.

Tan, S.T.M., Quill, T.J., Moser, M., Lecroy, G., Chen, X., Wu, Y., Takacs, C.J., Salleo, A. and Giovannitti, A. (2021). Redox-Active Polymers Designed for the Circular Economy of Energy

Storage Devices. *ACS Energy Letters*, **6(10):** 3450–3457. https://doi.org/10.1021/acsenergylett.1c01625.

Tawalbeh, M., Khan, H.A. and Al-Othman, A. (2023). Insights on the applications of metal oxide nanosheets in energy storage systems. *Journal of Energy Storage*, **60**(October 2022), 106656. https://doi.org/10.1016/j.est.2023.106656.

Wakchaure, V.C., Kottaichamy, A.R., Nidhankar, A.D., Ranjeesh, K.C., Nazrulla, M.A., Thotiyl, M.O, and Babu, S.S. (2020). Hexaaminobenzene Derived Two-Dimensional Polymer Supercapacitor with High Specific Capacitance and Energy Density. *ACS Applied Energy Materials*, **3(7):** 6352–6359. https://doi.org/10.1021/acsaem.0c00569.

Wang, C., Wang, Z., Zhao, D., Ren, J., Liu, S., Tang, H., Xu, P., Gao, F., Yue, X., Yang, H., Niu, C., Chu, W., Wang, D., Liu, X., Wang, Z., Wu, Y. and Zhang, Y. (2021). Core-Shell Co_2VO_4/Carbon Composite Anode for Highly Stable and Fast-Charging Sodium-Ion Batteries. *ACS Applied Materials and Interfaces*, **13(46):** 55020–55028. https://doi.org/10.1021/acsami.1c16035.

Wang, H., Xu, J. and Xie, Y. (2023). Bridge molecule thiourea enforced polyaniline growing on activated carbon fiber for wearable braidable-supercapacitor. *Applied Surface Science*, **637**(July), 157980. https://doi.org/10.1016/j.apsusc.2023.157980.

Wu, L., Shi, X. and Wu, Z.S. (2023). Recent Advancements and Perspectives of Biodegradable Polymers for Supercapacitors. *Advanced Functional Materials*, **33(16)**. https://doi.org/10.1002/adfm.202211454.

Xu, H., Yuan, S., Wang, Z., Zhao, Y., Fang, J. and Shi, L. (2014). Graphene anchored with ZrO_2 nanoparticles as anodes of lithium ion batteries with enhanced electrochemical performance. *RSC Advances*, **4(17):** 8472–8480. https://doi.org/10.1039/c3ra47653g.

Xu, J., Yin, Q., Li, X., Tan, X., Liu, Q., Lu, X., Cao, B., Yuan, X., Li, Y., Shen, L. and Lu, Y. (2022). Spheres of Graphene and Carbon Nanotubes Embedding Silicon as Mechanically Resilient Anodes for Lithium-Ion Batteries. *Nano Letters*, **22(7):** 3054–3061. https://doi.org/10.1021/acs.nanolett.2c00341.

Xu, Y.W., Zhang, B.H., Li, G.R., Liu, S. and Gao, X.P. (2020). Covalently Bonded Sulfur Anchored with Thiol-Modified Carbon Nanotube as a Cathode Material for Lithium-Sulfur Batteries. *ACS Applied Energy Materials*, **3(1):** 487–494. https://doi.org/10.1021/acsaem.9b01761.

Yang, G., Li, L., Lee, W. B. and Ng, M.C. (2018). Structure of graphene and its disorders: a review. *Science and Technology of Advanced Materials*, **19(1):** 613–648. https://doi.org/10.1080/14686996.2018.1494493.

Yang, H.X., Liu, Z.K., Wang, Y., Li, N.W. and Yu, L. (2023). Multiscale Structural Gel Polymer Electrolytes with Fast Li+ Transport for Long-Life Li Metal Batteries. *Advanced Functional Materials*, **33(1)**. https://doi.org/10.1002/adfm.202209837.

Ye, R., Henkensmeier, D., Yoon, S.J., Huang, Z., Kim, D.K., Chang, Z., Kim, S. and Chen, R. (2018). Redox Flow Batteries for Energy Storage: A Technology Review. *Journal of Electrochemical Energy Conversion and Storage*, **15(1):** 1–21. https://doi.org/10.1115/1.4037248.

Zhang, M. and Li, J. (2009). Carbon nanotube in different shapes. *Materials Today*, **12(6):** 12–18. https://doi.org/10.1016/S1369-7021(09)70176-2.

Zhao, H., Wu, Q., Hu, S., Xu, H. and Rasmussen, C.N. (2015). Review of energy storage system for wind power integration support. *Applied Energy*, **137:** 545–553. https://doi.org/10.1016/j.apenergy.2014.04.103.

Zhao, Z., Sun, L., Li, Y. and Feng, W. (2023). Polymer-derived carbon materials for energy storage devices: A mini review. *Carbon*, **210**(February). https://doi.org/10.1016/j.carbon.2023.118066.

Chapter 21
NaCMC-Based Biopolymer Composite Electrolytes for Supercapacitor Applications

Monarch Kumar, Himanshu Sharma, Nandini Nair, Anjali Verma, Jitesh Kumar, Mohd Suleman, Mohd Sadiq Anjani Kumar Singh, R. S. Maharia, B. L. Prashant, and Yogesh Kumar

1 Introduction

Metal-ion batteries have become an essential facilitator in transition from fossil fuels to sustainable forms of energy, using smartly built materials as the technical engine (Wang et al., 2020). With the lithium-ion battery serving as well-known example, this idea, sometimes known as a rocking-chair battery, has become extremely popular. Lithium-ion Battery (LIB), Sodium-ion Battery (SIB) (only in one instance where this battery has been termed NIB), and Potassium-ion battery (PIB) are the three secondary batteries based on metal ions that have received most attention to date (Balli et al., 2019). A well-designed electrolyte poses a broad voltage window, shows excellent electrochemical stability, and has high ionic concentration and conductivity, low toxicity, and little viscosity, all being important for supercapacitor performance (Govindarajan et al., 2022). Types of liquid electrolytes used can be differentiated as organic electrolytes, ionic liquids electrolytes, graphene quantum dots, and aqueous electrolytes to improve the electrochemical stability window for supercapacitors (Burke, 2000). Current conditions require a trade-off between satisfying

M. Kumar · H. Sharma · N. Nair · M. S. A. K. Singh (✉) · B. L. Prashant · Y. Kumar
Department of Physics, ARSD College, University of Delhi, New Delhi 110021, India
e-mail: sadiqphy@gmail.com

A. Verma · R. S. Maharia
Department of Chemistry, ARSD College, University of Delhi, New Delhi 110021, India

J. Kumar
Department of Physics and Electronics, Rajdhani College University of Delhi, New Delhi, India

M. Suleman
Department of Physics, Shibli National College, MSDS University Azamgarh, UP, Azamgarh, India

R. K. Kotnala et al. (eds.), *Advanced Functional Materials for Sustainable Environments*, https://doi.org/10.1007/978-3-031-62620-3_21

energy requirements and addressing environmental challenges, essential for attaining long-term viability. Due to their affordability and environmental friendliness, literature analysis shows biopolymer electrolytes are currently gaining a lot of attention as alternatives to synthetic polymers in electrochemical systems (Cingolani, 2013; Naderi et al., 2014). Solid biopolymer electrolytes have several advantages over liquid or gel-based electrolytes such as flexibility, lightweight, low self-discharge in batteries, good electrode surface contact, and leakage problem and are environmentally friendly. Solid biopolymer electrolytes have become the focus of interest for researchers in recent years due to their potential to enhance the performance of battery and energy storage devices. Due to their improved stability, increased ion conductivity, and better mechanical strength they have found applications in batteries, fuel cells, sensors, and supercapacitors (Guzmán-González et al., 2023; Nakamur et al., 2004). This study focuses on Sodium carboxymethyl cellulose (NaCMC) as a polymer host in a polymer electrolyte system. NaCMC is a derivative of cellulose, which is the most abundant natural polymer, found in plants and trees. NaCMC is produced by modifying the cellulose molecule by a process called carboxymethylation in which carboxymethyl groups (-CH_2-COOH) are introduced onto the cellulose backbone. It is an anionic, water-soluble biopolymer.

One of the unique properties of NaCMC is its ability to form gels in aqueous solutions. This property is because of the presence of carboxymethyl groups, which act as a negative charge to the polymer molecule. The negative charges cause the NaCMC chains to repel each other, leading to the formation of a 3-dimensional network structure that can trap water molecules and form a gel (Samsudin et al., 2012; Ahmad et al., 2016). This property makes NaCMC an effective thickener and stabilizer in food and pharmaceutical formulations. NaCMC also has excellent biocompatibility and biodegradability. It is non-toxic and is widely used in the food and pharmaceutical industries. Being biodegradable, it can be broken down by natural processes into harmless byproducts, making it an environmentally friendly material. NaCMC has also been investigated for its potential use in energy storage applications (Cyriac et al., 2022). NaCMC-based composites have shown high capacitance values and good stability, making them promising candidates for use in supercapacitor applications. The goal of the current effort is to create a solid polymer electrolyte layer with sufficient mechanical strength and ionic conductivity for energy storage applications. The ionic conductivity of SPEs of the sodium salt of carboxymethyl cellulose (NaCMC) doped with $NaNO_3$ was found to be around ~3×10^{-3} S/cm for the sample containing 30% $NaNO_3$ (Shetty et al., 2021). Plasticizers, due to their relatively small molecular size, are capable of infiltrating the gaps between neighboring polymer chains, thereby reducing the secondary attractive forces between neighboring polymer chains. This infiltration leads to an expansion of the available space within the polymer structure, consequently enhancing the mobility of polymer segments as well as the mobility of charge carriers (Ahmed et al., 2019). In the present study NaCMC-based solid biopolymer electrolytes were prepared by solution cast techniques. The ethylene glycol solution acts as plasticizer, Na-CMC as biopolymer doped with $NaNO_3$ as salt. The effect of glycerol plasticizers on the electrochemical parameters of the system was studied. Electrical impedance spectroscopy was used to examine the ionic conductivity of plasticized biopolymer electrolytes PEs generated at room temperature. The highest ionic conductivity, and their performance was assessed using LSV and CV.

2 Materials and Method

2.1 Materials

Ethylene glycol ($C_2H_6O_2$) AR, ≥99%, sodium carboxymethyl cellulose salt (NaCMC), sodium nitrate ($NaNO_3$) and deionized (DI) distilled water were materials purchased from an Indian company.

2.2 Method: Preparation of Biopolymer Electrolyte

Thin films of biopolymer NaCMC having 5 wt% ethylene glycol solution was prepared by solution casting method with varying $NaNO_3$ salt concentrations of 10 wt% (as C10), 20 wt% (as C20), 30 wt% (as C30) and 40 wt% (as C40) in DI distilled water. The sample preparation of biopolymer composite electrolytes is shown in Fig. 1. The solution was prepared using the traditional solution casting method in distilled water by stirring for 5 h until the solution became homogeneous. The samples were then poured into Petri dishes and were dehydrated by keeping samples at temperatures 70°C-90°C for over 24 hours until the samples were dry in an oven. The formed electrolyte films were then peeled off and prepared for characterization.

3 Results and Discussions

Electrochemical impedance spectroscopy showed the electrical properties of the electrolyte. In an EIS measurement, a small perturbation signal (typically a sinusoidal voltage or current) is applied to the system, and the resulting response is measured. By varying the frequency of the perturbation signal over a range of values, the impedance of the system can be determined as a function of frequency (Liew et al., 2015). The impedance spectra of all prepared biopolymer samples are shown in Fig. 2.

A graph between Z' and Z" was plotted from which the bulk ionic conductivity was calculated using the relation

$$\sigma = \frac{t}{R_b \times A}$$

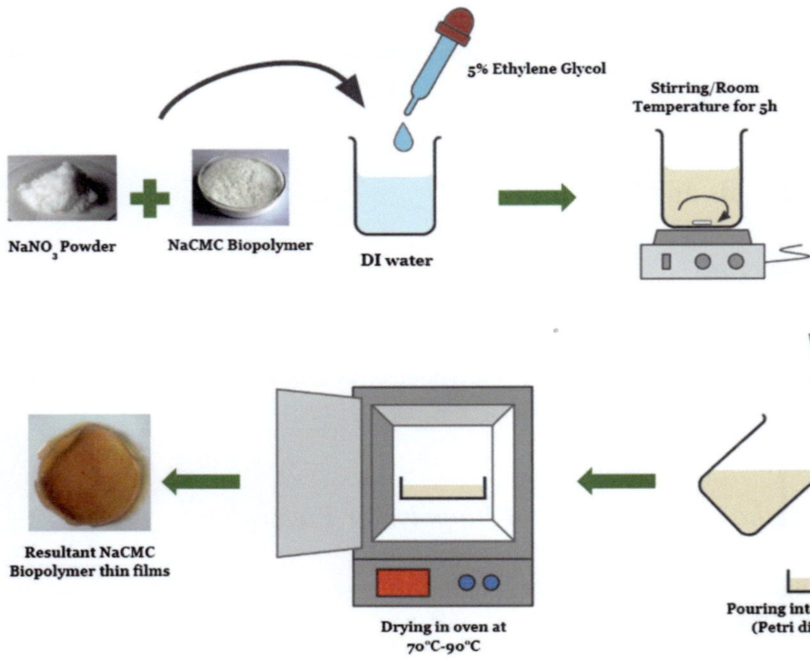

Fig. 1 Biopolymer composite electrolytes

where t is the thickness, R_b bulk resistance of the specimen and A the area of the electrodes (Arya et al., 2018). The ionic conductivity of all biopolymer composite samples is discussed in Table 1.

Chronoamperometry was used to study the kinetics of the electrolyte system by measuring the current as a function of time. In this technique, the direct current occurring at the electrode is monitored as a function of time when a step potential is applied (Sadiq et al., 2023). A step potential was applied to the system and the resulting current was measured. By analyzing the current as a function of time, the electronic transference number (t_e) and ion transference number (t_{ion}) were calculated by using

$$t_{ion} = \frac{I_i - I_f}{I_i}$$

and

$$t_e = \frac{I_f}{I_i}$$

where I_i is the total initial current and I_f the total final current.

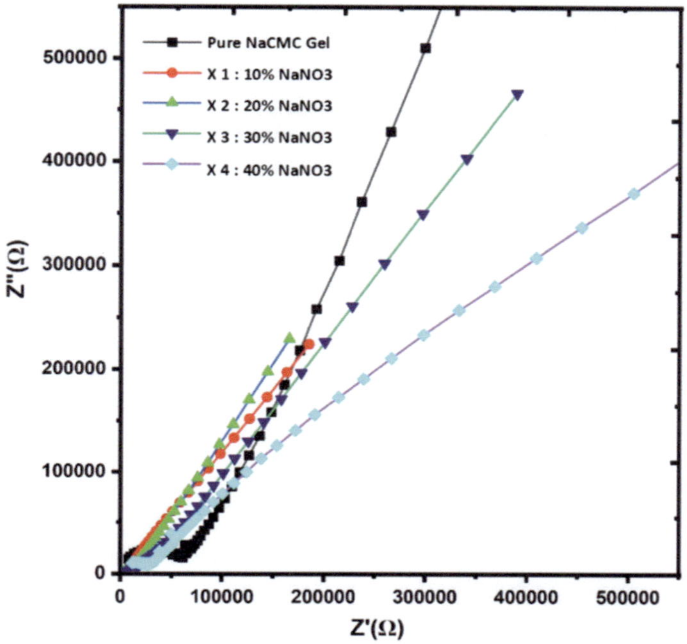

Fig. 2 Ionic conductivity of biopolymer electrolytes

Table 1 Stability window, conductivity, ion and electron transference and bulk resistance for samples in study

Conc. of $NaNO_3$ in NaCMC sample	*Conductivity (10^{-6} in S/m)*	*Chronoamperometry ion and electron transference*		*Bulk resistance (R_b) (Z' in Ω)*
		t_{ion}	t_e	
Pure (0%)	0.265	-	-	62,943.72
10%	1.861	0.943	0.0568	4342.67
20%	3.002	0.923	0.07195	4007.74
30%	2.525	0.951	0.04912	8111.18
40%	0.802	0.945	0.05565	30,451.12

The transference number quantifies the fraction of the total current carried by a specific ion in an electrolyte as shown in Fig. 3. The above relation was used for calculating ion transference number (t_{ion}) as well as electronic transference number (t_e) and found the most interesting results (Table 1). The ion transference number obtained was larger than the electronic transference number, therefore, concluding the fact that conductivity is majorly caused by ions and the distribution of resultant current is negligible due to electrons.

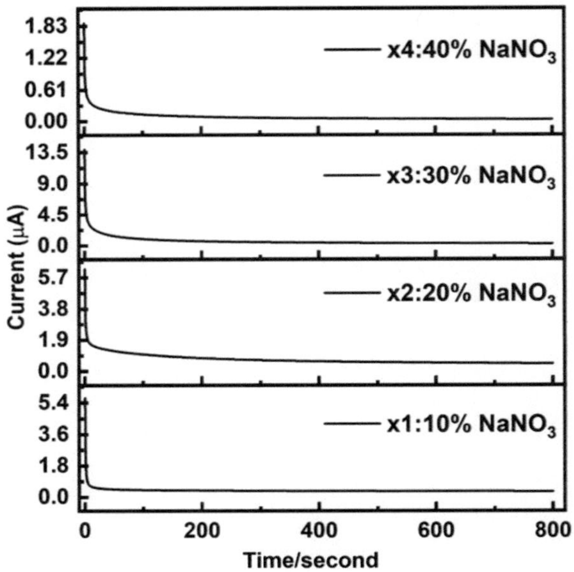

Fig. 3 Chronoamperometry biopolymer composite electrolytes samples

Linear sweep voltammetry (LSV) is a voltammetry technique that measures the current response of a sample as a function of a linearly varying potential. The current peaks or troughs at the onset potential of redox reactions of the sample indicate the oxidation or reduction of a species in LSV. Although the potential differential between the working electrode and the reference electrode utilized is linearly time-swept, the current at the electrode is measured (Aziz et al., 2023). Here LSV was used to calculate the stability window conclude that 20% $NaNO_3$ had the highest stability window of 3.71 vas compared to the pure sample with 1.57 V (Fig. 4).

Cyclic voltammetry was used to investigate the oxidation and reduction processes of the electrolyte system (Fig. 5). It involves application of a potential waveform to an electrochemical cell, and the measurement of the resulting current as a function of the applied potential (Sadiq et al., 2021). By analyzing the shape and magnitude of the resulting voltammogram information about the oxidation and reduction potentials of electrolytes and their respective stability windows was obtained.

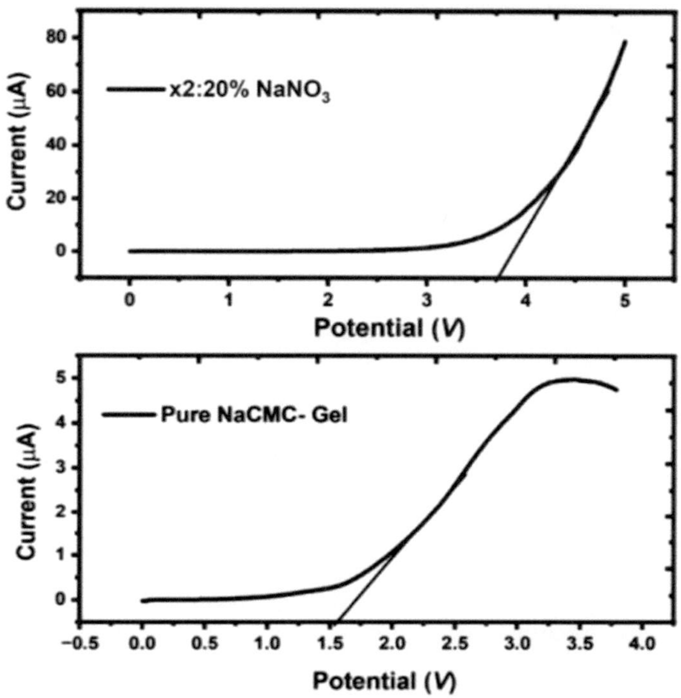

Fig. 4 Linear sweep volumetry (LSV) graphs for pure and 20% doped biopolymer samples

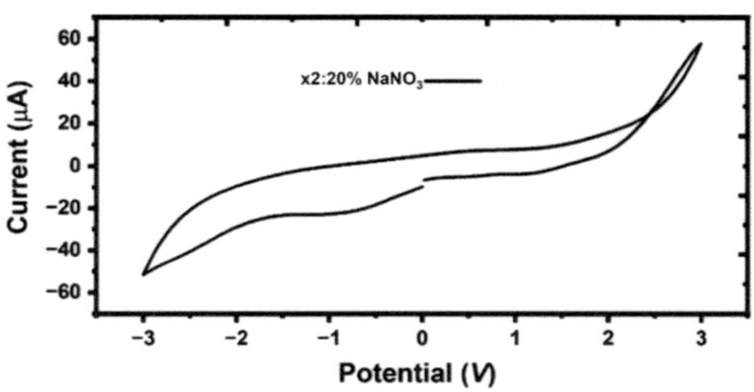

Fig. 5 Cyclic volumetry graph for various compositions of doped NaCMC polymer

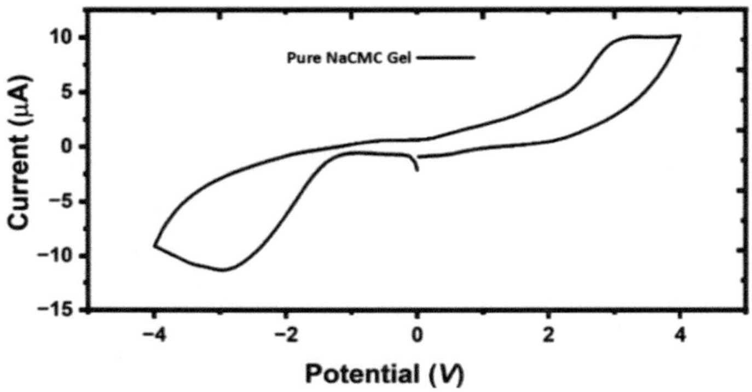

Fig. 5 (continued)

4 Conclusion

NaCMC-based solid biopolymer electrolyte was prepared by doping with $NaNO_3$ and ethylene glycol using the method of solution casting. Addition of ethylene glycol as a plasticizer enhanced the conductivity, flexibility, and other mechanical properties of the solid polymer electrolyte. Chronoamperometry and EIS techniques were used to calculate the bulk ionic conductivity and ion transference numbers. The transference number obtained for various solutions was found to be >0.9. Thus, we found that ions are the dominant charge carriers. The stability window which was calculated using linear sweep voltammetry was found to be greatest for the sample doped with 20 wt% of $NaNO_3$. The highest ionic conductivity was obtained for the host containing 20 wt% of $NaNO_3$ which was found to be around 3.002×10^{-6} S/cm compared with other doped films.

References

Arya, A. and Sharma, A.L. (2018). Structural, microstructural and electrochemical properties of dispersed-type polymer nanocomposite films. *Journal of Physics D: Applied Physics*, **51:** 045504.

Ahmad, N.H. and Isa, M.I.N. (2016). Characterization of un-plasticized and propylene carbonate plasticized carboxymethyl cellulose doped ammonium chloride solid biopolymer electrolytes. *Carbohydrate polymers*, **137:** 426–432.

Ahmed, H.T., Jalal, V.J., Tahir, D.A., Mohamad, A.H. and Abdullah, O.G. (2019). Effect of PEG as a plasticizer on the electrical and optical properties of polymer blend electrolyte MC-CH-$LiBF_4$ based films. *Results in Physics*, **15:**102735.

Aziz, S.B., Ahmed, M.J., Abdullah, O.G., Murad, A.R., Hamad, S.M. and Hadi, J.M. (2023). Magnesium Ion Conducting Biopolymer Blend-Based Electrolyte for Energy Storage Application: Electrochemical Characteristics. *Electrochimica Acta*, 142659.

Balli, B., Şavk, A. and Şen, F. (2019). Graphene and polymer composites for supercapacitor applications. In Nanocarbon and its Composites, Woodhead Publishing, pp. 123–151.

Burke, A. (2000). Ultracapacitors: why, how, and where is the technology. *Journal of Power Sources*, **91:** 37–50.

Cingolani, R. (2013). The road ahead. *Nature Nanotechnology*, **8:** 792–793.

Cyriac, V., Ismayil, Noor, I.M., Mishra, K., Chavan, C., Bhajantri, R.F. and Masti, S.P. (2022). Ionic conductivity enhancement of PVA: carboxymethyl cellulose poly-blend electrolyte films through the doping of NaI salt. *Cellulose*, **29:** 3271–3291.

Govindarajan, D. and Chinnakutti, K.K. (2022). Fundamentals, basic components and performance evaluation of energy storage and conversion devices. In Oxide Free Nanomaterials for Energy Storage and Conversion Applications. Elsevier, pp. 51–74.

Guzmán-González, G., Avila-Paredes, H.J. and Santos-Mendoza, I. (2023). The role of nanoceramic additives surface charge on the ionic transport of single lithium-ion conducting polymer electrolytes. *Journal of Solid State Electrochemistry*, **27:** 1–11.

Liew, C.W. and Ramesh, S. (2015). Electrical, structural, thermal and electrochemical properties of corn starch-based biopolymer electrolytes. *Carbohydrate polymers*, **124:** 222–228.

Naderi, A. and Lindström, T. (2014). Carboxymethylated nanofibrillated cellulose: effect of monovalent electrolytes on the rheological properties. *Cellulose*, **21:** 3507–3514.

Nakamur, K., Minagaw, Y., Hatakeyam, T. and Hatakeyama, H. (2004). DSC studies on bound water in carboxymethylcellulose–polylysine complexes. *Thermochimica acta*, **416:**135–140.

Samsudin, A.S., Khairul, W.M. and Isa, M.I.N. (2012). Characterization on the potential of carboxy methylcellulose for application as proton conducting biopolymer electrolytes. *Journal of Non-crystalline Solids*, **358:** 1104–1112.

Sadiq, M., Khan, M.A., Raza, M.M.H., Zulfequar, M. and Ali, J. (2023). Improved performance of biopolymer composite electrolyte based cellulose acetate/Zinc Oxide filler for supercapacitors. *Energy & Environment*, 0958305X231159443.

Sadiq, M., Raza, M.M.H., Chaurasia, S.K., Zulfequar, M. and Ali, J. (2021). Studies on flexible and highly stretchable sodium ion conducting blend polymer electrolytes with enhanced structural, thermal, optical, and electrochemical properties. *Journal of Materials Science: Materials in Electronics*, **32:** 19390–19411.

Wang, Q., Chen, S. and Zhang, D. (2020). CNT yarn-based supercapacitors. In Carbon nanotube fibers and yarn, Woodhead Publishing, pp. 243–270

Chapter 22
Biomass-Derived Carbon Coupled with $NiCo_2O_4$ as an Effective Electrode Material for High-Performance Supercapacitor

Brijeshkumar Yadav, Giriraj Dandekar, Rounak Atram, and Mahadev Sonawane

1 Introduction

The necessity for the development of clean energy is driven by the rising global energy demand and new ecological problems brought on by the burning of fossil fuels. Due to the difficult problems with non-renewable sources, such as growing global warming, air and water pollution, and a limited supply of natural re-sources, it is necessary to move towards renewable sources of energy. (Ferreira et al., 2016). By harnessing renewable energy sources like sun, water, wind, etc. the need for clean energy may be satisfied. Due to irregular energy generation, these environmentally friendly alternatives need effective energy storage system. Researchers are working hard to develop new energy storage technologies, such as supercapacitors, batteries, fuel cells, etc. by studying revolutionary materials (Gielen et al., 2019).

Supercapacitors have become a potential substitute for batteries since they possess superior performance in various aspects, making them a promising energy storage technology. These electrochemical devices have outstanding energy, power and capacity density and also offer quicker charging times and greater energy capacity compared to traditional batteries. This makes supercapacitors an attractive alternative for energy storage purposes. These energy storage devices may be identified by, among other characteristics, their rapid cycle rate, high power density, durability,

B. Yadav · M. Sonawane
Department of Physics, The Institute of Science, Dr. Homi Bhabha State University, 15, Madam Cama Road, Fort Mumbai, Mumbai, Maharashtra 400032, India

G. Dandekar · R. Atram (✉)
Department of Physics, M.M College of Arts, N.M. Institute of Science H.R.J College of Commerce, Bhavans College, Andheri-(W), Mumbai, Maharashtra 400058, India
e-mail: rounak.atram@bhavans.ac.in

R. K. Kotnala et al. (eds.), *Advanced Functional Materials for Sustainable Environments*, https://doi.org/10.1007/978-3-031-62620-3_22

extended life, and environmentally benign functioning (Satyajit and Samantara, 2018).

On the basis of charge storage mechanism supercapacitors are classified in two categories, viz. EDLC (electric double layer capacitors) and pseudocapacitors. The EDLCs store charge faradaically, while pseudocapacitors store charge non-faradaically. The EDLCs are mainly carbon materials like graphene, CNTs, CNF, biomass-derived carbon and others, whereas, pseudocapacitors are mostly conducting polymers and metal oxide materials. (Rajan et al., 2022; Frackowiak Beguin, 2001).

Carbon generated from biomass, such as biochar, can be utilized as a carbon source for supercapacitors (Jameel and Keshwani, 2017). Being a renewable and sustainable source of carbon as opposed to conventional fossil fuel-based sources, the use of carbon from biomass in supercapacitors is a desirable choice. Biomass may be converted thermochemically into a variety of energies, including heat, electricity and bio-fuels (Enock et al., 2017). Combustion, pyrolysis, carbonization, torrefaction, co-firing, gasification, and liquefaction are some of the thermochemical conversion processes. Pyrolysis, which includes all chemical procedures to produce char solid, bio-oil, and bio-gas under oxygen-free or oxygen-limiting circumstances, is regarded as the origin of all thermochemical conversion techniques (Amer and Elwardany, 2020). These thermochemical conversion processes are characterized by employing heat, which is often provided by external sources, to break down organic materials into richer sources of energy by destroying the hydro-carbonaceous bonds (Yaashikaa et al., 2020).

The performance of a supercapacitor heavily relies on the selection of an electrode material with high specific capacitance, superior chemical stability, and an optimum pore size distribution, resulting in extensive surface area, which in turn enables the efficient transport of ions and electrons across the electrode-electrolyte interface (Rajendran and Wang, 2018; Simon and Gogotsi, 2008; Abeykoon et al., 2017; Ray et al., 2019) (Fig. 1).

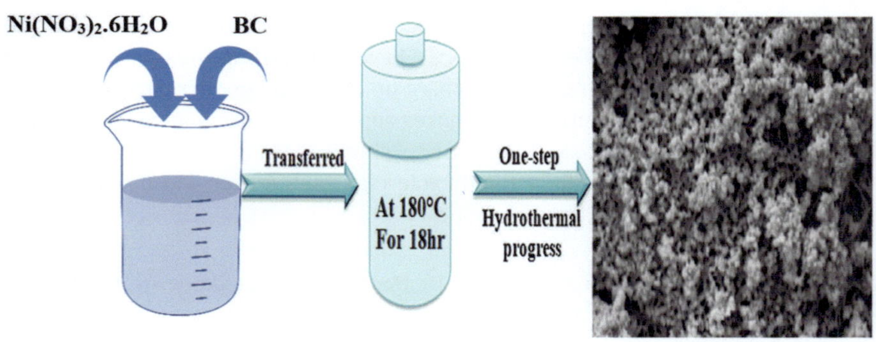

Fig. 1 Preparation of $NiCo_2O_4$/BC

This article describes, synthesized biocarbon (BC) from royal palm trees which is fibrous in nature. To create composite electrode materials for supercapacitors, transition metal oxide ($NiCo_2O_4$) was grown on BC that was produced from the petioles of royal palms by carbonization. High specific capacitance, high energy density, and outstanding performance were produced using an asymmetric electrode made of nickel oxide ($NiCo_2O_4$/BC).

2 Materials and Method

2.1 Materials

Materials used for synthesis are nickel nitrate hexahydrate ($Ni(NO_3)_2.6H_2O$), cobalt nitrate hexahydrate ($Co(NO_3)_2.6H_2O$), NMP, polyvinylpyrrolidone (PVP), acetone, potassium hydroxide (KOH) purchased from Loba Chem. All the chemicals were analytical grade and used without further purification.

2.2 Preparation of Biocarbon (BC)

Pieces of the Royal Palm (Roystonea regia) such as the petiole and leaf sheath were gathered from Bhavan's College in Andheri (W), Mumbai, India. It was cut into small pieces (~1 cm) and dried under the sun until all visible moisture had evaporated. Gathered biomass was washed and cleaned with distilled water to remove dust from the surface and further dried in hot air oven at 80 °C for 1 hr with further rise in temperature to 140 °C for 4 hours and cooled overnight (Hooda and Ghany, 2019). The biomass pieces were then placed in silica crucibles for torrefaction, sealed with a tight lid, and coated with several layers of aluminum foil to keep out atmospheric air except for air trapped inside vessel cavities and material being physically activated, which created a low oxygen environment (Sivaraj et al., 2010; Bergman and Kiel, 2005; Lu and Zhao, 2017).

These crucibles were kept in the muffle furnace at 200 °C for 2 hours. After cooling, the sample was crushed with a mortar and pestle and filtered through many layers of cloth. Following that, the samples were placed in tightly sealed crucibles, and the entire setup was placed in a muffle furnace at 400 °C for 2 hours for carbonization (Amer and Elwardany, 2020). After overnight natural cooling, the material was collected and stored. After rinsing with concentrated HNO_3 and washing with distilled water until the pH was neutralized, the sample was soaked in KOH to remove any residue before drying in the hot air oven for 6 hours (Isahak et al., 2013).

2.3 Preparation of $NiCo_2O_4$/BC

$NiCo_2O_4$/BC was prepared using hydrothermal synthesis method. First, 100 mg of BC derived from royal palm biomass was ultrasonicated for 2 h in 100 mL of deionized water. Later, 0.5 g of nickel nitrate hexahydrate and 0.5 g cobalt nitrate hexahydrate were dissolved in the same solution to make a homogenous mixture. Second, the solution was transferred into a 100 mL Teflon-lined autoclave and heated to 180 °C for 18 hours. Lastly, the autoclave was allowed to cool naturally to ambient temperature. The product was washed with distilled water, filtered with Whatman filter paper and dried overnight in a hot air oven (Fu et al., 2019; Guo et al., 2018).

3 Material Characterization

The morphology of the biocarbon (BC) and $NiCo_2O_4$/BC were examined by scanning electron microscopy (SEM EVO18, CARLZEISS). The phase structures of the as-synthesized biocarbon and $NiCo_2O_4$/BC were studied by XRD (RIGAKU) using Cu K_α radiation ($\lambda = 1.5406$) at scanning rate of 7 min^{-1} ranging from 10 to 80 °C. The elemental composition of the samples were studied using energy dispersive X-ray spectroscopy (EDS). Functional group study was done using Fourier transform infrared spectroscopy (FTIR).

4 Electrochemical Measurements

The electrochemical properties were studied on Metro ohm PGSTAT204 electrochemical workstation (Institute of Science, Nagpur, Maharashtra, India). The 1 M H_2SO_4 electrolyte were used to study electrochemical properties in a three-electrode system. Platinum and Ag/$AgCl_2$ were the counter and reference electrode, respectively. Synthesized biocarbon and $NiCo_2O_4$/BC were the working electrodes. The working electrodes were prepared as 85 wt.% of as-prepared Biocarbon and $NiCo_2O_4$/BC, 10 wt.% polyvinylidene fluoride (PVDF) and 5 wt.% carbon black and add few drop N-methyl-2-pyrrolidone (NMP) binder were mixed to obtain homogeneous slurry. The slurry was pasted on graphite electrode using brush and dried in hot air vacuum at 70 °C for 12 h. The electrochemical analysis was done by cyclic voltammetry (CV), galvanostatic charge-discharge (GCD) and electrochemical impedance spectroscopy (EIS) techniques with range from 0.01 MHz to 100 kHz. The specific capacitance (C_s) of the electrodes were calculated from the galvanostatic charge-discharge (GCD) curves by following equation:

$$C_S = \frac{I \times \Delta t}{\Delta V \times m} \tag{1}$$

where C_s is the specific capacitance (Fg^{-1}), I the current during the charge-discharge period (A), Δt the discharge time during the measurement (sec), ΔV the operating voltage window (V) and m is the mass of corresponding electrode materials (mg).

The energy density (ED) and power density (PD) of the $NiCo_2O_4$/BC electrode were calculated according to equations:

$$ED = \frac{1}{2 \times 3.6} \times \Delta V^2 \times C_S \quad (2)$$

$$PD = \frac{3600 \times ED}{\Delta t} \quad (3)$$

where C_S is the specific capacitance in (Fg^{-1}).

5 Result and Discussion

5.1 *Scanning Electron Microscope and Energy Dispersive Spectroscopy*

The surface morphology of the as-synthesized BC and $NiCo_2O_4$/BC composite material was studied using scanning electron microscope (SEM). In Fig. 2 (a) the SEM image show rough surfaces with irregular shapes and pores of various sizes, randomly distributed all over. The porous structure in the material contributes to enhancing the surface area of the sample. Porosity also allows deep penetration of electrolytes in the material which activates various sites to store the ions resulting in higher electrochemical results. The micrograph was captured at 2 μm and a magnification of 11.32 KX. The SEM image also reveals that the torrefaction and carbonization processes create microporous carbon after activation. Figure 2 (b) reveals the randomized distribution of $NiCo_2O_4$ nanoparticle all over the surface of biocarbon. The SEM image also shows nanopores of the biocarbon visible in few parts of the images. The growth of $NiCo_2O_4$ nanoparticles is visible on the surface of biocarbon. The micrograph was captured at 2 μm and a magnification of 12.74 KX.

Figure 3 (a) and (b) shows the elemental composition of biocarbon and $NiCo_2O_4$/BC. Elemental composition was done using energy dispersive spectroscopy (EDS). The EDS spectra of $NiCo_2O_4$/BC reveals the peaks of carbon, oxygen, nitrogen, cobalt, nickel which are in proportion to our stoichiometric ratio and confirms the presence of these elements in our material.

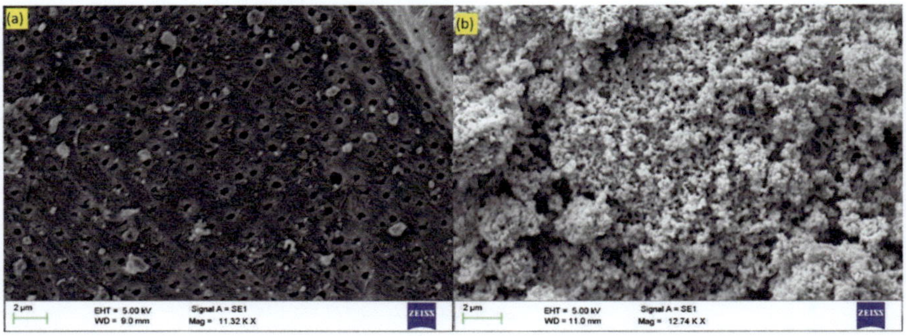

Fig. 2 The SEM images of: (a) biocarbon (BC) and (b) $NiCo_2O_4$/BC

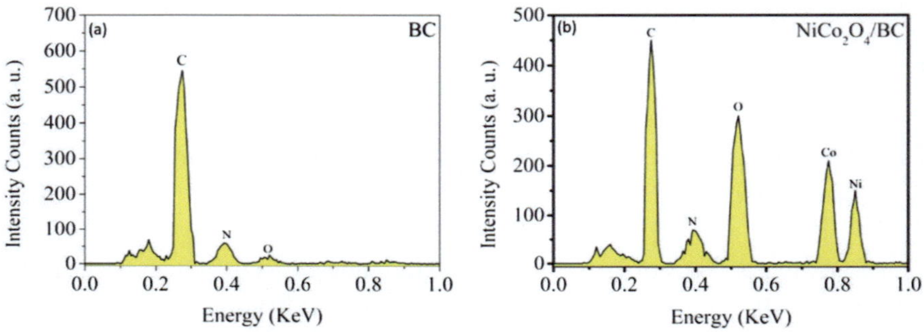

Fig. 3 The EDS spectra of biocarbon and $NiCo_2O_4$/BC

5.2 *X-Ray Diffraction (XRD)*

As shown in Fig. 4, XRD patterns are used to analyze the phase and crystallinity of composites made of $NiCo_2O_4$ and BC generated from biomass. The maximum peak is used to standardize all peaks. The crystalline character of the nanocomposite is indicated by the strong peaks. When the amount of $NiCo_2O_4$ nanoparticle on BC is increased, the $NiCo_2O_4$/BC diffraction peak grows stronger, demonstrating the presence of the $NiCo_2O_4$ phase in the $NiCo_2O_4$/BC composites. The peaks could be indexed to the $NiCo_2O_4$/BC (220), (311) and (400) spinel phase diffraction peaks obtained at 32°, 36.7° and 46° (JCPDS Card no. 20-0781) (Yang and Park, 2020; Kamble et al., 2021).

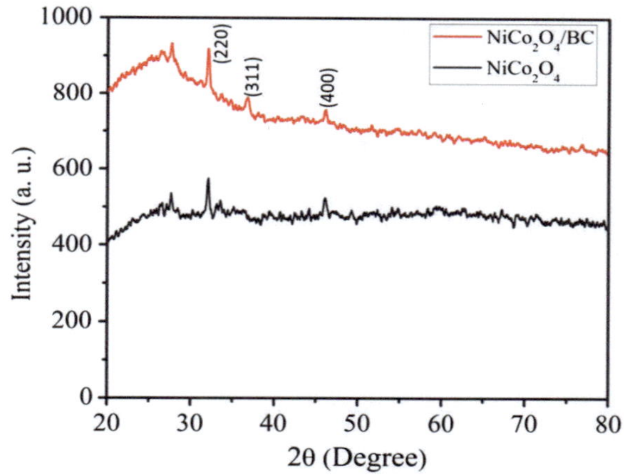

Fig. 4 The XRD pattern of biocarbon and $NiCo_2O_4$/BC

5.3 Fourier Transform Infrared Spectroscopy (FTIR)

The FTIR spectrum of $NiCo_2O_4$/BC is presented in Fig. 5. It was noticed that there were over four peaks in the FTIR band. FTIR spectra displayed bands at 795, 875, 1098 and 1412 cm^{-1}, which could be assigned to the C-C stretching, C-O stretch secondary alcohol which is ~1100 and C-N (nitro compounds) respectively. In this work, nickel and cobalt nitrate hexahydrate were used as a precursor. The characteristic peak of N-O stretching vibrations attributed at 1047 cm^{-1} of $-NO_3$ group are absent in the curve which indicates that reactants was converted to product of $NiCo_2O_4$.

The peaks observed at 795 and 875 cm^{-1} represents C-C stretching vibrations which confirms the formation of biocarbon (BC) (Kamble et al., 2021; Nandiyanto et al., 2019).

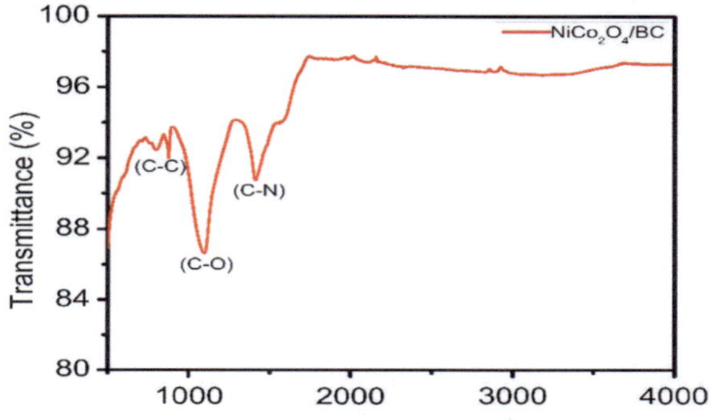

Fig. 5 The FTIR spectra of $NiCo_2O_4$/BC

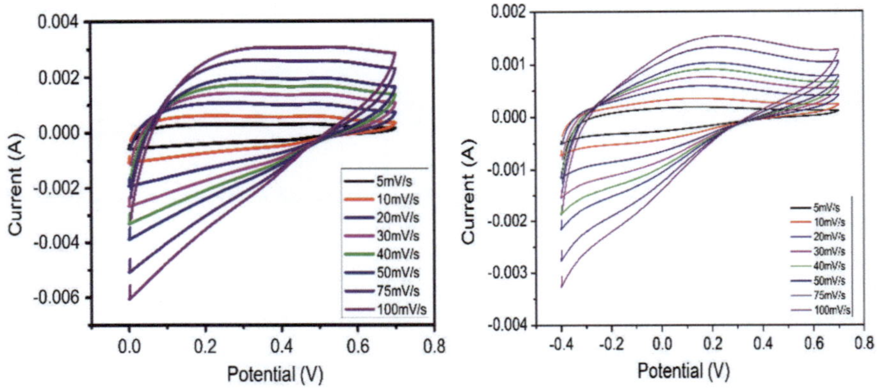

Fig. 6 Cyclic voltammetry of biocarbon and $NiCo_2O_4$/BC

5.4 Cyclic Voltammetry

The electrochemical properties of biocarbon (BC) and $NiCo_2O_4$/BC composite electrode material were studied using cyclic voltammetry. The CV for biocarbon (BC) and $NiCo_2O_4$/BC electrodes were studied within the potential window of 0 to 0.5 V (vs. Ag/AgCl) and −0.4 to 0.7 V (vs. Ag/AgCl), respectively, with varying scan rate from 5 to 100 mVs^{-1}.

Figure 6 (a) shows CV curves with a quasi-rectangular form exhibit typical electrical double-layer capacitive behavior. With low scan rates, voltammograms have a rectangular curve approximating the ideal capacitor behavior (Fig. 6). A fast scan rate, alters the ideal rectangular CV behavior. The CV curves remains unchanged even at higher scan rates for Biocarbon (BC).

Figure 6 (b) shows the CV curves of the $NiCo_2O_4$/BC which exhibits quasi-rectangular behavior at lower scan rates. This reveals ideal capacitor phenomenon of EDLC. The CV curve also reveals shifting of anodic peaks towards positive potential and cathodic peaks towards negative potential. This is simply due to ion diffusion rate as it limits the need of electronic neutralization during the redox reaction (Guo et al., 2018 and Wang et al., 2018).

5.5 Galvanostatic Charge-Discharge

Galvanostatic charge-discharge were studied to determine charge storage capability of the bio-carbon and tsynthesized electrode material. The GCD curves for biocarbon are shown in Fig. 7, performed at various current densities varying from 0.25 Ag^{-1} to 3 Ag^{-1}. The GCD curve showed a perfect triangular form and almost symmetric features with little IR drop which imply optimal EDLC behaviors. The triangular

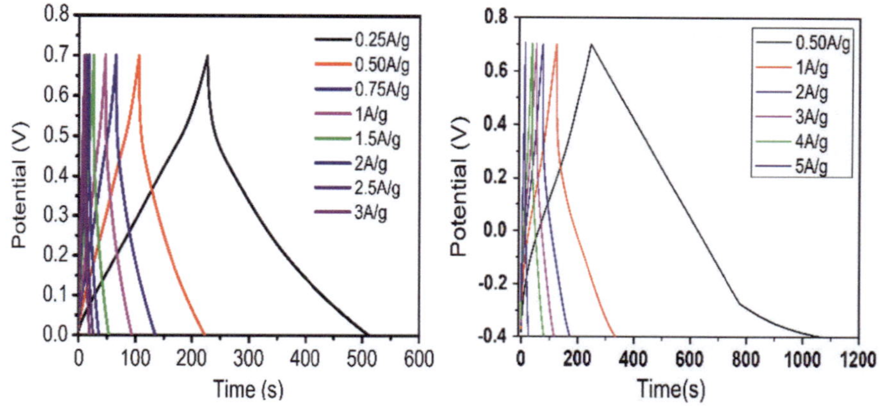

Fig. 7 Galvanostatic charge-discharge of biocarbon and $NiCo_2O_4$/BC

shape in the GCD curve also indicates the reversible nature of the material with chemical stability. The specific capacitance of biocarbon was calculated using the GCD curves. Biocarbon achieved a higher specific capacitance of 128 Fg^{-1} at a lower current density of 0.25 Ag^{-1}. For various charge-discharge current densities varying from 0.25, 0.5, 0.75, 1, 1.5, 2, 2.5 and 3 Ag^{-1} biocarbon achieved a specific capacitance of 128. 105, 94, 86, 74, 66, 59 and 54 Fg^{-1}, respectively. With increasing current densities of biocarbon, specific capacitance starts decreasing.

Compared with $NiCo_2O_4$ coupled with biocarbon galvanostatic charge–discharge (GCD) curves shown in Fig. 7, GCD tests were performed at various current densities varying from 0.5 to 5 Ag^{-1} which showed a perfect triangular form and almost symmetric features with no IR drop which imply optimal EDLC behavior. With the increasing current density discharge time decreases because ions in the electrolytes faces time limitation which controls ion diffusion on the surface of electrode at higher current densities. The $NiCo_2O_4$/BC achieved a higher specific capacitance of 377.5 Fg^{-1} at lower current density 0.5 Ag^{-1}. For various current densities varying from 0.5, 1, 2, 3, 4 and 5 Ag^{-1}, the $NiCo_2O_4$/BC achieved a specific capacitance of 377.5, 193.6, 181.8, 176.8, 160.5 and 63.6 Fg^{-1}, respectively. With increasing current densities of $NiCo_2O_4$/BC, specific capacitance starts decreasing (Khalid et al., 2016; Wang et al., 2016).

5.6 Electrochemical Impedance Spectroscopy

The ion transport property within the biocarbon (BC) and $NiCo_2O_4$/BC device was investigated using EIS. The EIS measurements were done in the frequency range of 0.01 to 1 MHz on the BC and $NiCo_2O_4$/BC electrodes. Supercapacitors show resistive and capacitive behavior at high and low frequencies, respectively. The variation of the

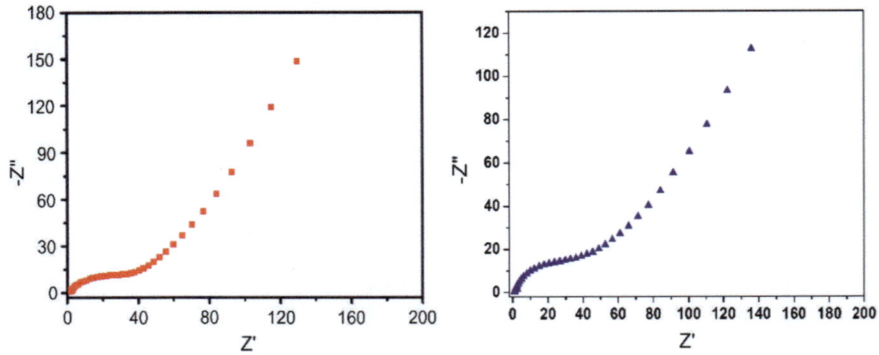

Fig. 8 Electrochemical impedance spectroscopy of biocarbon and $NiCo_2O_4$/BC

real and imaginary components of the impedance data (Nyquist plots) are presented in Fig. 8. To study the interface behavior between $NiCo_2O_4$/BC electrodes with electrolyte solution, EIS measurements were performed.

The Nyquist plot is divided into three frequency regions of low, middle and high. The low frequency range shows contact resistance (R_s) and mid frequency range shows charge transfer resistance (R_{ct}). Semicircle area in the Nyquist plot represents the charge transfer resistance R_{ct} of the material. The small kinetic (semicircle) arc in the high-frequency region depicts the charge transfer resistance (~35 Ω) between electrode-electrolyte interface and high slope in the low frequency regime indicates small diffusive resistance of electrodes owing to the fast ion diffusion within the nanopores layers during redox processes. The variation of the real and imaginary components of the impedance data (Nyquist plots) are presented in Fig. 8 (Parveen et al., 2019).

5.7 *Ragone Plot*

The relation between energy density and power density can be well explained by the Ragone plot. Biocarbon achieved energy densities of 2.17, 1.78, 1.59, 1.46, 1.25, 1.11, 1.0 and 0.91 Wh-kg^{-1} at power densities of 28, 54, 82, 110, 164, 217, 273 and 328 W-kg^{-1} (Fig. 9).

Ragone plot shows how much energy and power the device can hold. At varying current density from 0.5 to 5 Ag^{-1}, the energy density was recorded as 15.86, 8.14, 7.60, 7.39, 6.74, 2.67 Wh-kg^{-1} at power density of 68.75, 137.49, 273.60, 410.41, 549.94, 687.52 W-kg^{-1}, respectively. The $NiCo_2O_4$/BC showed highest energy density of 15.86 Wh-kg^{-1} at the power density of 68.75 W-kg^{-1}. Even at higher power density of 687.52 W-kg^{-1} the $NiCo_2O_4$/BC holds a higher energy density of 2.67 Wh-kg^{-1}. These results reveal that $NiCo_2O_4$/BC is a potential energy storage material which can be useful for energy storage.

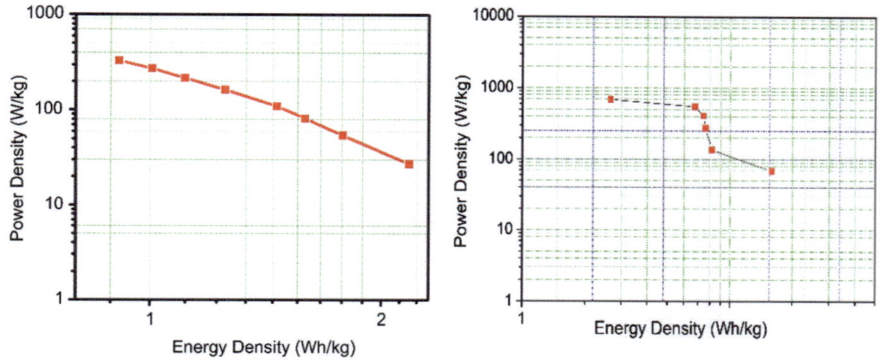

Fig. 9 Ragone plot of biocarbon and $NiCo_2O_4$/BC

5.8 *Capacitive Curve*

The relation between specific capacitance and current densities is shown by the capacitive curve. Biocarbon achieved the highest specific capacitance of 128 Fg^{-1} at the lowest current densities of 0.25 Ag^{-1}. Biocarbon achieved specific capacitance of 128, 105, 94, 86, 74, 66, 59 and 54 at current densities of 0.25, 0.5, 0.75, 1, 1.5, 2, 2.5 and 3 Ag^{-1}.

$NiCo_2O_4$/BC achieved the highest specific capacitance of 377.5 Fg^{-1} at the lowest current densities of 0.5 Ag^{-1}. The $NiCo_2O_4$/BC achieved specific capacitance of 377.50, 193.64, 181.81, 176.85, 160.46 and 63.64 Fg^{-1} at current densities of 0.5, 1, 2, 3, 4 and 5 Ag^{-1}. The gradual decrease in specific capacitance is basically due to limited ion diffusion at higher current density. At higher current densities charge transfer on the electrode surface is decreased resulting in drop in the specific capacitance. This behaviour is well explained and shown in the capacitive curve in Fig. 10.

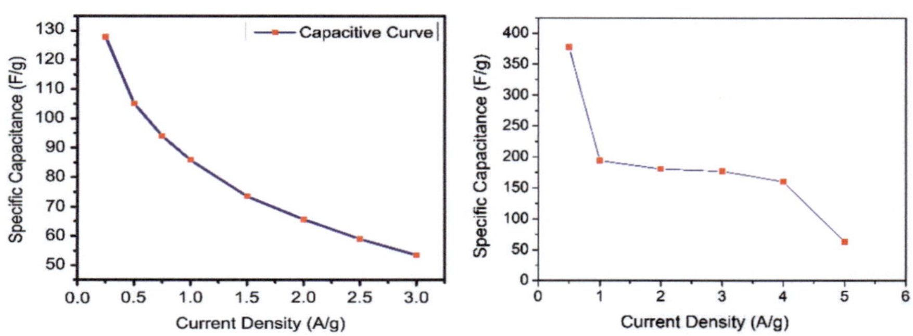

Fig. 10 Capacitive curve of biocarbon and $NiCo_2O_4$/BC

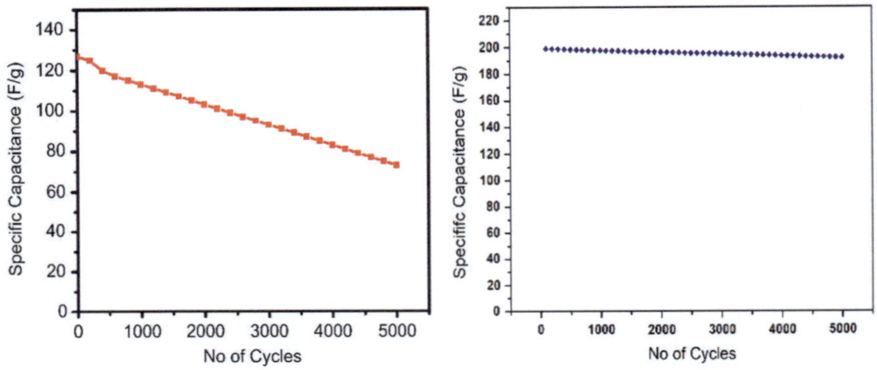

Fig. 11 Cycling stability of biocarbon and $NiCo_2O_4$/BC

5.9 Cycling Stability

Figure 11 shows another critical aspect of a supercapacitor, which is cyclic stability. Cycling stability was studied using 5000 GCD cycles to determine the endurance of the supercapacitor cycles. The sample demonstrated a fine cycling capability and displayed nearly 60% retention of the capacitance after 5000 cycles.

In electrochemical analysis of supercapacitor, cycling performance is a very vital characteristic. It is used to determine the capacitive behavior of the electrode material in the practical application. Figure 11 shows the cycling performance of $NiCo_2O_4$/BC at the current density of 1 Ag^{-1}. The $NiCo_2O_4$/BC displayed excellent cycling performance with over 95% of capacitance retention in the initial 1000 cycles.

The capacitance for $NiCo_2O_4$/BC gradually drops with increasing cycling numbers. Even after 5000 cycles and from the cycling stability curve we can confirm that $NiCo_2O_4$/BC holds the capacitance retention above 90%. The good stability of this material is due to its unique porous structure which helps to lessen volume changes during charging and dis-charging. The overall result of electrochemical results proves that $NiCo_2O_4$/BC is one of the potential electrode materials for supercapacitor application.

$NiCo_2O_4$ is one of the widely studied electrode material for supercapacitor. Kong et al. (2015) studied $NiCo_2O_4$ with carbon textile and PPY and achieved ultra-high specific capacitance of 2244 Fg^{-1}. Lei et al. (2017) also studied electrochemical properties of $NiCo_2O_4$ with carbon nano fibers (CNF) which reported specific capacitance if 1757 Fg^{-1}. In both the cases high specific capacitance was achieved but with the use of lot of chemicals which are harmful for environment. In this paper we have reported $NiCo_2O_4$ coupled with carbon derived from biomass without using a single chemical which is environmentally friendly and a novel work. The $NiCo_2O_4$/BC also achieved specific capacitance of 377.50 Fg^{-1} at lower scan rate which is comparatively low to the above two mentioned cases. The specific capacitance of $NiCo_2O_4$/BC can be improved by coupling with conducting polymers.

6 Conclusion

In summary, utilizing two-step-facile carbonizations, carbon was effectively synthesized from royal palm biomass. Low-temperature carbonization was used to physically activate the sample in the presence of air. The sample has a high specific capacitance of 128 Fg^{-1} at a current density of 0.25 for optimal low-temperature (400 °C) carbonization. The stability line of cyclic stability illustrates that it has excellent performance (~80%) even after 1000 cycles and retained 60% after 5000 cycles. SEM examination revealed that carbonized biochar developed more pores than predicted. In future work, this biocarbon's morphology can be studied more deeply using TEM. $NiCo_2O_4$ nanoparticles were grown on the biocarbon to prepare $NiCo_2O_4$/BC composite by hydrothermal method. The morphological, compositional and structural study using SEM, EDS, XRD and FTIR confirms the formation of $NiCo_2O_4$/BC composite material. Electrochemical results revealed that $NiCo_2O_4$/BC exhibits a higher specific capacitance of 377.50 Fg^{-1} at the current density of 0.5 Ag^{-1} and higher energy density of 15.86 Wh kg^{-1} at power density of 68.75 W-kg^{-1}. The $NiCo_2O_4$/BC showed a capacitive retention of ~90% making it a potential electrode material.

Acknowledgements Authors would like to thank Director, Institute of Science, Mumbai, HOD, Department of Physics, RTM Nagpur University, Director, Institute of Science, Nagpur, Principal and Management of M.M. College of Arts, N.M. Institute of Science, H.R.J. College of Commerce, Bhavan's College (Autonomous), Andheri (W), India for providing research facilities and their constant encouragement and support.

References

Abeykoon, N.C., Garcia, V., Jayawickramage, R.A., Perera, W., Cure, J., Chabal, Y.J., ... and Ferraris, J.P. (2017). Novel binder-free electrode materials for supercapacitors utilizing high surface area carbon nanofibers derived from immiscible polymer blends of PBI/6FDA-DAM: DABA RSC advances, **7(34):** 20947–20959.

Amer, M. and Elwardany, A. (2020). Biomass Carbonization Renewable Energy–Resources, Challenges and Applications, IntechOpen.

Bergman, P.C. and Kiel, J.H. (2005). Torrefaction for biomass upgrading. *In*: Proc. 14th European Biomass Conference, pp. 17–21, Paris.

Enock, T., King'ondu, C.K., Pogrebnoi, A. and Jande (2017). Status of biomass-derived carbon materials for supercapacitor application. International Journal of Electrochemistry.

Ferreira, P., Araújo, M. and Hens, L. (2016). Energy and environment: bringing together engineering and economics. Environment, Development, and Sustainability, **18(5):** 1275–1277.

Frackowiak, E. and Béguin, F. (2001). Carbon Materials for the Electrochemical Storage of Energy in Capacitors. *Carbon*, **39(6):** 937–950.

Fu, M., Chen, W., Ding, J., Zhu, X. and Liu, Q. (2019). Biomass waste derived multi-hierarchical porous carbon combined with $CoFe_2O_4$ as advanced electrode materials for supercapacitors, *Journal of Alloys and Compounds*, **782:** 952–960.

Gielen, D., Boshell, F., Saygin, D., Bazilian, M., Wagner, N. and Gorini, R. (2019). The role of renewable energy in the global energy transformation. *Energy Strategy Reviews*, **24:** 38–50.

Guo, D., Li, Z., Song, X., Tan, L., Ma, H., Jiao, J., Zhua D. and Lia F. (2018). $NiCo_2O_4$ nanosheets grown on interconnected honeycomb-like porous biomass carbon for high performance asymmetric supercapacitor. *New J. Chem.*, **42:** 8478–8484.

Hooda, M. and Ghany, A. (2019). Production of a new activated carbon prepared from palm fronds by thermal activation. *Int. J. Eng. Technol. Manag. Res*, **6:** 34–43.

Isahak, W.N.R.W., Hisham, M.W.M. and Yarmo, M.A. (2013). Highly porous carbon materials from biomass by chemical and carbonization method: A comparison study. *Journal of Chemistry*, **2013:** 1–6.

Jameel, H. and Keshwani, D. (2017). Biomass to Renewable Energy Processes. 2nd edn. CRC, Boca Raton.

Kamble, G., Kashale, A., Rasal, A., Mane, S., Chavan, R., Chang, J., Ling, Y., Kolekar, S. and Ghule, A. (2021). Marigold micro-flower like $NiCo_2O_4$ grown on flexible stainless-steel mesh as an electrode for supercapacitors. *RSC Adv.*, **11:** 3666.

Khalid, S., Cao, C. and Wang, L. (2016). Microwave Assisted Synthesis of Porous $NiCo_2O_4$ Microspheres: Application as High Performance Asymmetric and Symmetric Supercapacitors with Large Areal Capacitance. *Sci Rep* **6:** 22699.

Lu, H. and Zhao, X.S. (2017). Biomass-derived carbon electrode materials for supercapacitors. *Sustainable Energy & Fuels,* **1(6):** 1265–1281.

Nandiyanto, A. Oktiani, R. and Ragadhita, R. (2019). How to read and interpret ftir spectroscope of organic material; *Indonesian Journal of Science & Technology* **4(1):** 97–118.

Parveen, N., Jaafari, A. and Han, J. (2019). Robust cyclic stability and high-rate asymmetric supercapacitor based on orange peel-derived nitrogen-doped porous carbon and intercrossed interlinked urchin-like $NiCo_2O_4$@3DNF framework. *Electrochimica Acta*, **293:** 84–96

Rajan, L., Kumar, R., Sahoo, P., Kumar, S. and Soam, A. (2022). Application of Iron Oxide in Supercapacitor. Iron Oxide Nanoparticles.

Rajendran, R. and Wang, F. (2018). Electrochemical Capacitor Performance: Influence of Aqueous Electrolytes. Supercapacitors-Theoretical and Practical Solutions, InTech.

Ray, A., Roy A., Saha, S. and Das, S. (2019). Transition Metal Oxide-Based Nano-materials for Energy Storage Application. Science, Technology and Advanced Application of Supercapacitors, IntechOpen.

Satyajit, R. and Samantara, A. (2018). Supercapacitor: Instrumentation, Measurement, and Performance Evaluation Techniques. 1st edn. Springer Singapore.

Simon, P. and Gogotsi, Y. (2008). Materials for electrochemical capacitors. *Nature Materials*, **7(11):** 845–854.

Sivaraj, R., Rajendran, V. and Gunalan, G. S. (2010). Preparation and characterization of activated carbons from parthenium biomass by physical and chemical activation techniques. *E-Journal of Chemistry*, **7(4):** 1314–1319.

Wang, J., Zhang, Y., Ye, J., Wei, H., Hao, J., Mu, J. and Hussain, S. (2016). Facile synthesis of three-dimensional $NiCo_2O_4$ with different morphology for supercapacitors. *RSC Advances*, **6(74):** 70077–70084.

Wang, W., Qi, J., Sui, Y., He, Y., Meng, Q., Wei, F. and Jin, Y. (2018). An Asymmetric Supercapacitor Based on Activated Porous Carbon Derived from Walnut Shells and $NiCo_2O_4$ Nanoneedle Arrays Electrodes. *Journal of Nanoscience and Nanotechnology*, **18(8):** 5600–5608(9).

Yaashikaa, P.R., Kumar, P.S., Varjani, S. and Saravanan A. (2020). A critical review on the biochar production techniques, characterization, stability and applications for a circular bioeconomy. *Biotechnology Reports*, **28:** e00570.

Yang, G. and Park, S. (2020). Nanoflower-like $NiCo_2O_4$ grown on biomass carbon coated nickel foam for asymmetric supercapacitor. *Journal of Alloys and Compounds*, **835:** 155270.

Chapter 23
Hydrothermally Prepared $NiCo_2O_4$/Biocarbon Composite Nanoparticles: Application of Energy Storage Devices

Brijesh Kumar Yadav, Rounak Atram, Mahadev Sonawane Jitesh Kumar, and Ashish Kumar Yadav

1 Introduction

The COVID-19 epidemic brought attention to the problem of managing energy consumption globally, where lack of infrastructure and resources may prevent the use of renewable energy sources (Sherpa et al., 2022). We may still heavily rely on traditional and fossil fuels. However, there is a global shift towards renewable energy options like solar, geothermal, and ocean energy driven by environmental concerns and the need to reduce dependency on finite resources. Additionally, there is an increase in energy efficiency and conservation measures (Gielen et al., 2019). Due to its ability to charge fast, long lifespan, compact size and light weight, supercapacitors offer an affordable alternative for storing energy from clean sources such as wind and solar power (Gautham Prasad et al., 2019; Iro et al., 2016). Supercapacitors are used in various applications including energy storage systems, electric vehicles, and portable electronics. Supercapacitor is classified into two categories, viz. EDLC and pseudocapacitors based on their charge storage mechanism. EDLCs store energy on

B. K. Yadav
Department of Physics, The Institute of Science, Dr. Homi Bhabha State University, 15, Madam Cama Road, Fort Mumbai, Mumbai, Maharashtra 400032, India

R. Atram (✉)
Department of Physics, M.M College of Arts, N.M. Institute of Science H.R.J College Of Commerce, Bhavans College, Andheri-(W), Mumbai, Maharashtra 400058, India
e-mail: rounak.atram@bhavans.ac.in

M. S. J. Kumar
Department of Physics and Electronics, Rajdhani College (University of Delhi), New Delhi 110015, India

A. K. Yadav
Department of Physics, R.J College of Arts, Science and Commerce Ghatkopar-(W), Mumbai, Maharashtra 400086, India

R. K. Kotnala et al. (eds.), *Advanced Functional Materials for Sustainable Environments*, https://doi.org/10.1007/978-3-031-62620-3_23

the surface of the electrodes, and pseudocapacitors using a chemical process to store energy. EDLCs use a non-faradic process which means they do not use chemical reactions to store energy, while pseudocapacitors use a faradic process, ie. they use chemical reactions to store energy (Iro et al., 2016; Zhang et al., 2015).

Carbonaceous materials such as graphene, CNT, CNF, biomass derived carbon, rGO and others are primary electrode materials for ELDC (Karaman et al., 2021; Zhang et al., 2019). The active component in the electrodes of pseudocapacitors is frequently mixed metal oxides, mixed metal sulfide, and conducting polymers (Mohd Abdah et al., 2020; Patil et al., 2022; Shaikh et al., 2022).

Biomass-derived carbon, like biochar, can be used as a source of carbon for supercapacitors (Gao et al., 2017; Rawat et al., 2022). Biomass-derived carbon are useful in energy storage and is an attractive option as it is a cost-effective, renewable and sustainable source of carbon, compared to traditional fossil fuel-based sources. Furthermore, the use of carbon produced from biomass in supercapacitors has the potential to mitigate greenhouse gas emissions through the atmospheric sequestration of carbon (Yang et al., 2019).

Thermochemical conversion, a method of converting biomass into various forms of energy, such as heat, electricity and biofuels, involves the following processes: gasification, liquefaction, co-firing, carbonization, pyrolysis, combustion, and torrefaction. Since pyrolysis includes all chemical reactions that produce char solid, bio-oil, and bio-gas in environments with low or no oxygen, it is regarded as the foundation for all thermochemical conversion techniques (Amer and Elwardany, 2020; Anon, 2017; Ghany, 2020; Isahak et al., 2013; Kwaku Armah et al., 2023; Patel et al., 2016; Sivaraj et al., 2010; Yaashikaa et al., 2020). These thermochemical conversion methods are characterized using heat, which is usually supplied by external sources, to degrade the organic materials, shattering the hydro-carbonaceous bonds, and transforming them into a richer source of energy (Anon, 2017; Ghany, 2020; Isahak et al., 2013; Patel et al., 2016; Sivaraj et al., 2010).

This article presents, synthesized biocarbon (BC) from royal palm trees which is fibrous in nature. This BC derived from petioles of royal palm by torrefaction, carbonization process and mixed transition metal oxide ($NiCo_2O_4$) grown on biomass derived BC to develop composite electrode materials for supercapacitors. The high specific capacitance, energy density, and superior performance were provided by an asymmetric electrode ($NiCo_2O_4$/BC).

2 Experimental Study

2.1 Materials

Nickel nitrate hexahydrate ($Ni(NO_3)_2.6H_2O$), polyvinylpyrrolidone (PVP), cobalt nitrate hexahydrate ($CO(NO_3)_2.6H_2O$), NMP, acetone, potassium hydroxide (KOH) were purchased from Loba Chem. There was no need for further purification because all compounds were analytical grade.

2.2 Synthesis of BC and $NiCo_2O_4$/BC

Preparation BC

Parts like petiole and leaf sheath of royal palm (Roystonea regia) were collected from the Bhavans College, Andheri (W), Mumbai, India. It was sliced into tiny (0.5–1 cm) pieces and sun-dried. In order to eliminate any remaining soil and dust, the harvested raw biomass was carefully cleaned with distilled water and dried for 4 hrs at 80 °C and 1 hr at 140 °C, and allowed to cool overnight (Gale et al., 2021). Then, the biomass pieces were placed inside silica crucibles for torrefaction. The lids were tightened, and the pieces were wrapped in several layers of aluminum foil to prevent any air from entering the vessels or the material being physically activated. This resulted in the creation of an atmosphere with low oxygen levels. These crucibles were placed in the muffle furnace at 200 °C for 2 hrs. Subsequently, the sample was grinded using a mortar and pestle and filtered through layers of cloth. Then these samples were filled in tightly sealed crucibles and this setup was put into muffle furnace at 600 °C for 2 hrs for carbonization (Amer and Elwardany, 2020; Ghany, 2020; Isahak et al., 2013; Luo et al., 2022; Patel et al., 2016; Sivaraj et al., 2010). After overnight natural cooling, the material was collected and stored. The sample was rinsed with conc. HNO_3 and cleaned multiple time with DI water until the pH was neutralized then immersed in KOH to remove any residue before drying in the hot air oven for 6 hrs (Guo et al., 2018; Parveen et al., 2019; Yang and Park, 2020).

2.3 Preparation of $NiCo_2O_4$/BC

Synthesis process of $NiCo_2O_4$/BC is shown in Schematic 1. The $NiCo_2O_4$/BC waa prepared by a facile hydrothermal synthesis method. Initially, 100 mg of BC obtained from royal palm biomass was kept in 100 mL of deionized water for ultrasonication for 2 h and later precursors of nickel (0.5 g) and cobalt (0.5 g) were dissolved in same solution under continuous magnetic string for 0.5 h to form a uniform mixture. This mixture was transferred into 100 ml of Teflon autoclave and kept for heating at

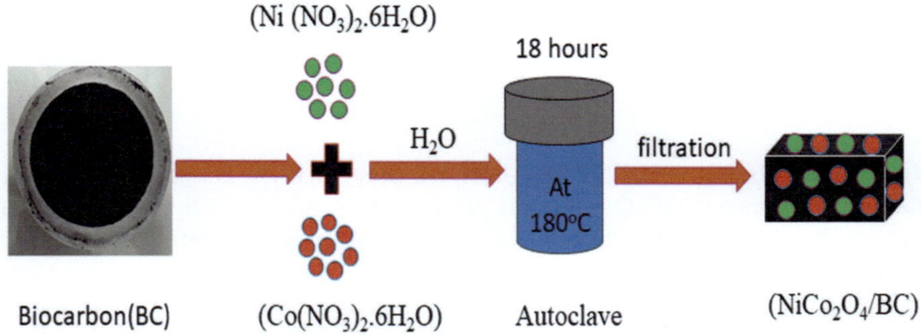

Schematic 1. Synthesis process of $NiCo_2O_4$/BC

180 °C for 18 hrs. After natural cooling the sample was washed with the help of DI water and dried overnight at 80 °C (Kamble et al., 2021).

2.4 Material Characterization

SEM (scanning electron microscopy) (SEM EVO18, CARLZEISS) was used to analyze the morphology of the $NiCo_2O_4$/BC. The phase structures of the as-synthesized $NiCo_2O_4$/BC were studied by the powder XRD (XRD, RIGAKU) using Cu Kα radiation (λ = 1.5406) at scanning rate of 7 per min ranging from 10 to 80 °C. The presence of elements in as-synthesized $NiCo_2O_4$/BC were investigated by EDS analysis. FTIR was performed to study functional group.

2.5 Electrochemical Measurements

The electrochemical properties of as-prepared materials were carried out on a Metro ohm PGSTAT204 electrochemical workstation (Institute of Science, Nagpur, Maharashtra, India) with 1 M H_2SO_4 aqueous as the electrolyte in a 3-electrode system. The three-electrode system consist of platinum and Ag/$AgCl_2$ as counter and reference electrode, respectively. The working electrodes were prepared as 85 wt.% of as-prepared $NiCo_2O_4$/BC, 10 wt. % polyvinylidene fluoride and 5 wt.% carbon black and add few drops of N-methyl-2-pyrrolidone binder were mixed to obtain homogeneous slurry. The slurry was pasted on graphite electrode using brush and dried in hot air vacuum at 70 °C for 12 hrs. The electrochemical analysis of prepared materials was characterized by CV, GCD and EIS with range from 0.01 MHz to 100 kHz. The specific capacitance (C_s) of $NiCo_2O_4$/BC electrode were calculated from the galvanostatic charge-discharge curves by following equation:

$$C_S = \frac{I \times \Delta t}{\Delta V \times m} \tag{1}$$

where, C_s is the specific capacitance, I the current, Δt is the discharge time, ΔV the operating voltage window (V) and m is the mass of corresponding electrode materials (mg).

The energy density (ED) and power density (PD) of the $NiCo_2O_4$/BC electrode were calculated according to following equations:

$$ED = \frac{1}{2 \times 1.36} \times \Delta V^2 \times C_S \tag{2}$$

$$PD = \frac{3600 \times ED}{\Delta t} \tag{3}$$

3 Result and Discussion

3.1 Scanning Electron Microscope and Energy Dispersive Spectroscopy

Using a scanning electron microscope, the surface morphology of the as-synthesized $NiCo_2O_4$/BC nano composite was examined. Figure 1 shows the SEM images of (a) BC and (b) $NiCo_2O_4$/BC. Figure 1 (a) reveals rough surfaces with pores of various sizes, randomly distributed all over surface of BC. The porous structure in the material enhances the surface area of the material. Porosity also allows deep penetration of electrolytes in the materials which activates various sites to store the ions resulting in higher electrochemical results.

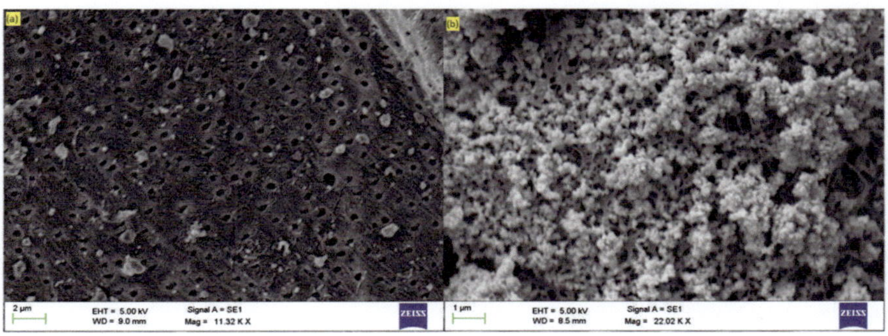

Fig. 1 The SEM images of: (a) BC and (b) $NiCo_2O_4$/BC

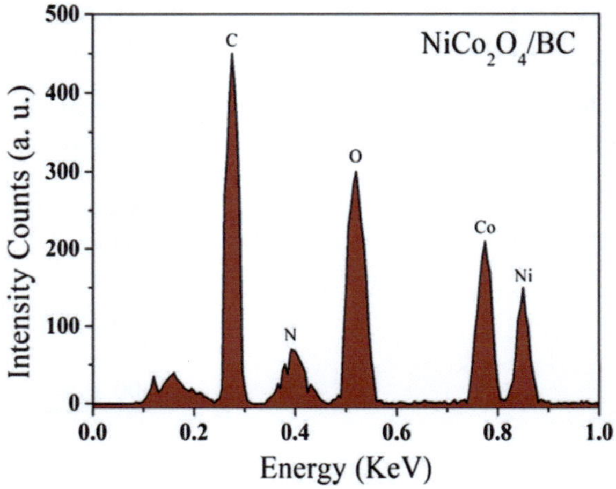

Fig. 2 EDS spectra of $NiCo_2O_4$/BC

Figure 1 (b) reveals the randomised distribution of $NiCo_2O_4$ nanoparticle all over the surface of biocarbon. SEM image also reveals nanopores of the Biocarbon visible in few parts of the images. The growth of $NiCo_2O_4$ nanoparticles on the surface of Biocarbon results in enhanced electrochemical of $NiCo_2O_4$/BC composite.

The elemental composition of $NiCo_2O_4$/BC was carried out using energy dispersive spectroscopy (EDS). The EDS spectra of $NiCo_2O_4$/BC reveals the peaks of carbon, oxygen, nitrogen, cobalt, nickel which are in proportion to our stoichiometric ration and confirms the presence of these elements in our material. Figure 2 shows elemental composition of $NiCo_2O_4$/BC using energy dispersive spectroscopy (EDS).

3.2 The X-Ray Diffraction (XRD)

The phase and crystallinity of the as-synthesized of the $NiCo_2O_4$/BC composites are evaluated by X-ray diffraction patterns as shown in Fig. 3. According to the highest peak, all peaks are normalized. The sharp peaks indicate the crystalline nature of the nanocomposite. Diffraction peaks are seen for the $NiCo_2O_4$/BC composites, and when the concentration of $NiCo_2O_4$ nanoparticles on BC increases, the strength of the $NiCo_2O_4$/BC diffraction peak increases, suggesting the presence of the $NiCo_2O_4$ phase in the materials. The peaks could be indexed to the diffraction peaks at 32°, 36.7° and 46° collected from $NiCo_2O_4$/BC and can be well indexed with (220) (311) and (400) spinel phase of $NiCo_2O_4$ (JCPDS Card no. 20-0781) (Khalid et al., 2016; Wang et al., 2018).

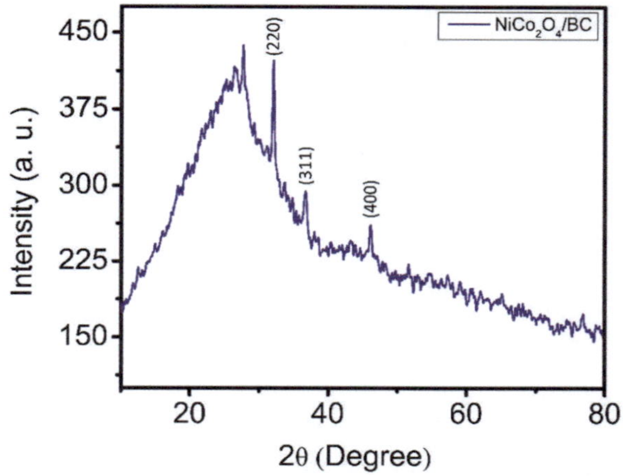

Fig. 3 The XRD pattern of $NiCo_2O_4$/BC

3.3 The FTIR

The FTIR spectrum of $NiCo_2O_4$/BC is presented in Fig. 4. It was noticed that there were over 4 peaks in the FTIR band. FTIR spectra displayed bands at 795,875,1098 and 1412 cm^{-1}, which could be assigned to the C-C stretching, C-O stretch secondary alcohol which is ~1100 and C-N (nitro compounds), respectively. The reaction product, $NiCo_2O_4$, was produced by converting the reactant to the distinctive peaks of the -NO_3 group at 1047 cm^{-1}, which are absent from the nickel and cobalt nitrate hexahydrate that we utilized as a precursor. The synthesis of BC is confirmed by the peaks seen at 795 and 875 cm^{-1} correspond to C–C stretching vibrations (Khalid et al., 2016; Wang et al., 2016).

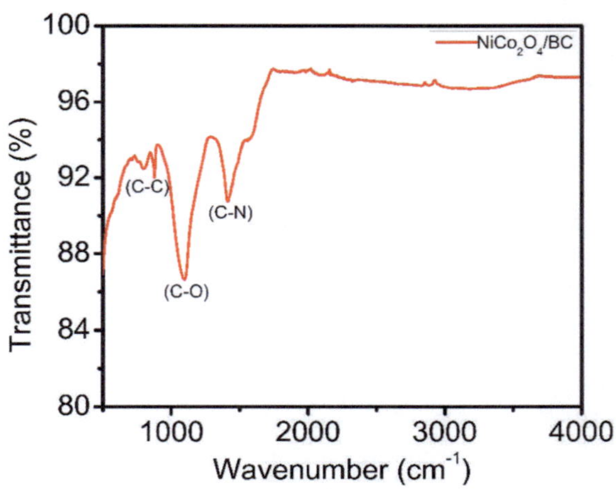

Fig. 4 The FTIR spectra of $NiCo_2O_4$/BC

Fig. 5 Cyclic voltammetry

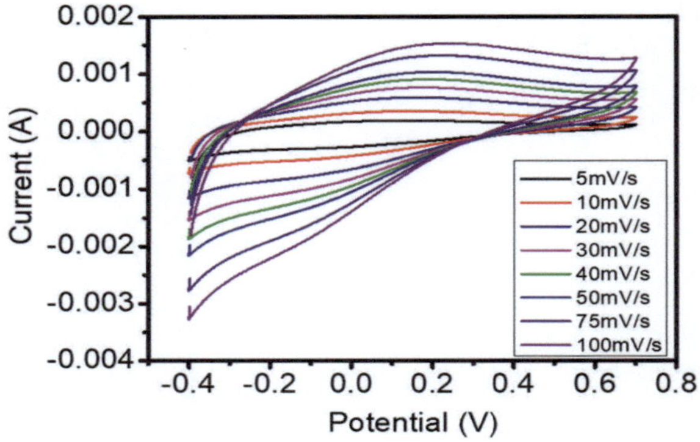

3.4 Cyclic Voltammetry (CV)

The $NiCo_2O_4$/BC electrode materials electrochemical characteristics were investigated by using cyclic voltammetry at various scan rates in the potential range of −0.4 to 0.7 V. Figure 5 shows CV curves of the $NiCo_2O_4$/BC at various scan rates ranging from 5 to 100 mV s^{-1}. At lower scan rates cyclic voltammetry curves showed a pure quasi-rectangular behavior of EDLC which is approximating the ideal capacitor behavior. As a result of the enclosed region growing and maintaining its form with increased scan rate, they also exhibit good reversibility and remarkable charge-discharge characteristics. The anodic peaks on the CV curve are moving in the direction of the positive potential, while the cathodic peaks are moving in the direction of the negative potential. This is only a result of the ion diffusion rate, which reduces the requirement for electronic neutralization throughout the redox process (Kamble et al., 2021; Wang et al., 2016, 2018).

3.5 Galvanostatic Charge-Discharge (GCD)

The GCD curves for $NiCo_2O_4$/BC are shown in Fig. 6. The GCD tests were performed at various current densities varying from 0.5 to 5 Ag^{-1}. The GCD curve showed a perfect triangular form and almost symmetric features with NO IR drop which imply optimal EDLC behaviour.

When the current density increases, the electrolyte ions encounter temporal constraints that regulate their diffusion at the active electrode surface, causing the discharge time to decrease. The most crucial factor in assessing the electrochemical performance of supercapacitors is their specific capacitance (Kamble et al., 2021; Khalid et al., 2016; Parveen et al., 2019; Wang et al., 2016; Yang and Park, 2020). The $NiCo_2O_4$/BC achieved a higher specific capacitance of 377.5 Fg^{-1} at lower current density 0.5 Ag^{-1}. The specific capacitance of $NiCo_2O_4$/BC was 377.5, 193.6, 181.8,

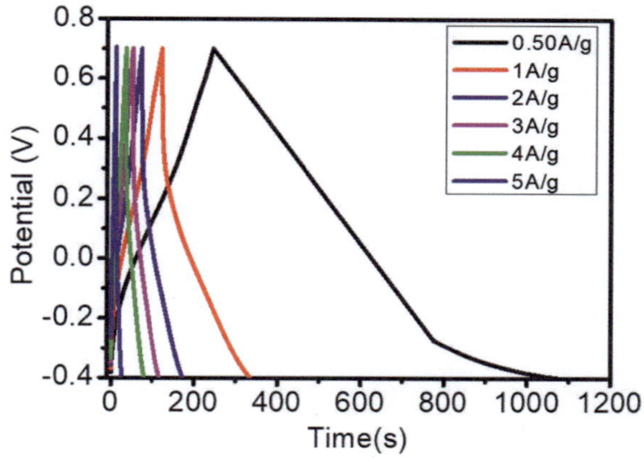

Fig. 6 Galvanostatic charge-discharge

176.8, 160.5, and 63.6 Fg^{-1} for different current densities ranging from 0.5, 1, 2, 3, 4 and 5 Ag^{-1}.

3.6 *Electrochemical Impedance Spectroscopy*

The interface study between electrode and electrolyte is carried out with the help of electrochemical impedance spectroscopy (EIS). Impedance curves in the Nyquist plots are distinguished by low, mid and high frequency ranges. In low frequency region electrode works as ideal capacitor. Low frequency region represents surface resistance (R_s) which is approx. 2 Ω. Mid frequency range in Nyquist plot represents the charge transfer resistance (R_{ct}) which is nearly 40 Ω.

The material's charge transfer resistance (R_{ct}), is shown as a semicircle in the Nyquist plot. The frequency range of 0.01 to 1 MHz was used for the $NiCo_2O_4$/BC electrode electrochemical impedance spectroscopy studies. Supercapacitors have a well-known tendency to behave as capacitive devices at low frequencies and as resistive (blocking) devices at high frequencies (Kamble et al., 2021; Khalid et al., 2016; Wang et al., 2016; Yang and Park, 2020). Figure 7 displays variations of the impedance data's real and imaginary components (Nyquist plots).

3.7 *Ragone Plot*

A Ragone plot illustrates the relationship between power density and energy density and shows how much energy and power the device can hold. At varying current density from 0.5 Ag^{-1} to 5 Ag^{-1}, the energy density was recorded as 15.86, 8.14, 7.60, 7.39, 6.74, 2.67 $Whkg^{-1}$ at power density of 68.75, 137.49, 273.60, 410.41, 549.94, 687.52 Wkg^{-1}, respectively. The $NiCo_2O_4$/BC showed highest energy density of

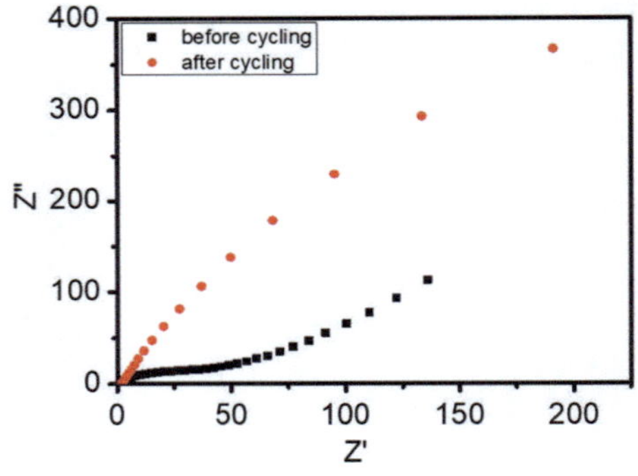

Fig. 7 Electrochemical impedance spectroscopy

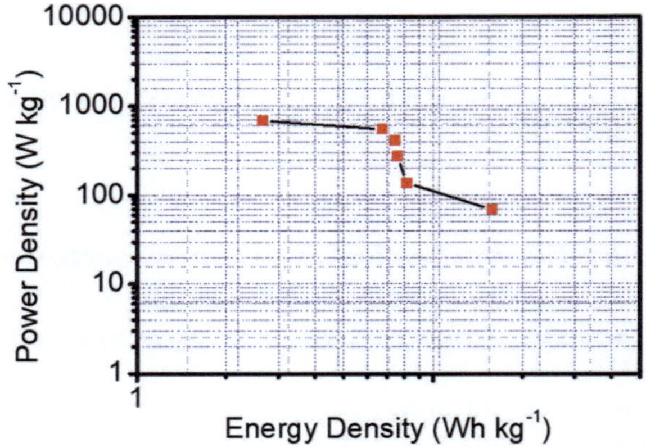

Fig. 8 Ragone plot

15.86 Whkg^{-1} at the power density of 68.75 Wkg^{-1} (Fig. 8). Even at higher power density of 687.52 Wkg^{-1} $NiCo_2O_4$/BC holds a higher energy density of 2.67 Whkg^{-1}. These results reveal that $NiCo_2O_4$/BC is a potential energy storage material which can be useful for energy storage (Khalid et al., 2016; Nandiyanto et al., 2019; Wang et al., 2018).

3.8 Capacitive Curve

The relation between specific capacitance and current densities is represented by the capacitive curve. The $NiCo_2O_4$/BC achieved the highest specific capacitance of 377.5 Fg^{-1} at the lowest current densities of 0.5 Ag^{-1}. The $NiCo_2O_4$/BC achieved specific capacitance of 377.50, 193.64, 181.81, 176.85, 160.46 and 63.64 Fg^{-1} at current densities of 0.5, 1, 2, 3, 4 and 5 Ag^{-1}. The gradual decrease in specific capacitance is basically due to limited ion diffusion at higher current density. At higher current densities charge transfer on the electrode surface is decreased resulting in drop in

Fig. 9 Capacitive curve

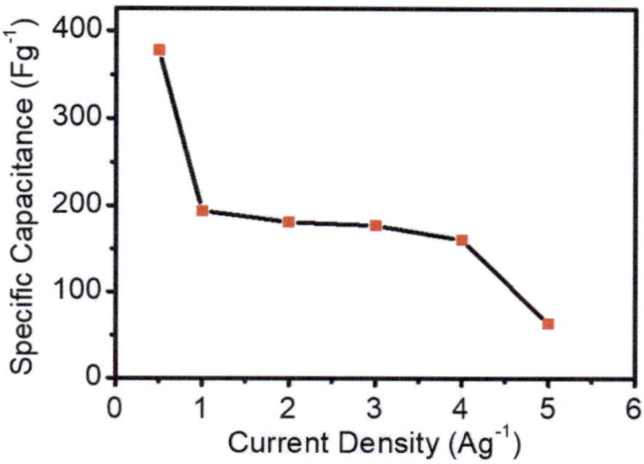

the specific capacitance (Kamble et al., 2021; Nandiyanto et al., 2019; Wang et al., 2016). This behaviour is well explained and shown in the capacitive curve in Figure 9.

3.9 *Cycling Stability*

In electrochemical analysis of supercapacitor, cycling performance is a vital characteristic. It is used to determine the capacitive behaviour of the electrode material in the practical application. Figure 10 shows the cycling performance of $NiCo_2O_4$/BC at the current density of 1 Ag^{-1}. The $NiCo_2O_4$/BC showed excellent capacitive behavior in the cycling test as in the initial 1000 cycles $NiCo_2O_4$/BC has over 95% of capacitance retention. The capacitance for $NiCo_2O_4$/BC gradually drops with increasing cycling numbers. Even after 5000 cycles and from the cycling stability curve one can confirm that $NiCo_2O_4$/BC holds the capacitance retention above 90%. The good stability of this material is due to its special porous construction, which minimizes changes in volume during charging and discharging. The overall result of electrochemical results proves that $NiCo_2O_4$/BC is one of the potential electrode materials for supercapacitor application (Guo et al., 2018; Luo et al., 2022; Parveen et al., 2019).

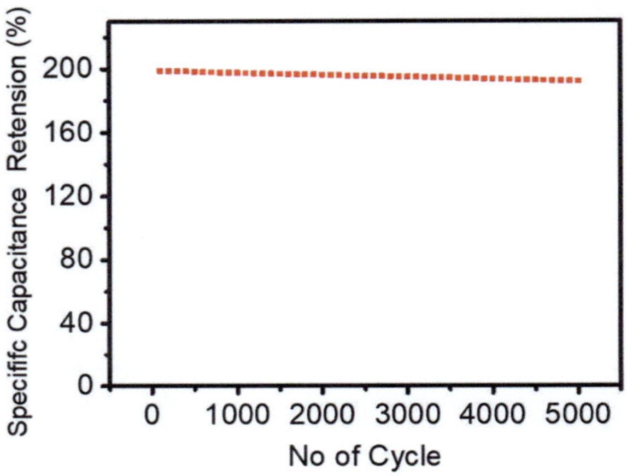

Fig. 10 Cycling stability

4 Conclusion

Torrefaction, carbonization and a straightforward hydrothermal process were used to create the $NiCo_2O_4$/BC nano composite. The morphological, compositional and structural study using SEM, EDS, XRD and FTIR confirms the formation of $NiCo_2O_4$/BC composite material. Electrochemical results revealed that $NiCo_2O_4$/BC exhibits a greater specific capacitance of 377.50 Fg^{-1} at the current density of 0.5 Ag^{-1} and higher energy density of 15.86 Wh kg^{-1} at power density of 68.75 Wkg^{-1}. The $NiCo_2O_4$/BC showed a capacitive retention of ~90% making it a potential electrode material.

Acknowledgements Authors would like to thank Director, Institute of Science, Mumbai, HOD, Department of Physics, RTM Nagpur University, Director, Institute of Science, Nagpur, Principal and Management of M.M. College of Arts, N.M. Institute of Science, H.R.J. College of Commerce, Bhavan's College (Autonomous), Andheri (W), India for providing the research facilities and their constant encouragement and support.

References

Amer, Mahmoud and Ahmed Elwardany (2020). Biomass Carbonization. in Renewable Energy - Resources, Challenges and Applications. IntechOpen.

Anon (2017). Biomass to Renewable Energy Processes, Second Edition. CRC Press.

Gale, Mark, Tu Nguyen, Marissa Moreno and Kandis Leslie Gilliard-AbdulAziz (2021). Physiochemical Properties of Biochar and Activated Carbon from Biomass Residue: Influence of Process Conditions to Adsorbent Properties. *ACS Omega,* **6(15):** 10224–10233.

Gao, Zan, Yunya Zhang, Ningning Song and Xiaodong Li (2017). Biomass-Derived Renewable Carbon Materials for Electrochemical Energy Storage. *Materials Research Letters,* **5(2):** 69–88.

Gautham Prasad, G., Nidheesh Shetty, Simran Thakur, Rakshitha and Bommegowda, K.B. (2019). Supercapacitor Technology and Its Applications: A Review. *IOP Conference Series: Materials Science and Engineering,* **561(1):** 012105.

Ghany, Hoda M. Abdel. (2020). Production of a new activated carbon prepared from palm fronds by thermal activation. *International Journal of Engineering Technologies and Management Research,* **6(4):** 34–43.

Gielen, Dolf, Francisco Boshell, Deger Saygin, Morgan D. Bazilian, Nicholas Wagner and Ricardo Gorini (2019). The Role of Renewable Energy in the Global Energy Transformation. *Energy Strategy Reviews,* **24:** 38-50.

Guo, Dongxuan, Li Zhang, Xiumei Song, Lichao Tan, Huiyuan Ma, Jia Jiao, Di Zhu and Fengbo Li (2018). $NiCo_2O_4$ Nanosheets Grown on Interconnected Honeycomb-like Porous Biomass Carbon for High Performance Asymmetric Supercapacitors. *New Journal of Chemistry,* **42(11):** 8478-8484.

Iro, Zaharaddeen S., Subramani, C. and Dash, S.S. (2016). A Brief Review on Electrode Materials for Supercapacitor. *International Journal of Electrochemical Science,* **11(12):** 10628–10643.

Isahak, Wan Nor Roslam Wan, Mohamed Wahab Mahamed Hisham and Mohd Ambar Yarmo (2013). Highly Porous Carbon Materials from Biomass by Chemical and Carbonization Method: A Comparison Study. *Journal of Chemistry,* **2013:** 1–6.

Kamble, Gokul P., Anil A. Kashale, Akash S. Rasal, Seema A. Mane, Rutuja A. Chavan, Jia-Yaw Chang, Yong-Chien Ling, Sanjay S. Kolekar and Anil V. Ghule (2021). Marigold Micro-Flower like $NiCo_2O_4$ Grown on Flexible Stainless-Steel Mesh as an Electrode for Supercapacitors. *RSC Advances,* **11(6):** 3666–3672.

Karaman, Ceren, Onur Karaman, Necip Atar and Mehmet Lütfi Yola (2021). Sustainable Electrode Material for High-Energy Supercapacitor: Biomass-Derived Graphene-like Porous Carbon with Three-Dimensional Hierarchically Ordered Ion Highways. *Physical Chemistry Chemical Physics,* **23(22):** 12807–12821.

Khalid, Syed, Chuanbao Cao, Lin Wang and Youqi Zhu (2016). Microwave Assisted Synthesis of Porous $NiCo_2O_4$ Microspheres: Application as High Performance Asymmetric and Symmetric Supercapacitors with Large Areal Capacitance. *Scientific Reports,* **6(1):** 22699.

Kwaku Armah, Edward, Maggie Chetty, Jeremiah Adebisi Adedeji, Denzil Erwin Estrice, Boldwin Mutsvene, Nikita Singh and Zikhona Tshemese (2023). Biochar: Production, Application and the Future. in *Biochar - Productive Technologies, Properties and Applications.* IntechOpen.

Luo, Lu, Yuling Lan, Qianqian Zhang, Jianping Deng, Lingcong Luo, Qinzhi Zeng, Haili Gao and Weigang Zhao (2022). A Review on Biomass-Derived Activated Carbon as Electrode Materials for Energy Storage Supercapacitors. *Journal of Energy Storage,* **55:** 105839.

Mohd Abdah, Muhammad Amirul Aizat, Nur Hawa Nabilah Azman, Shalini Kulandaivalu and Yusran Sulaiman (2020). Review of the Use of Transition-Metal-Oxide and Conducting Polymer-Based Fibres for High-Performance Supercapacitors. *Materials & Design,* **186:** 108199.

Nandiyanto, Asep Bayu Dani, Rosi Oktiani and Risti Ragadhita (2019). How to Read and Interpret FTIR Spectroscope of Organic Material. *Indonesian Journal of Science and Technology,* **4(1)**: 97.

Parthenium Biomass by Physical and Chemical Activation Techniques. *E-Journal of Chemistry,* **7(4):** 1314–1319.

Parveen, Nazish, A. Ibrahim Al-Jaafari and Jeong In Han (2019). Robust Cyclic Stability and High-Rate Asymmetric Supercapacitor Based on Orange Peel-Derived Nitrogen-Doped Porous Carbon and Intercrossed Interlinked Urchin-like $NiCo_2O_4$@3DNF Framework. *Electrochimica Acta,* **293:** 84–96.

Patel, Madhumita, Xiaolei Zhang and Amit Kumar (2016). Techno-Economic and Life Cycle Assessment on Lignocellulosic Biomass Thermochemical Conversion Technologies: A Review. *Renewable and Sustainable Energy Reviews,* **53:** 1486–99.

Patil, Pranoti H., Vidya V. Kulkarni and Sushilkumar A. Jadhav (2022). An Overview of Recent Advancements in Conducting Polymer–Metal Oxide Nanocomposites for Supercapacitor Application. *Journal of Composites Science,* **6(12):** 363.

Rawat, Shivam, Rakesh K. Mishra and Thallada Bhaskar (2022). Biomass Derived Functional Carbon Materials for Supercapacitor Applications. *Chemosphere,* **286:** 131961.

Shaikh, Navajsharif S., Shivaji B. Ubale, Vikas J. Mane, Jasmin S. Shaikh, Vaibhav C. Lokhande, Supareak Praserthdam, Chandrakant D. Lokhande and Pongsakorn Kanjanaboos (2022). Novel Electrodes for Supercapacitor: Conducting Polymers, Metal Oxides, Chalcogenides, Carbides, Nitrides, MXenes, and Their Composites with Graphene. *Journal of Alloys and Compounds,* **893:** 161998.

Sherpa, Knawang Chhunji, Gour Gopal Satpati, Navonil Mal, Agatha Sylvia Khalko and Rajiv Chandra Rajak (2022). Effect of the COVID-19 on Access to Affordable and Clean Energy. pp. 79–104 in COVID-19 and the Sustainable Development Goals. Elsevier.

Sivaraj, Rajeshwari, Venckatesh Rajendran and G. Sangeetha Gunalan (2010). Preparation and Characterization of Activated Carbons from Parthenium Biomass by Physical and Chemical Activation Techniques. *E-Journal of Chemistry,* **7(4):** 1314–1319.

Wang, Jinxing, Yangyang Zhang, Junhua Ye, Hongmei Wei, Jinghua Hao, Juyi Mu, Shuoqing Zhao and Shahid Hussain (2016). Facile Synthesis of Three-Dimensional $NiCo_2O_4$ with Different Morphology for Supercapacitors. *RSC Advances,* **6(74):** 70077–70084.

Wang, Wei, Jiqiu Qi, Yanwei Sui, Yezeng He, Qingkun Meng, Fuxiang Wei and Yunxue Jin (2018). An Asymmetric Supercapacitor Based on Activated Porous Carbon Derived from Walnut Shells and $NiCo_2O_4$ Nanoneedle Arrays Electrodes. *Journal of Nanoscience and Nanotechnology,* **18(8):** 5600–5608.

Yaashikaa, P.R., Senthil Kumar, P. Varjani, Sunita and Saravanan, A. (2020). A Critical Review on the Biochar Production Techniques, Characterization, Stability and Applications for Circular Bioeconomy. *Biotechnology Reports,* **28:** e00570.

Yang, Guijun and Soo-Jin Park (2020). Nanoflower-like $NiCo_2O_4$ Grown on Biomass Carbon Coated Nickel Foam for Asymmetric Supercapacitor. *Journal of Alloys and Compounds,* **835:** 155270.

Yang, Hui, Shewen Ye, Jiaming Zhou and Tongxiang Liang (2019). Biomass-Derived Porous Carbon Materials for Supercapacitor. *Frontiers in Chemistry,* 7.

Zhang, Lulu, Xiumei Song, Boxin Xiao, Lichao Tan, Huiyuan Ma, Xinming Wang, Bonan Li, Dongxuan Guo and Dawei Chu (2019). Highly Graphitized and N, O Co-Doped Porous Carbon Derived from Leaves of Viburnum Sargenti with Outstanding Electrochemical Performance for Effective Supercapacitors. *Colloids and Surfaces A: Physicochemical and Engineering Aspects,* **580:** 123721.

Zhang, Yi-Zhou, Yang Wang, Tao Cheng, Wen-Yong Lai, Huan Pang and Wei Huang (2015). Flexible Supercapacitors Based on Paper Substrates: A New Paradigm for Low-Cost Energy Storage. *Chemical Society Reviews,* **44(15):** 5181–5199.

Chapter 24
Carbon Nanotubes and Its Composites in Sensing of Drugs and Pesticides

Anirudh Pratap Singh Raman, Sandeep Yadav Madhur Babu Singh, Indra Bahadur, Nagendra Kumar Kaushik, Prashant Singh, and Kamlesh Kumari

1 Introduction

Carbon nanotubes are cylindrically shaped with diameters as tiny as one nanometer. Significant research has been carried out in last two decades and have intriguing electrical and mechanical properties, dimensional solids and hold great promise for use as components in a variety of real-world nanoscale technologies. It has been shown that nanotubes can be used as components in macroscopic composites and as materials for nanoscale electrochemical, electrical, mechanical and other devices. The study and use of nanotubes on a fundamental level have been made easier by advancements in nanotube development. Scientists are diligently seeking resolutions to intricate challenges in nanotube growth, recognizing its pivotal role in the forthcoming advancement of CNTs (Dai, 2002; Kumar et al., 2017). Among several carbon-based nanomaterials, carbon nanotubes (CNTs) have emerged as a potential subgroup. While boron and molybdenum-based nanotubes have also been explored by many research groups, but CNTs are now the most significantly explored material.

A. P. S. Raman · S. Y. M. B. Singh · P. Singh (✉)
Department of Chemistry, Atma Ram Sanatan Dharma College University of Delhi, New Delhi, India
e-mail: psingh@arsd.du.ac.in

I. Bahadur
Department of Chemistry, Faculty of Natural and Agricultural Sciences, North-West University, Mmabatho, South Africa

N. K. Kaushik (✉)
Plasma Bioscience Research Center, Department of Electrical and Biological Physics, Kwangwoon University, Seoul 01897, South Korea
e-mail: kaushik.nagendra@kw.ac.kr

K. Kumari
Department of Zoology, University of Delhi, New Delhi, India

R. K. Kotnala et al. (eds.), *Advanced Functional Materials for Sustainable Environments*, https://doi.org/10.1007/978-3-031-62620-3_24

CNTs are made up of one or more concentric layers that resemble graphite and have a diameter between 0.4 and 10 nm. The discovery of CNTs can be attributed to Iijima in 1991 (Kimberlin, 2018) who observed their presence through transmission electron microscopy (TEM) and subsequently developed methods for synthesizing large quantities of nanotubes (Mamalis et al., 2004). The CNTs can be conceptualized as cylindrical structures formed by rolling up graphite sheets and are considered derivatives of carbon fibres and fullerene, with their molecular structures of carbon atoms arranged in specific tubular configurations. There are primarily two types of CNTs, differentiated based on layers, viz. SWCNTs and MWCNTs. One layer of graphene when rolled into a cylindrical shape is known as SWCNTs and their sizes range from 0.4 to 2 nm, and they frequently form hexagonally packed bundles. MWCNTs are made of more than one cylinder (made of graphene sheets) and their diameter ranges from 1 to 3 nm (He et al., 2013).

1.1 Single-Walled Carbon Nanotubes (SWCNTs)

The SWCNTs are one-dimensional structures composed of rolled-up graphene sheets, representing a unique form of carbon allotrope. These nanostructures can be envisioned as hollow cylinders with diameters ranging from 0.4 to 5 nm. The SWCNTs exhibit remarkable properties in terms of thermal conductivity, optical behaviour, and mechanical strength, leading to extensive research worldwide (Koziol et al., 2017; Pop et al., 2006; Kataura et al., 1999; Star et al., 2004). Different classifications include armchair ($n = m$), zig-zag ($m = 0$), and chiral nanotubes that do not conform to these specific rules. The chiral vector significantly influences various properties of SWCNTs, particularly their electrical behaviour. On considering the value of n and m as a multiple of three, the bands of SWCNTs is above the Fermi level, and it results in obtaining the metallic or quasi-metallic abilities. On other side, SWCNTs exhibit moderate semiconducting behaviour, with exceptions observed in small-diameter SWCNTs due to curvature effects (Matsuda et al., 2010). The chiral vector also plays a crucial role in the optical properties of SWCNTs. The configuration of carbon atoms in the nanotube has direct impact on various phenomenon like electroluminescence, photoluminescence and other optical (Lu and Chen, 2005). The chiral vector determines the possible optical transitions and has a significant impact on the observed optical properties. Raman spectroscopy has proven to be a valuable tool for providing insight into the electrical character of SWCNTs. Furthermore, the chirality of SWCNTs can manipulate the thermal conductivity as well as mechanical properties (slightly) (Janas et al., 2014). A significant development in SWCNT research occurred when methods for sorting individual nanotubes were discovered. Sorting allows the separation of SWCNTs based on their diameter, length, electrical behaviour, chirality and handedness. Over the years, the techniques for sorting SWCNTs have become more advance and diverse, enabling researchers to obtain highly purified samples of nanotubes with specific properties. This development has significantly expanded the possibilities for studying and utilizing SWCNTs in various applications (Telg et al., 2004; Oyama et al., 2006). The SWCNTs exhibit remarkable properties that are highly dependent on their chiral vector. Their electrical, optical, thermal and mechanical characteristics are strongly influenced by the arrangement of carbon atoms within the nanotube. Sorting methods have opened up new avenues for research, allowing for the isolation of SWCNTs with desired properties. The continued advancement of sorting techniques will further contribute to the understanding and utilization of SWCNTs in numerous scientific and technological fields (Janas, 2018; Liu et al., 2017; Zheng, 2017).

1.2 Multi-walled Carbon Nanotubes (MWCNTs)

The MWCNTs are fabricated of many concentric layers of graphene rolled into tubes with widths varying from 1 to 100 nm and offer distinct features that have led to a wide range of applications in materials science, electronics, energy storage and biomedical engineering (Zhang et al., 2009). The MWCNTs have a structure made up of many concentric graphene layers that form a tube within a tube (Fig. 1). Each layer is generally one atom thick, although the distance between levels might vary. On changing the number of layers in MWCNTs, the size also varies (Devgan et al., 2021; Hirlekar et al., 2009). The top and bottom layers are frequently topped by hemispherical fullerene-like structures. MWCNTs are classified into different types depending on variables such as layer arrangement, interlayer spacing, and the presence of defects. Concentric MWCNTs have layers that are well-aligned and concentrically organised, giving them outstanding mechanical characteristics and structural stability (Khan et al., 2017). Non-concentric MWCNTs feature misaligned or helically wrapped layers, which results in unusual characteristics due to their twisted or helical structure. MWCNTs with bamboo-like characteristics have alternating segments of bigger and smaller diameter tubes, similar to the pattern observed in bamboo stalks (Sun et al., 2021). Due open-ended MWCNTs lack fullerene-like caps, they can be used as nanoscale containers or templates for encapsulating other materials. Because of their peculiar structure MWCNTs have unique features. They have high mechanical strength and stiffness, making them excellent for composite reinforcement. The inclusion of numerous layers improves deformation resistance (Bharti et al., 2020). Due to the delocalized-electrons along the layers, MWCNTs offer great electrical conductivity and high thermal conductivity for efficient heat dissipation and are also chemically stable, resistant to environmental deterioration, and have intriguing optical features relating to light absorption and emission with diverse applications (Lephuthing et al., 2021). It strengthen mechanical properties in composite materials by acting as reinforcing agents, enhance energy storage in batteries and supercapacitors, serves as catalyst supports for various chemical processes, and show promise in biomedical engineering for drug delivery, biosensing and tissue engineering (Turkani et al., 2019). However, challenges remain in areas like large-scale production, structural property management, and understanding possible health and environmental implications (Donaldson et al., 2013).

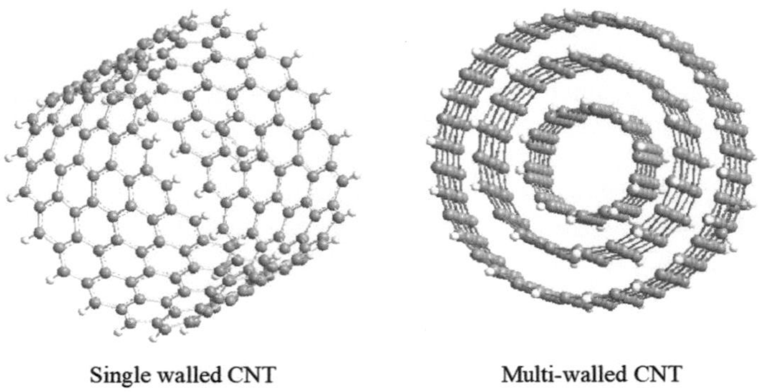

Fig. 1 Single-walled carbon nanotubes and multi-walled carbon nanotubes

1.3 The CNT-Based Polymer Composites

Due to simplicity in manufacturing, replication, and capacity to form solid-state structures, polymers are frequently used materials in a range of applications, including coatings, high-strength composites for the aerospace and automotive industries, thermal and electrical insulation, electronic devices. Polymer composites (PMCs) are made by combining several components, especially additives and polymer matrices, resulting in distinctive characteristics. PMCs, on the other hand, have intrinsic restrictions that have been resolved with the discovery of carbon nanotubes (CNTs) (Ajayan et al., 1994; Soni et al., 2020). Polymer nanocomposites (PNCs), which combine polymer matrices with CNTs, have received considerable interest and have emerged as a focus area for nanocomposite researchers. PNCs have the benefit of being easy to fabricate while retaining the tubular structure of CNTs. Researchers have been drawn to Ajayan et al. (1994) pioneering work in fabrication PNCs by the interplay between their structure and properties, which include mechanical, electrical, and thermal qualities, as well as its potential applications in a range of areas. Furthermore, with the arrival of cost-effective, large-scale manufacturing of nanotubes at the beginning of the twenty-first century, research efforts in this areas have increased. The strength, electrical conductivity, and durability of composites can be greatly enhanced by CNTs. Certain CNTs have directed only along one axis of the composite and it results in high electrical conductivity and open up new possibilities for the creation of conducting polymers (Grimes et al., 2000, 2001). However, a uniform dispersion of CNTs throughout the matrix is necessary to use them in composites effectively while preserving structural integrity. There should be a robust interfacial interaction among the matrix and CNTs in order to transfer load across the CNT-matrix contact to enhance the mechanical properties of the composite. Insufficient adhesion between the matrix and the CNTs compromises the benefits of the nanotubes' high tensile strength (Bal and Samal, 2007). The interfacial stress caused by shear among the fibers and the matrix regulates the amount of load is transmitted. Smaller load transfer channels are feasible with high interfacial shear stress and vice-versa (Andrews et al., 2002). The combination of polymers with carbon nanotubes in polymer nanocomposites has enormous potential for improving different characteristics and opening up new applications (Schadler et al., 1998).

1.4 The CNT-Based Metal Matrix Composites Composite

In recent years, sectors like automotive, aerospace, defence, sports goods, and medical equipment have shifted towards lightweight metal matrix composites (MMCs). These composites have excellent features such as high strength, low weight, stiffness, modulus, electrical conductivity, wear and corrosion resistance. There is a desire for even stronger and better performing advanced composites in some applications where cost is not a barrier. This has resulted in the development of MMCs reinforced with nanomaterials, notably CNTs, which have demonstrated exceptional mechanical and thermal characteristics augmentation. Researchers want to develop components that are incredibly rigid, strong, light in weight, and have exceptional heat, physical, and electrical qualities (Agarwal et al., 2019; Soni and Thomas, 2020; Radhamani et al., 2018). Cha et al. (2005) developed a unique CNT-alumina nanocomposites production technique to have high hardness. It is due to CNT load transfer and higher fracture toughness due to CNT bridging during crack propagation. Deng et al. (2007) have reported the synthesis of 1 wt.% CNT-reinforced Al composites for better tensile strength. They revealed that the CNTs were dispersed uniformly and that the Al matrix and CNTs were well bonded. Chen et al. (2000) have used electroless plating to deposit cobalt nanoparticles onto CNT surfaces, creating a coating of cobalt nanoparticles. The coating layer was enhanced by heat treatment. Uddin et al. (2010) studied the mechanical and electrical characteristics of Cu and alloy of Cu enhanced with CNTs. The inclusion of CNTs increased the composites' hardness and electrical conductivity (Uddin et al., 2010). Pang et al. (2009) synthesised multi-walled CNTs-Fe_3Al composites and discovered that they had soft magnetic characteristics equivalent to Fe_3Al. Aung et al. (2010) have investigated the inhibition of corrosion using composites of CNT and Mg and reported that the corrosion inhibition was enhanced.

1.5 The CNT-Based Sensors for Drug Sensing

Analytical chemistry development revealed that electrochemical sensors emerged as the primary and rapidly advancing class of chemical sensors. They provided continuous data concerning the presence of chemicals in their environments. A perfect chemical sensor would have a particular reaction that was related to the concentration of an individual chemical species. The essential characteristics of electrochemical sensors are sensitivity and selectivity. CNT-based sensors can be designed to interact with specific drug molecules, enabling rapid and sensitive detection. Moreover, functionalizing CNTs with specific receptors or antibodies can enhance their selectivity for drugs (Table 1). The CNT-modified electrodes using the nanomaterials and polymers to develop new nanocomposites with exceptional conductivity, great electron transfer ability, and exceptional stability for the development of sensor materials (Meskher et al., 2023). In a novel approach, the amplified voltametric sensor has been utilized to sense or detect anticancer drugs (doxorubicin (DOX) and 5-fluorouracil (FU)). The oxidation process of DOX in aqueous solutions was greatly performed by modified electrode. The oxidation of doxorubicin and 5-fluorouracil observe at potentials of approx. 680 and 970 mV, respectively (Fig. 2) (Fouladgar, 2018).

Superparamagnetic iron oxide nanoparticles (SPIONs), produced by a simple hydrothermal method, were used to make SPIONs-CNT nanocomposite material. The resultant nanocomposite undergone electrochemical investigations have been carried out to determine its compatibility with the anti-TB medication Rifampicin (Pinto et al., 2017). Foroughi et al. (2019) previously developed a promising electrochemical sensor to detect two molecules (tramadol and acetaminophen). Balram et al. (2020) employed a zinc oxide decorated MWCNTs modified GCE to fabricate a promising and selective electrochemical sensor to detect 4-nitrophenol. Sadanandhan et al. (2017) developed a transducer for serotonin detection, which involved a modified electrode (PEDOTNTs/rGO/Ag NPs) as shown in (Fig. 3).

Researchers have successfully constructed an electrochemical sensor for detecting serotonin (SE) at physiological pH using GCE/MWCNT-metal oxide nanocomposite electrodes. When compared to other evaluated electrodes, these

Table 1 Electrochemical application of CNTs

Electrode	*Drug*	*Method*	*Linear Range*	*References*
NH_2fMWCNT	Entacapone	CV, DPV	1–5 nM	Aftab et al., 2019
RGO-CB-CTS	Paracetamol	SWV, CV	3.2–32 μM	Baccarin et al., 2017
MWCNT	Tadalafil	LSW	0.85–8.9 μM	Zambianco et al., 2020
CNTs/ILC/CNTs/ FeNPs	Daclatasvir	EIS	3×10^{-9}-15×10^{-6} M	Atta et al., 2018
ZnO/functionalized MWCNT and Ag/ functionalized MWCNT	Dopamine	DPV	0.2–5 μM	Kumar et al., 2019
MWCNTs@MIP/ rGO/CKM/GCE	Chloramphenicol	DPV	5×10^{-9}-5×10^{-7} mol/L	Yang and Zhao, 2015
MMWCNTs@MIP/ CPE	Ciprofloxacin	DPV	0.005–0.85 × 10^{-6} mol/L	Bagheri et al., 2016

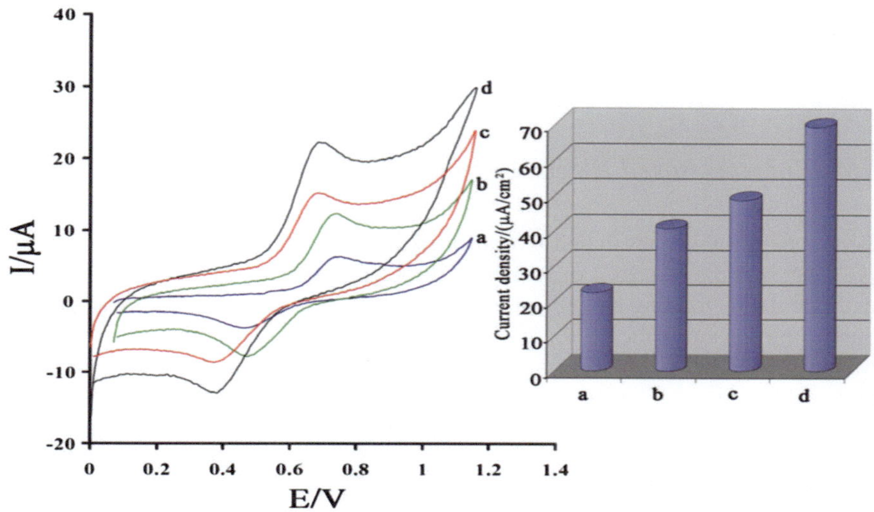

Fig. 2 Cyclic voltammogram on: (a) CPE, (b) CuO/CNT/CPE, (c) IL/CPE and (d) L/CuO/CNT/CPE in the presence of 500 μmol L^{-1} DOX at pH 5.0 (Fouladgar, 2018)

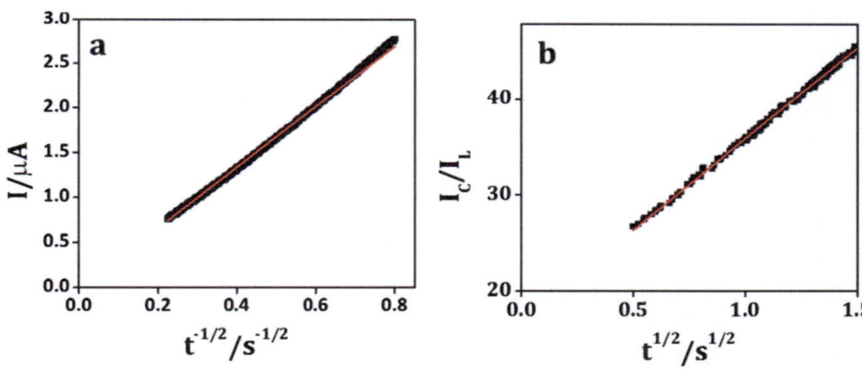

Fig. 3 (a) Plot of: I vs. $t^{-1/2}$ using chronoamperogram and (b) I_C / I_L/ vs. $t^{1/2}$ / using chronoamperogram (Sadanandhan et al., 2017)

modified electrodes showed great electrocatalytic activity for the oxidation of serotonin, resulting in significant increase in peak currents and reduced oxidation potentials. The limits of detection for the GCE/MWCNT-metal oxide electrodes were determined to be 129 nM for GCE/MWCNT-ZnO sensors (Fayemi et al., 2017). In the study, clomipramine, a significant antidepressant drug with limited redox activity, was efficiently electro-catalyzed on functionalised GCE (f-MWCNTs/GCE). This process produced an anodic peak at 5.5 pH approx. 1.120 V. The achieved limits of detection (LOD) for clomipramine were 1.315×10^{-8} g/mL respectively (M.S. et al., 2019).

1.6 Enzyme-Based Electrochemical Biosensors

Electrochemical enzyme-based biosensors work either detecting directly electroactive species at the electrode or indirectly through a medium. This method has been instrumental in creating an enzyme immunoassay system and the highly acclaimed glucose biosensor (Kilic et al., 2023). Utilising CV, the working electrodes were evaluated. Ability of the biosensors were evaluated using the chronoamperometric method by understanding the relationship between substrate's concentration and current. The electrochemical efficacy in detecting hydroquinone was assessed using both these methods (Lou et al., 2020). Amperometric biosensors were developed for the determination of dopamine using GCE with GO and MWCNTs (Vlamidis et al., 2017). Researchers have developed a novel biosensor that utilizes a biocomposite consisting of NPs of iron oxide, and tyrosinase to detect or sense methimazole. By producing thioquinone and binding copper at the tyrosinase active site, it inhibits the Tyr-based mechanism (Kurbanoglu et al., 2015). A photoelectrochemical sensor based on the heterojunction of g-C3N4 and CNT has been created for the precise sensing detection of GA. The photocurrent of g-C3N4@CNT is seven times more than that of g-C3N4 alone (Li et al., 2022). A nanocomposite of CNT-muc through an amperometric enzyme-electrode was developed. By comparing several compositions, the biosensor was optimised, and the analytical qualities were assessed. The effectiveness of the generated enzyme matrix for the assessment of glucose in blood plasma from humans was next assessed (Comba et al., 2018).

An enzymatic biosensor designed with D-amino acid oxidase (DAAO) immobilised on the GCE modified with a Au-NF and MWCNTs was used to study the detection of D-alanine. The potentiostatic approach was used to electro-deposit Au-NF onto the MWCNTs immobilised on the GCE surface during the synthesis of the Au-NF/MWCNT nanocomposite (Shoja et al., 2017). Recent studies have revealed that biosensors based on CNTs offer high selectivity and sensitivity for measuring blood glucose levels (Same and Samee, 2018). Kolahi-Ahari et al. (2020) quantitatively measured tramadol (TRA) utilizing a pencil graphite electrode (PGE) modified using AuNPs & MWNTs. Tramadol's electroanalysis was carried out using DPV and CV methods. The interfacial electron transport characteristics of electrodes during various modification stages have been examined using electron impedance spectroscopy as shown in Fig. 4.

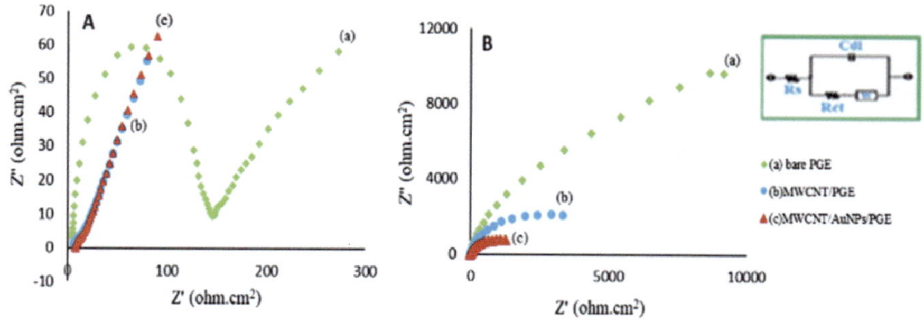

Fig. 4 Electrochemical impedance spectra of: (a) bare PGE, (b) f-MWCNTs/PGE and (c) f-MWCNTs/AuNps/PGE in (A) 5 mM [Fe (CN) 6]- 3/-4 (1: 1) solution containing 0.1 M of KCl and (B) 1 mM TRA in PBS (pH 7.0) (Kolahi-Ahari et al., 2020)

1.7 Pesticide Sensing by Carbon Nanotubes

Pesticide sensing and monitoring are highly important in protecting human health, environment and agricultural product safety. These methods accurately detect and quantify pesticides, aiding in risk assessment and reduction of pesticide exposure. They also identify contaminated areas, enforce maximum residue limits (MRLs) and implement mitigation measures. Furthermore, they support sustainable agriculture through optimized pesticide use, integrated pest management, and reduced environmental impact. Technological advancements continuously enhance our ability to monitor and mitigate pesticide impacts (Wong et al., 2017).

The limitations of conventional pesticide sensing methods, such as sensitivity, selectivity, and sample complexity, can be overcome with the use of CNTs that have a high surface-to-volume ratio, which greatly enhances their sensitivity to detect even trace amounts of pesticides. By functionalizing CNTs with specific groups or polymers, their selectivity can be improved as they can selectively interact with targeted pesticides. Furthermore, CNT-based sensors are compatible with complex sample matrices, eliminating the need for extensive sample preparation. This simplifies the analysis process and enables rapid on-site detection. Additionally, CNT-based sensors can be designed to be portable and cost-effective, making them accessible in various environments. Overall, CNTs provide a promising solution to overcome the limitations of traditional pesticide sensing methods, offering improved sensitivity, selectivity and ease of use (Pathak et al., 2022).

The CNTs can be used for detecting pesticides through different sensing methods, including electrical, optical and electrochemical principles. In electrical sensing, CNTs act as conductive channels that can be modified to respond to pesticides. Changes in conductivity revealed the amount of pesticide present that relied on its optical properties. Interactions with pesticides cause detectable changes in light absorption, fluorescence, or Raman scattering (Fang et al., 2021). Electrochemical sensing involves chemical reactions between CNTs and pesticides on the electrode surface. Binding events create measurable changes in current, potential, or impedance. Each sensing method has its advantages, like high sensitivity, real-time monitoring (electrical), compatibility with different setups (optical), or quick response (electrochemical). The choice of sensing method depends on the specific pesticide, desired sensitivity, and the properties of CNTs used in the sensor. By combining or customizing these methods, CNT-based sensors can be optimized for pesticide detection (Wang et al., 2020).

1.7.1 Electrochemical Sensing

Pesticide sensing using electrical sensors made of CNTs and their composites has shown great promise. CNTs have appreciable electrical properties and a high surface-to-volume ratio, making them ideal for detecting and measuring pesticides with great sensitivity and selectivity. These sensors work by detecting changes in electrical conductivity or impedance when they come in contact with the target pesticides. By incorporating CNTs into sensor devices, real-time monitoring, fast response times, and the potential for making them small in size are possible. This makes them useful in various areas like agriculture, environmental monitoring, and ensuring food safety (Wang et al., 2020).

Zhao et al. (2023) developed an electrochemical biosensor with high sensitivity for detecting organophosphorus pesticides (OPs). The biosensor utilized a unique electrode, that is, -αFe_2O_3NFs-AuNPs-CNTs for understanding loading the acetylcholinesterase (AChE). The electrical conductivity of the biosensor was improved significantly on incorporating the AuNPs and CNTs. The biosensor showed a straight-line correlation between the signal detected and the pesticide's (omethoate) concentration (10^{-13} to 10^{-7} M). It also exhibited stability, and repeatability, making it a valuable tool for pesticide sensing in environmental applications.

Xu et al. (2019) designed an AChE biosensor for detecting OPs by employing AuNPs on vertical nitrogen-doped VNSWCNTs. The combined actions of AuNPs and VNSWCNTs allowed the improved electrode to exhibit outstanding electron transfer capacity. The created biosensor showed a good linearity connection. The AChE biosensor demonstrated a desired level of reproducibility, sensitivity, and stability.

Nagabooshanam et al. (2020) have created an electrochemical sensor for detecting OPs in real samples. The sensor utilized polyaniline (PANI) and carboxyl functionalized (fMWCNT) to achieve in Fig. 5. The sensitivity of designed electrochemical sensor was determined to be 0.41 mA/ng/L/cm^2 for CPF and 0.58 mA/ng/L/cm^2 for MP. Tests conducted on cucumber samples showed an average recovery rate of 98.05% for CPF and 96.63% for MP. The sensor demonstrated stability over a 30-day period and showed minimal interference (<10%) from heavy metals. The creation of this sensor offers a lot of potential for continuous monitoring of food items, which will improve food safety.

Chen et al. (2017) developed a biosensor for detecting chlorpyrifos using a combination of dicyclohexyl phthalate (DCHP) and SPE-MWCNTs. Modified electrode enhanced electron transfer and enabled efficient electrochemical oxidation of

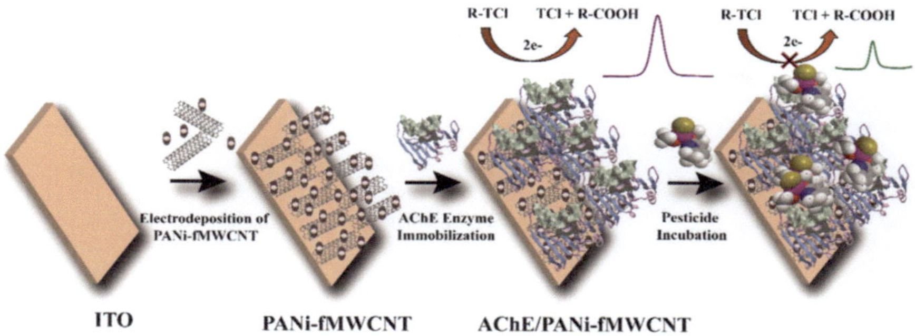

Fig. 5 Fabrication and working of the carboxyl functionalized multi-walled carbon nanotube sensor for detecting organophosphates (Nagabooshanam et al., 2020)

acetylthiocholine. By immobilizing AChE on the nanocomposite film, a biosensor specifically has been fabricated for pesticide detection. The biosensor exhibited a wide detection range, a low limit of detection (0.05 μg/L), good reproducibility, and acceptable stability. It successfully detected low concentrations of pesticides in actual vegetable samples, demonstrating its potential for practical applications.

To build a sensor for diazinon (DZN) detection, modified CNT electrospun nanofibers was used. The electrochemical efficacy of the redesigned electrode was assessed by optimizing factors like duration of time, pH and scanning rate. The sensor operated throughout a straight-line concentration that ranged from 3 to 100 nM and displayed an impressive limit of detection of 2.888 nM. It demonstrated outstanding results. Furthermore, the sensor sensed DZN in tomato juice with a good rate of recovery. This promising sensor made of CNTs has a great deal of potential for correctly determining the extent and concentration of DZN in actual samples (Topsoy et al., 2022).

Gholivand et al. (2018) introduced an electrochemical sensor using a modified pencil graphite electrode with a hollow fiber (HF) containing a composite of MWCNTs, ionic liquid (IL), and CUO NPs (Fig. 6) for the detection of glyphosate (GPS) by using the differential pulse voltammetry (DPV). Optimal conditions for GPS extraction and preconcentration were examined and a direct relation is observed between current and concentration. The technique was successfully applied to real sample analysis and compared to a standard method (Gholivand et al., 2018).

1.7.2 Optical Sensing

The CNT-based optical sensors have become a promising technology for pesticide detection. These sensors utilize the unique properties of CNTs, such as absorption, fluorescence, and raman scattering, to detect and quantify pesticide molecules. By functionalizing CNTs with specific groups or polymers, selectivity towards targeted pesticides can be enhanced. CNTs' high ratio of surface area to volume enables the

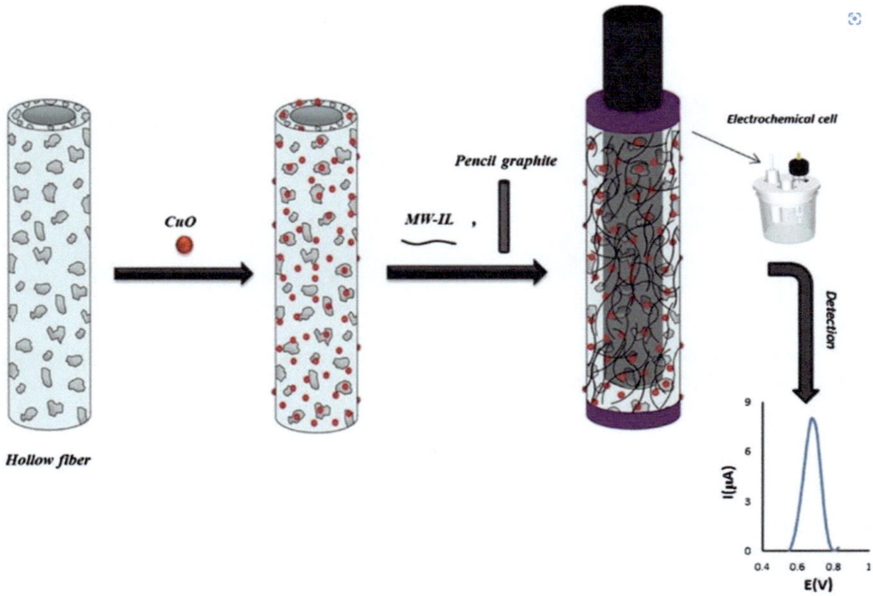

Fig. 6 Fabrication and working of electrochemical biosensor for detection of glyphosate (Gholivand et al., 2018)

sensitive detection of pesticides even at low levels. Moreover, CNT-based optical sensors offer advantages such as real-time monitoring, compatibility with various detection setups, and rapid response time. With ongoing advancements in CNT synthesis and modification techniques, optical sensors based on CNTs continue to hold great potential for sensitive and selective pesticide detection in various applications. Cost-effective and highly sensitive metamaterial-free THz sensors are gaining is significant as they overcome limitations associated with metal's optical properties (Fang et al., 2021).

Zhang et al. (2022) proposed a flexible THz sensor for detecting nereistoxin-related insecticides (NRIs) using AuNPS and N-CNTs. The accuracy and selectivity of the sensor was confirmed by achieving high recovery rates in spiked food samples. Uses for these results in environmental monitoring and food safety are possible. Abnous et al. (2017) developed a fluorometric assay for acetamiprid detection using an aptamer-based approach. It involves releasing a fluorescein-labelled complementary strand (CS) from the aptamer/CS conjugate (dsDNA) in the presence of acetamiprid. AuNPs, SWNTs, and streptavidin-coated SiNPs are utilized. The assay demonstrates high selectivity with a low limit of detection (LOD) of 127 pM.

A heterostructure of CdS/SnS_2/CNTs was formed by immobilizing CdS nanoparticles onto a flower-like SnS_2 structure using a simple mixing technique. CNTs were added as photosensitizers, creating synergistic effects. Coupled with HQ detector, the modified photoelectrodes showed increased photocurrent signal through oxidation of HQ holes. The photoelectrochemical sensor featured an outstanding detection limit of 0.1 μM and a broad detection range for HQ of 0.2–100 μM. Optimized boron nitride modified multiwalled carbon nanotubes (BN@MWCNTs) were used as a coating for solid-phase microextraction (SPME) fibers to extract organochlorine pesticides (OCPs) from fruits and vegetables (Huang et al., 2023).

Zhang et al. (2021) employed a combination of solid-phase microextraction (SPME) and gas chromatography-electron capture detection (GC-ECD). The method demonstrated a linear response within the concentration range of 0.03–200 ng g^{-1}. The technique provided a trustworthy method for pesticide analysis by effectively detecting organochlorine pesticides (OCPs) in samples of actual fruits and vegetables.

2 Conclusion

This chapter highlights the significant potential of CNTs and their composites as potential sensors for the detection or sensing of drugs as well as pesticides. Extraordinary properties of CNTs, including their high surface-to-volume ratio, excellent electrical conductivity, and chemical stability, make them ideal for sensing applications. The research in this chapter showcases the advancements in functionalizing CNTs to enhance their sensing performance and selectivity. The development of CNT-based sensors has demonstrated improved sensitivity, selectivity and detection limits, surpassing those of conventional sensors. However, challenges remain, including the optimization of sensor designs for long-term stability and the exploration of new applications. Despite these challenges, the future of CNT-based sensors in drug detection and pesticide monitoring appears promising. Ongoing research efforts and advancements in sensor fabrication techniques will contribute to the fabrication of extremely sensitive and selective/ precise CNT-based sensors, thereby advancing the field of sensor technology and benefiting various industries, including healthcare and agriculture.

References

Abnous, K., Danesh, N.M., Ramezani, M., Alibolandi, M., Lavaee, P. and Taghdisi, S.M. (2017). Aptamer based fluorometric acetamiprid assay using three kinds of nanoparticles for powerful signal amplification, *Microchimica Acta*, **184(1)**: 81–90 Springer-Verlag Wien.

Aftab, S., Ozcelikay, G., Kurbanoglu, S., Shah, A., Iftikhar, F.J. and Ozkan, S.A. (2019). A novel electrochemical nanosensor based on NH2-functionalized multi walled carbon nanotubes for the determination of catechol-orto-methyltransferase inhibitor entacapone, *Journal of Pharmaceutical and Biomedical Analysis*, **165:** 73-81.

Agarwal, P., Kishore, A., Kumar, V., Soni, S.K. and Thomas, B. (2019). Fabrication and machinability analysis of squeeze cast Al 7075/h-BN/graphene hybrid nanocomposite, *Engineering Research Express*, IOP Publishing, **1(1):** 15004.

Ajayan, P.M., Stephan, O., Colliex, C. and Trauth, D. (1994). Aligned Carbon Nanotube Arrays Formed by Cutting a Polymer Resin Nanotube Composite, *Science*, **265(5176):** 1212-1214.

Andrews, R., Jacques, D., Minot, M. and Rantell, T. (2002). Fabrication of carbon multiwall nanotube/polymer composites by shear mixing, *Macromolecular Materials and Engineering*, Wiley Online Library, **287(6):** 395–403.

Atta, N.F., Ahmed, Y.M. and Galal, A. (2018). Nano-magnetite/ionic liquid crystal modifiers of carbon nanotubes composite electrode for ultrasensitive determination of a new anti-hepatitis C drug in human serum, *Journal of Electroanalytical Chemistry*, **823:** 296-306.

Aung, N.N., Zhou, W., Goh, C.S., Nai, S.M.L. and Wei, J. (2010). Effect of carbon nanotubes on corrosion of Mg–CNT composites, *Corrosion Science*, Elsevier, **52(5):**1551–1553.

Baccarin, M., Santos, F.A., Vicentini, F.C., Zucolotto, V., Janegitz, B.C. and Fatibello-Filho, O. (2017). Electrochemical sensor based on reduced graphene oxide/carbon black/chitosan composite for the simultaneous determination of dopamine and paracetamol concentrations in urine samples, *Journal of Electroanalytical Chemistry*, **799:** 436-443.

Bagheri, H., Khoshsafar, H., Amidi, S. and Hosseinzadeh Ardakani, Y. (2016). Fabrication of an electrochemical sensor based on magnetic multi-walled carbon nanotubes for the determination of ciprofloxacin, *Anal. Methods*, The Royal Society of Chemistry, **8(16):** 3383–3390.

Bal, S. and Samal, S.S. (2007). Carbon nanotube reinforced polymer composites—a state of the art, *Bulletin of Materials Science*, Springer, **30:** 379–386.

Balram, D., Lian, K.-Y. and Sebastian, N. (2020). Ultrasound-assisted synthesis of 3D flower-like zinc oxide decorated fMWCNTs for sensitive detection of toxic environmental pollutant 4-nitrophenol, *Ultrasonics Sonochemistry*, **60:** 104798.

Bharti, M., Singh, A., Singh, B.P., Dhakate, S.R., Saini, G., Bhattacharya, S., Debnath, A.K., et al. (2020). Free-standing flexible multiwalled carbon nanotubes paper for wearable thermoelectric power generator, *Journal of Power Sources*, **449:** 227493.

Cha, S.I., Kim, K.T., Lee, K.H., Mo, C.B. and Hong, S.H. (2005). Strengthening and toughening of carbon nanotube reinforced alumina nanocomposite fabricated by molecular level mixing process, *Scripta Materialia*, Elsevier, **53(7):** 793–797.

Chen, D., Liu, Z., Fu, J., Guo, Y., Sun, X., Yang, Q. and Wang, X. (2017). Electrochemical acetylcholinesterase biosensor based on multi-walled carbon nanotubes/dicyclohexyl phthalate modified screen-printed electrode for detection of chlorpyrifos, *Journal of Electroanalytical Chemistry*, Elsevier, **801:** 185–191.

Chen, X., Xia, J., Peng, J., Li, W. and Xie, S. (2000). Carbon-nanotube metal-matrix composites prepared by electroless plating, *Composites Science and Technology*, Elsevier, **60(2):** 301–306.

Comba, F.N., Romero, M.R., Garay, F.S. and Baruzzi, A.M. (2018). Mucin and carbon nanotube-based biosensor for detection of glucose in human plasma, *Analytical Biochemistry*, **550:** 34-40.

Dai, H. (2002). Carbon nanotubes: opportunities and challenges, *Surface Science*, **500(1):** 218-241.

Deng, C., Zhang, X., Wang, D., Lin, Q. and Li, A. (2007). Preparation and characterization of carbon nanotubes/aluminum matrix composites, *Materials Letters*, Elsevier, **61(8–9):** 1725–1728.

Devgan, S., Mahajan, A. and Sidhu, S.S. (2021). Multi-walled carbon nanotubes in powder mixed electrical discharge machining: an experimental study, state of the art and feasibility prospect, *Applied Physics A*, **127(11):** 806.

Donaldson, K., Poland, C.A., Murphy, F.A., MacFarlane, M., Chernova, T. and Schinwald, A. (2013). Pulmonary toxicity of carbon nanotubes and asbestos—similarities and differences, *Advanced Drug Delivery Reviews*, Elsevier, **65(15):** 2078–2086.

Fang, L., Jia, M., Zhao, H., Kang, L., Shi, L., Zhou, L. and Kong, W. (2021). Molecularly imprinted polymer-based optical sensors for pesticides in foods: Recent advances and future trends, *Trends in Food Science & Technology*, Elsevier, **116:** 387–404.

Fayemi, O.E., Adekunle, A.S. and Ebenso, E.E. (2017). Electrochemical determination of serotonin in urine samples based on metal oxide nanoparticles/MWCNT on modified glassy carbon electrode, *Sensing and Bio-Sensing Research*, **13:** 17-27.

Foroughi, M.M., Jahani, S. and Hassani Nadiki, H. (2019). Lanthanum doped fern-like CuO nanoleaves/MWCNTs modified glassy carbon electrode for simultaneous determination of tramadol and acetaminophen, *Sensors and Actuators B: Chemical*, **285:** 562-570.

Fouladgar, M. (2018). CuO-CNT Nanocomposite/Ionic Liquid Modified Sensor as New Breast Anticancer Approach for Determination of Doxorubicin and 5-Fluorouracil Drugs, *Journal of The Electrochemical Society*, **165(13):** B559-B564.

Gholivand, M.B., Akbari, A. and Norouzi, L. (2018). Development of a novel hollow fiber- pencil graphite modified electrochemical sensor for the ultra-trace analysis of glyphosate, *Sensors and Actuators B: Chemical*, Elsevier, **272:** 415–424.

Grimes, C.A., Dickey, E.C., Mungle, C., Ong, K.G. and Qian, D. (2001). Effect of purification of the electrical conductivity and complex permittivity of multiwall carbon nanotubes, *Journal of Applied Physics*, American Institute of Physics, **90(8):** 4134–4137.

Grimes, C.A., Mungle, C., Kouzoudis, D., Fang, S. and Eklund, P.C. (2000). The 500 MHz to 5.50 GHz complex permittivity spectra of single-wall carbon nanotube-loaded polymer composites, *Chemical Physics Letters*, Elsevier, **319(5–6):** 460–464.

He, H., Pham-Huy, L.A., Dramou, P., Xiao, D., Zuo, P. and Pham-Huy, C. (2013). Carbon Nanotubes: Applications in Pharmacy and Medicine, edited by Noda, T. *BioMed Research International*, Hindawi Publishing Corporation, **2013:** 578290.

Hirlekar, R., Yamagar, M., Garse, H., Vij, M. and Kadam, V. (2009). Carbon nanotubes and its applications: a review, *Asian Journal of Pharmaceutical and Clinical Research*, **2(4):** 17-27.

Huang, L., Wang, C., Yang, Y., Wang, Y., Li, C., Xie, Y., Zhao, P., et al. (2023). A light-driven photoelectrochemical sensor for highly selective detection of hydroquinone based on type-II heterojunction formed by carbon nanotubes immobilized in 3D honeycomb CdS/SnS2, *Journal of Colloid and Interface Science*, Academic Press, **643:** 585–599.

Janas, D. (2018). Towards monochiral carbon nanotubes: a review of progress in the sorting of single-walled carbon nanotubes, *Mater. Chem. Front.*, The Royal Society of Chemistry, **2(1):** 36–63.

Janas, D., Czechowski, N., Mackowski, S. and Koziol, K.K. (2014). Direct evidence of delayed electroluminescence from carbon nanotubes on the macroscale, *Applied Physics Letters*, **104(26):** 261107.

Kataura, H., Kumazawa, Y., Maniwa, Y., Umezu, I., Suzuki, S., Ohtsuka, Y. and Achiba, Y. (1999). Optical properties of single-wall carbon nanotubes, *Synthetic Metals*, **103(1):** 2555-2558.

Khan, S.A., Gao, M., Zhu, Y., Yan, Z. and Lin, Y. (2017). MWCNTs based flexible and stretchable strain sensors, *Journal of Semiconductors*, Chinese Institute of Electronics, **38(5):** 1674-4926.

Kilic, N.M., Singh, S., Keles, G., Cinti, S., Kurbanoglu, S. and Odaci, D. (2023). Novel Approaches to Enzyme-Based Electrochemical Nanobiosensors, *Biosensors*, **13(6):** 622.

Kimberlin, D.W. (2018). Antiviral Agents, Principles and Practice of Pediatric Infectious Diseases, Fifth Edit., Elsevier Inc., https://doi.org/10.1016/B978-0-323-40181-4.00295-4.

Kolahi-Ahari, S., Deiminiat, B. and Rounaghi, G.H. (2020). Modification of a pencil graphite electrode with multiwalled carbon nanotubes capped gold nanoparticles for electrochemical determination of tramadol, *Journal of Electroanalytical Chemistry*, **862:** 113996.

Koziol, K.K., Janas, D., Brown, E. and Hao, L. (2017). Thermal properties of continuously spun carbon nanotube fibres, *Physica E: Low-Dimensional Systems and Nanostructures*, **88:** 104-108.

Kumar, M., Kumara Swamy, B.E., Reddy, S., Zhao, W., Chetana, S. and Gowrav Kumar, V. (2019). ZnO/functionalized MWCNT and Ag/functionalized MWCNT modified carbon paste electrodes for the determination of dopamine, paracetamol and folic acid, *Journal of Electroanalytical Chemistry*, **835:** 96-105.

Kumar, S., Rani, R., Dilbaghi, N., Tankeshwar, K. and Kim, K.-H. (2017). Carbon nanotubes: a novel material for multifaceted applications in human healthcare, *Chem. Soc. Rev.*, The Royal Society of Chemistry, **46(1):** 158–196.

Kurbanoglu, S., Mayorga-Martinez, C.C., Medina-Sánchez, M., Rivas, L., Ozkan, S.A. and Merkoçi, A. (2015). Antithyroid drug detection using an enzyme cascade blocking in a nanoparticle-based lab-on-a-chip system, *Biosensors and Bioelectronics*, **67:** 670-676.

Lephuthing, S.S., Okoro, A.M., Mohlala, M., Malgas, N., Ige, O. and Apata Olubambi, P. (2021). Effect of Multi-walled Carbon Nanotubes Additions on Its Dispersion Characteristics in Titanium Oxides BT–Trends in Manufacturing and Engineering Management, in Vijayan, S., Subramanian, N. and Sankaranarayanasamy, K. (Eds.), Springer Singapore, pp. 135–143.

Li, J., Wang, C., Chen, X., Huang, M., Fu, Q., Li, R., Wang, Y., et al. (2022). A non-enzymatic photoelectrochemical sensor based on g-C3N4@CNT heterojunction for sensitive detection of antioxidant gallic acid in food, *Food Chemistry*, **389:** 133086.

Liu, B., Wu, F., Gui, H., Zheng, M. and Zhou, C. (2017). Chirality-Controlled Synthesis and Applications of Single-Wall Carbon Nanotubes, *ACS Nano*, American Chemical Society, **11(1):** 31–53.

Lou, C., Jing, T., Zhou, J., Tian, J., Zheng, Y., Wang, C., Zhao, Z., et al. (2020). Laccase immobilized polyaniline/magnetic graphene composite electrode for detecting hydroquinone, *International Journal of Biological Macromolecules*, **149:** 1130-1138.

Lu, X. and Chen, Z. (2005). Curved Pi-Conjugation, Aromaticity, and the Related Chemistry of Small Fullerenes (<C60) and Single-Walled Carbon Nanotubes, *Chemical Reviews*, American Chemical Society, **105(10)**: 3643–3696.

Mamalis, A.G., Vogtländer, L.O.G. and Markopoulos, A. (2004). Nanotechnology and nanostructured materials: trends in carbon nanotubes, *Precision Engineering*, **28(1):** 16-30.

Matsuda, Y., Tahir-Kheli, J. and Goddard, W.A.I.I.I. (2010). Definitive Band Gaps for Single-Wall Carbon Nanotubes, *The Journal of Physical Chemistry Letters*, American Chemical Society, **1(19):** 2946-2950.

Meskher, H., Ragdi, T., Thakur, A.K., Ha, S., Khelfaoui, I., Sathyamurthy, R., Sharshir, S.W., et al. (2023). A Review on CNTs-Based Electrochemical Sensors and Biosensors: Unique Properties and Potential Applications, *Critical Reviews in Analytical Chemistry*, pp. 1–24.

M.S., J., K., M., K.K., J. and D.K., S. (2019). Sensitive electro-chemical determination of antidepressant drug clomipramine at f-mwcnts nano clusters modified glassy carbon electrode, *Journal of Pharmaceutical Sciences and Research*, **11(3):** 700–707.

Nagabooshanam, S., John, A.T., Wadhwa, S., Mathur, A., Krishnamurthy, S. and Bharadwaj, L.M. (2020). Electro-deposited nano-webbed structures based on polyaniline/multi walled carbon nanotubes for enzymatic detection of organophosphates, *Food Chemistry*, Elsevier, **323:** 126784.

Oyama, Y., Saito, R., Sato, K., Jiang, J., Samsonidze, Ge.G., Grüneis, A., Miyauchi, Y., et al. (2006). Photoluminescence intensity of single-wall carbon nanotubes, *Carbon*, **44(5):** 873–879.

Pang, L.X., Xing, D.J., Zhang, A.Q., Xu, J. and Sun, S.S. (2009). Magnetic properties of MWNTs-Fe3Al composite, *Advanced Materials Research*, Trans Tech Publ, **79:** 1351–1354.

Pathak, V.M., Verma, V.K., Rawat, B.S., Kaur, B., Babu, N., Sharma, A., Dewali, S., et al. (2022). Current status of pesticide effects on environment, human health and it's eco-friendly management as bioremediation: A comprehensive review, *Frontiers in Microbiology*, Frontiers Media SA., **13**.

Pinto, T., Cai, L., Wang, C. and Tan, X. (2017). CNT-based sensor arrays for local strain measurements in soft pneumatic actuators, *International Journal of Intelligent Robotics and Applications*, Springer Singapore, **1(2):** 157–166.

Pop, E., Mann, D., Wang, Q., Goodson, K. and Dai, H. (2006). Thermal Conductance of an Individual Single-Wall Carbon Nanotube above Room Temperature, *Nano Letters*, American Chemical Society, **6(1):** 96–100.

Radhamani, A.V, Lau, H.C. and Ramakrishna, S. (2018). CNT-reinforced metal and steel nanocomposites: A comprehensive assessment of progress and future directions, *Composites Part A: Applied Science and Manufacturing*, **114:** 170-187.

Sadanandhan, N.K., Cheriyathuchenaaramvalli, M., Devaki, S.J. and Ravindranatha Menon, A.R. (2017). PEDOT-reduced graphene oxide-silver hybrid nanocomposite modified transducer for the detection of serotonin, *Journal of Electroanalytical Chemistry*, **794:** 244-253.

Same, S. and Samee, G. (2018). Carbon nanotube biosensor for diabetes disease, *Crescent Journal of Medical and Biological Sciences*, **5(1):** 1-6.

Schadler, L.S., Giannaris, S.C. and and Ajayan, P.M. (1998). Load transfer in carbon nanotube epoxy composites, *Applied Physics Letters*, American Institute of Physics, **73(26):** 3842-3844.

Shoja, Y., Rafati, A.A. and Ghodsi, J. (2017). Enzymatic biosensor based on entrapment of d-amino acid oxidase on gold nanofilm/MWCNTs nanocomposite modified glassy carbon electrode by sol-gel network: Analytical applications for d-alanine in human serum, *Enzyme and Microbial Technology*, **100:** 20–27, https://doi.org/10.1016/j.enzmictec.2017.02.001.

Soni, S.K. and Thomas, B. (2020). Microstructure, mechanical properties and machinability studies of Al7075/SiC/h-BN hybrid nanocomposite fabricated via ultrasonic-assisted squeeze casting, *FME Transactions*, **48(3):** 532-542.

Soni, S.K., Thomas, B. and Kar, V.R. (2020). A Comprehensive Review on CNTs and CNT-Reinforced Composites: Syntheses, Characteristics and Applications, *Materials Today Communications*, **25:** 101546.

Star, A., Lu, Y., Bradley, K. and Grüner, G. (2004). Nanotube Optoelectronic Memory Devices, *Nano Letters*, American Chemical Society, **4(9):** 1587–1591.

Sun, W., Hui, L., Yang, Q. and Zhao, G. (2021). Nanofiltration filter paper based on multi-walled carbon nanotubes and cellulose filter papers, *RSC Adv.*, The Royal Society of Chemistry, **11(2):** 1194–1199.

Telg, H., Maultzsch, J., Reich, S., Hennrich, F. and Thomsen, C. (2004). Chirality Distribution and Transition Energies of Carbon Nanotubes, *Physical Review Letters*, American Physical Society, **93(17):** 177401.

Topsoy, O.K., Muhammad, F., Kolak, S., Ulu, A., Güngör, Ö., Şimşek, M., Köytepe, S., et al. (2022). Fabrication of electrospun polycaprolactone/chitosan nanofiber-modified screen-printed electrode for highly sensitive detection of diazinon in food analysis, *Measurement*, Elsevier, **187:** 110250.

Turkani, V.S., Maddipatla, D., Narakathu, B.B., Saeed, T.S., Obare, S.O., Bazuin, B.J. and Atashbar, M.Z. (2019). A highly sensitive printed humidity sensor based on a functionalized MWCNT/HEC composite for flexible electronics application, *Nanoscale Adv.*, RSC, **1(6):** 2311–2322.

Uddin, S.M., Mahmud, T., Wolf, C., Glanz, C., Kolaric, I., Volkmer, C., Höller, H., et al. (2010). Effect of size and shape of metal particles to improve hardness and electrical properties of carbon nanotube reinforced copper and copper alloy composites, *Composites Science and Technology*, Elsevier, **70(16):** 2253–2257.

Vlamidis, Y., Gualandi, I. and Tonelli, D. (2017). Amperometric biosensors based on reduced GO and MWCNTs composite for polyphenols detection in fruit juices, *Journal of Electroanalytical Chemistry*, **799:** 285-292.

Wang, W., Wang, X., Cheng, N., Luo, Y., Lin, Y., Xu, W. and Du, D. (2020). Recent advances in nanomaterials-based electrochemical (bio)sensors for pesticides detection, *TrAC Trends in Analytical Chemistry*, Elsevier, **132:** 116041.

Wong, A., Silva, T.A., Caetano, F.R., Bergamini, M.F., Marcolino-Junior, L.H., Fatibello-Filho, O. and Janegitz, B.C. (2017). An Overview of Pesticide Monitoring at Environmental Samples Using Carbon Nanotubes-Based Electrochemical Sensors, *C 2017*, Multidisciplinary Digital Publishing Institute, **3(1):** 8.

Xu, M., Jiang, S., Jiang, B. and Zheng, J. (2019). Organophosphorus pesticides detection using acetylcholinesterase biosensor based on gold nanoparticles constructed by electroless plating on vertical nitrogen-doped single-walled carbon nanotubes, *International Journal of Environmental Analytical Chemistry*, Taylor & Francis, **99(10):** 913–927.

Yang, G. and Zhao, F. (2015). Electrochemical sensor for chloramphenicol based on novel multi-walled carbon nanotubes@molecularly imprinted polymer, *Biosensors and Bioelectronics*, **64:** 416-422.

Zambianco, N.A., da Silva, V.A.O.P., Orzari, L.O., Corat, E.J., Zanin, H.G., Silva, T.A., Buller, G.A., et al. (2020). Determination of tadalafil in pharmaceutical samples by vertically oriented multi-walled carbon nanotube electrochemical sensing device, *Journal of Electroanalytical Chemistry*, **877:** 114501.

Zang, X., Chang, Q., Pang, Y., Wang, L., Zhang, S., Wang, C. and Wang, Z. (2021). Solid-phase microextraction of eleven organochlorine pesticides from fruit and vegetable samples by a coated fiber with boron nitride modified multiwalled carbon nanotubes, *Food Chemistry*, Elsevier, **359:** 129984.

Zhang, H.-B., Liang, X.-L., Dong, X., Li, H.-Y. and Lin, G.-D. (2009). Multi-Walled Carbon Nanotubes as a Novel Promoter of Catalysts for CO/CO2 Hydrogenation to Alcohols, *Catalysis Surveys from Asia*, **13(1):** 41-58.

Zhang, Y., Xu, Y., Liu, H. and Sun, B. (2022). Ultrahigh sensitivity nitrogen-doping carbon nanotubes-based metamaterial-free flexible terahertz sensing platform for insecticides detection, *Food Chemistry*, Elsevier, **394:** 133467.

Zhao, C., Zhao, Y., Wang, L., Lu, K., Guo, W., Lu, X. and Gao, F. (2023). AuNPs and CNTs embellished three-dimensional bloom-like α-Fe2O3 nanocomposites for highly sensitive electrochemical pesticides detection, *Microchemical Journal*, Elsevier, **191:** 108762.

Zheng, M. (2017). Sorting Carbon Nanotubes, *Topics in Current Chemistry*, **375(1):** 13.

Part VI
Materials for EMI Shielding

Chapter 25
Ku-Band Radiation Shielding Response of $La_{0.7}Sr_{0.3}MnO_3$ Thin Film

Hilal Ahmad Reshi, Shreeja Pillai, Avanish Pratap Singh, Aijaz A. Wani, S. K. Dhawan, and Vilas Shelke

1 Introduction

The overwhelming progress in the development of electronic devices and communication equipment in recent times produced a pollution generally known as electromagnetic interference (EMI). This EMI is ideally considered to be a serious issue as it adversely affected various aspects such as life-time, safety operation and efficiency of electronic devices (Gao et al., 2020a, b). To address this problem, it needs to explore superior EMI shielding materials to eliminate or decrease (Park et al., 2006; Zhang et al., 2018; Lv et al., 2018). The technology works on the materials which are able to absorb electromagnetic radiation and make electromagnetic interference free environment (Kimura et al., 2007). There are defence, commercial, scientific electronic devices as well as communication instruments that were widely used which increased the interest in shielding against electromagnetic interference (Singh et al., 2013; Wang et al., 2013; Liu et al., 2012). In past, various shielding materials have been used to protect humans and scientific instruments as well. However, conventional materials were the first choice till the last few decades. Traditional metallic materials have excellent electromagnetic shielding performance. However,

H. A. Reshi (✉) · S. Pillai · V. Shelke
Novel Materials Research Laboratory, Department of Physics, Barkatullah University, Bhopal 462026, India
e-mail: hilal.phy@gmail.com

A. P. Singh
Department of Physics, Atma Ram Sanatan Dharma College, Dhaula Kuan, New Delhi 110021, India

A. A. Wani
Department of Physics, Govt. Degree College, Kulgam, Jammu and Kashmir 192231, India

S. K. Dhawan
Polymeric and Soft Materials Section, CSIR-National Physical Laboratory, Dr. K. S. Krishnan Road, New Delhi 440012, India

R. K. Kotnala et al. (eds.), *Advanced Functional Materials for Sustainable Environments*, https://doi.org/10.1007/978-3-031-62620-3_25

they suffer from many disadvantages such as propensity to corrosion, high density, high weight and high processing cost. Generally, electromagnetic wave experiences reflection, absorption and transmission during its encounter with a material. The EMI shielding efficiency depends on intrinsic properties such as conductivity, dielectric constant, frequency of source, and distance between the shield and source. It is highly imperative to develop thinner and more effective EMI shielding materials. Widespread investigations are paying attention to develop materials of high electromagnetic matching. Still, the materials fail to fulfill the required conditions like low density, high EM absorbance, less brittleness, corrosion free and ease of fabrication process and hampered their integration in practical applications (Zou et al., 2006).

Metal-oxide compounds reveal some good functional properties that conventional metallic elements or semiconductors do not possess. These properties are superconductivity, ferro-electricity, colossal magneto-resistance (CMR), multiferroicity etc. In rare-earth thin films of manganites having mixed valence perovskite configuration, usually the transport properties are mostly dependent on deposition method, processing, and substrate on which the film is deposited. These CMR oxide thin films could be attractive for various advanced applications (Haghiri-Gosnet and Renard, 2003). The spin polarization and magneto-transport properties have fascinated the attention in manganites in recent time (Yang et al., 2010; Yan et al., 2000; Wang et al., 2006). These LSMO films are considered to be best option for bottom electrode ferroelectric films as functional component (Vaz et al., 2010a, b; Zhang et al., 2010). Several physical and EMI shielding properties of oxide materials in bulk and nanoscale form have been studied (Reshi et al., 2013, 2014, 2016; Shelke et al., 2011, 2012; Mazumdar et al., 2010; Reshi et al., 2015a, b). The LSMO with a composition $La_{0.7}Sr_{0.3}MnO_3$ is an engrossing material due to its exclusive combination of electric and magnetic properties. There is still no systematic exploration on EMI shielding effects of the materials in thin film for the microwave frequency range. LSMO with strong transport properties at room temperature makes it a potential candidate for microwave absorption (Li et al., 2002). The LSMO thin film is well recognized and can be obtained by a variety of techniques. LSMO films can show significant EMI shielding behavior as they possess tunable magnetic and electrical properties.

EMI shielding properties of manganite based LSMO thin film deposited on quartz via spin coating method was investigated. The observed shielding effectiveness of LSMO thin film of high frequency region (Ku-band) presents an insight and interesting approach for LSMO thin film absorbing materials.

2 Materials and Method

The LSMO thin film was grown on quartz substrate by sol-gel based spin coating method. The chemicals used as starting materials for solution preparation were acetate salts of lanthanum, strontium, and manganese [La $(CH_3COO)_3$. XH_2O, Sr $(CH_3COO)_2$, Mn $(CH_3COO)_2$.XH_2O]. Distilled water and 30 ml acetic acid solvents were used to prepare the lanthanum, strontium and manganese precursors.

Lanthanum and strontium solutions were poured drop-wise with the help of burettes taking manganese precursor as base. The quartz substrates were pretreated for much cleaner and fresher substrate surface. The preliminary step was to drench the substrate in the ethanol for about 20 min, and then in distilled water for 20 min and finally in acetone for 20 min. Each step was followed by an ultrasonic cleaning. Finally, substrates were annealed at 300 °C for 10 min to evaporate the organics absorbed by surface. The homogeneous solution of LSMO was spin coated on the cleaned quartz substrate. The seed layer was deposited at 2000 rpm for 5 s followed by 4000 rpm for the same time. After each layer deposition, the sample was pyrolyzed at 250 °C for 10 min to evaporate the solvent out of the film. The process was repeated 12 times to achieve desired thickness of film. Finally the film was annealed in air at 600C$^{\infty}$ for 60 min.

The crystalline configuration of samples was determined by X-Ray method employing D8-ADVANCE BRUKER instrument. Diffraction spectrum was recorded using CuK_{α} radiation source. The complex and shielding measurements were done using E8362B Vector Network Analyzer in Ku-band frequency region. The film deposited substrate was cut in the prescribed dimensions of Ku-band of 15.8 mm × 7.9 mm loaded into a sample holder connected between the waveguide flanges of Vector Network Analyzer.

3 Results and Discussion

The diffraction pattern of LSMO thin film on quartz is shown in Fig. 1. It reveals polycrystalline phase formation of LSMO compound on given substrate (Andres et al., 1999). The XRD pattern depicts a single phase without any kind of impurity.

The development of a pure phase on the film is evidenced by the peak reflections identified to $La_{0.7}Sr_{0.3}MnO_3$ and by the lack of appearance of any impurity. The indexing of diffraction data was done for rhombohedral symmetry. This observation recommends that rhombohedral structure La_{1-x} Sr_xMnO_3 is valid for strontium doping level up-to 30% (Zarbali et al., 2012). A big hump around 23 degree angle corresponds to the substrate.

The electromagnetic (EM) wave absorption is related to complex parameters such as permeability, permittivity and loss constants. The electromagnetic radiation absorption features of manganites are dominant by magnetic and dielectric losses as well (Pawar et al., 2012).

The relative complex properties i.e. permittivity ($\varepsilon^* = \varepsilon' - i\varepsilon\gg$) and permeability ($\mu^* = \mu> - i\mu\gg$) of LSMO film was tested within Ku-band frequency region at room temperature. The variation of real, imaginary and loss parts with frequency is shown in Fig. 2. Scattering parameters (S_{11} and S_{21}) in line with Nicholson-Ross and Weir algorithms have been used to calculate complex parameters (Nicholson et al., 1970). These relative complex properties describe electromagnetic wave propagation through a medium. The dielectric constant (ε') and (ε") generally explain the amount of polarization that occurs in the material and energy dissipation. The materials

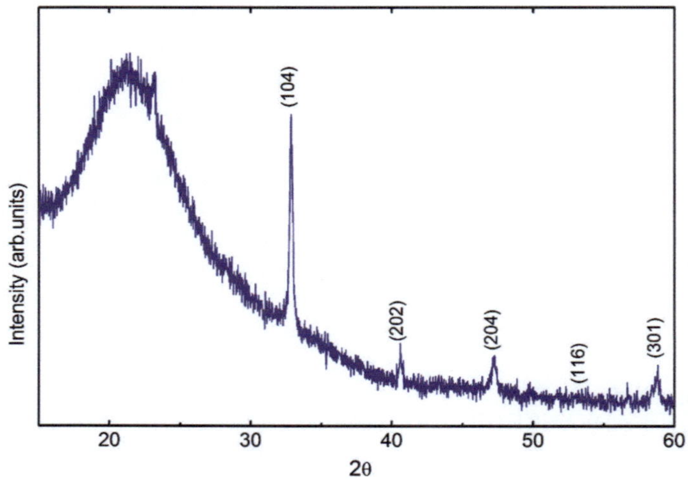

Fig. 1 Diffraction pattern for $La_{0.7}Sr_{0.3}MnO_3$ thin film on quartz substrate annealed at 600 °C

dielectric performance is suggested by ionic, orientational, electronic, and space charge polarization. The presence of dipoles in a material divulge the orientation polarization.

The (ε>) value is almost constant between 12 and15 GHz. The value increases slightly for frequencies above 15 to 17.5 GHz which implies that orientational polarization is the most possible mechanism in the said frequency region (Kim et al., 2004b, a). The fluctuating nature above 17.5 GHz depicts a resonance behavior as expected in high conducting region in the material. The dielectric loss (ε") value increases from 0.1 to 2.2 with increase in frequency from 12.4 to 17.5 GHz. As a whole, dielectric constant and dielectric loss are considered to be very low. The values may be improved by improving the different parameters like increasing number of layers on substrate.

The real and imaginary permeability of LSMO film is revealed in Fig. 2 (d, e). It is seen that both μ' and μ" high at low frequency region and the values decrease in a linear manner as the frequency is increased. The declining trend with increase in frequency may be due to low resonance behavior in this frequency region. The permeability of film play a prominent role in improving the microwave absorption effects by converting electromagnetic energy into heat (Zou et al., 2010). The maximum value of μ' observed Fig. 2 is 0.95 which decreases linearly with increasing frequency. Interestingly, the imaginary part (μ") is negative in measured range and the value varies from –0.05 to –0.6. The positive region in imaginary permeability (μ≫) evidently hints towards magnetic energy loss. Perversely, a negative region of imaginary μ≫ delineate magnetic energy is radiated out from the surface of film. The gradual decrease in μ' with increasing frequency implies that magnetic loss improves as frequency increases due to hysteresis of magnetism (Yang et al., 2009). Magnetic loss occurs due to time lag of magnetization vector behind the magnetic

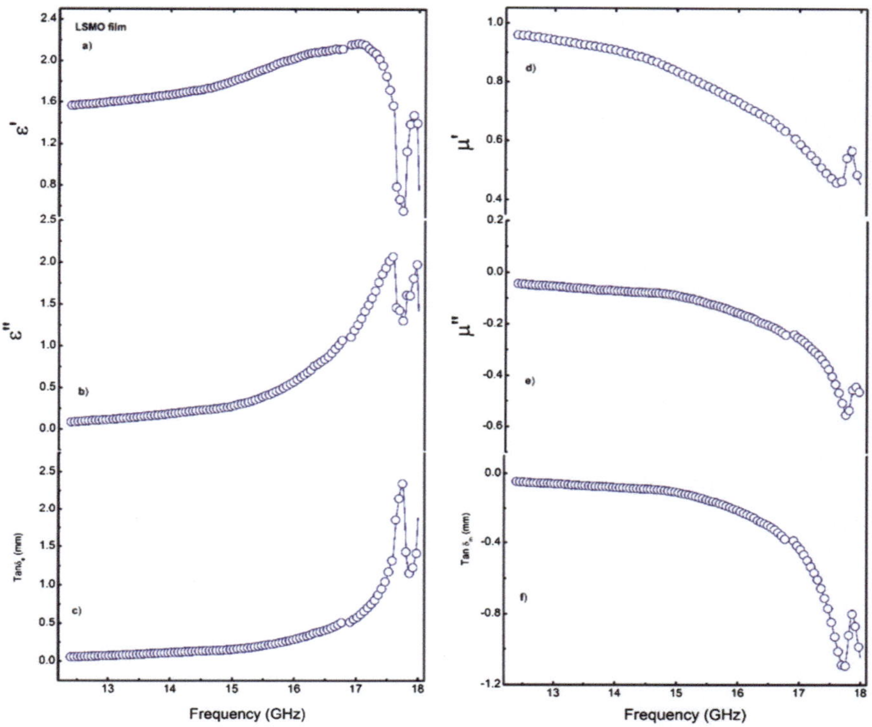

Fig. 2 Frequency-dependent complex parameters: (a) permittivity, (b) imaginary permittivity, (c) dielectric tangent loss, (d) permeability, (e) imaginary permeability and (f) magnetic tangent loss

field vector. The magnetic rotation and domain wall displacement bring the change in magnetization vector. These motions delay due to change in magnetic field and support to magnetic loss (μ''). The number of atoms with baggy bonds, surface area, and unsaturated coordination on the surface leads to the interfacial polarization along with multiple scatterings, helpful for the absorption of electromagnetic waves (Zhang et al., 2006).

The imaginary complex permittivity indicates the films capability of absorb EM waves. The tangent loss (tan δ) entails the materials ability to convert energy storage into heat. Therefore higher value of loss factor would indicate better EM wave absorbing property (Yussuf et al., 2005). Materials of higher value of tan δ; ie. >1 are good conductors and those with<1 are weak conductors. In former case, the conductivity (σ) plays an important role rather than (ε_r) while in later case electromagnetic dissipation occur due to permeability. Therefore, the value of tan δ ~1 is considered to be optimum in which σ_{ac} and ε_r contribute to the EMI shielding. Figure 2 (c, f) shows the tan δ_e and tan δ_m values of the LSMO thin film. The values are in the range of 0.03 to 2.27 and –0.03 to –1.12 repectively. The tan δ is around 0.03 throughout measured frequency range which indicates that dielectric loss is present in entire frequency region (Singh et al., 2014). The tan δ_m hints the effect of eddy current, anisotropy energy and natural resonances of material. Also, a steady dielectric loss in the whole frequency region proves that LSMO film allow electromagnetic wave absorption due to their steady dielectric loss capability (Cheng et al., 2009).

The EMI shielding creates an obstacle containing electrical conductivity matrix which blocks radiated EM energy via reflection as well as absorption. Interference shielding means attenuation due to absorption and/or reflection (Das et al. 2000). The shielding effectiveness (SE) of any material is articulated as a ratio of incident power to the transmitted power and can be in logarithmic form expressed as (Joo et al., 2000, 1994).

$$SE_T(dB) = 10\log\{P_T/P_I\} = SE_R + SE_A + SE_M \quad (1)$$

where P_I and P_T represent incident and transmitted power of electromagnetic energies (Seo et al., 2008) while SE_R, SE_A and SE_M derive the contribution of total shielding effectiveness from reflection, absorption and internal multiple reflections of EM wave.

The scattering parameters for transmission (T) and reflection (R) S_{11} or S_{22} and S_{21} or S_{12} of a two port Vector Network Analyzer can be related with the reflectance (R); and transmittance (T) as

$$T = |P_T/P_I|^2 = |S_{21}|^2 = |S_{12}|^2 \quad (2)$$

$$R = |P_R/P_I| = |S_{11}|^2 = |S_{22}|^2 \quad (3)$$

The coefficient of absorption (A) is equal to 1–R–T.

The suitable way to express the reflection and effective absorption in dB form respectively which gives reflection loss (SE_R), absorption loss (SE_A) and multiple internal reflections (SE_M) as

$$SE_R(dB) = -10\log(1 - R) \quad (4)$$

$$SE_A(dB) = -10\log(1 - A_{eff}) = -10\log(T/1 - R) \quad (5)$$

$$SE_M = -20\log\left(1 - 10^{-SEA/10}\right) \quad (6)$$

The materials effectiveness (A_{eff}) as a shield can be expressed as

$$A_{eff} = (1 - R - T)/(1 - R)$$

Figure 3 shows the deviation of radiation shielding effectiveness of LSMO film in the frequency region from 12.2–18.1 GHz. The effectiveness due to reflection (SE_R) as well as absorption (SE_A) varied from 0.82 to 1.05 dB and 0.08 to 0.75 dB while as the shielding effectiveness under multiple internal reflections (SE_M) varied from 0.16 to 1.41 dB, respectively. Thus, the total shielding effectiveness achieved for LSMO thin film is 3.01 dB which is very low. The reason being a very thin film was deposited and was annealed at low temperature. The shielding effectiveness of thin film is dominated by reflection while as absorptive and multiple reflective shielding contributes very little. The quartz substrate is transparent to electromagnetic radiations. Therefore, the attenuation originates mainly from the interface of deposited LSMO layers on-to the quartz substrate. The LSMO film forms a conductive percolating path. The electrical conduction may take place either via electron transport or by hopping/tunneling of electrons, thereby contributing to EMI shielding. Annealing at low temperature and less film thickness leads to a relatively less attenuation. The grain boundary barriers for electrons or holes are accompanied by the increase in density of surface, thereby reducing the overall conductivity of the LSMO. The absorption and multiple reflections depend upon the film thickness and magnetic behaviour of material. The LSMO film may be used to attenuate the electromagnetic waves, however, the film requires high annealing temperature and film thickness. The shielding effectiveness may be improved further by simply increasing film thickness (Kim et al., 2004a; b).

Variation of conductivity ($\sigma_{ac} = \omega\varepsilon_o\varepsilon\gg$) (Park et al., 2009) with the frequency is shown in Fig. 4 (a). The σ_{ac} plays a prominent role to enhance the reflection

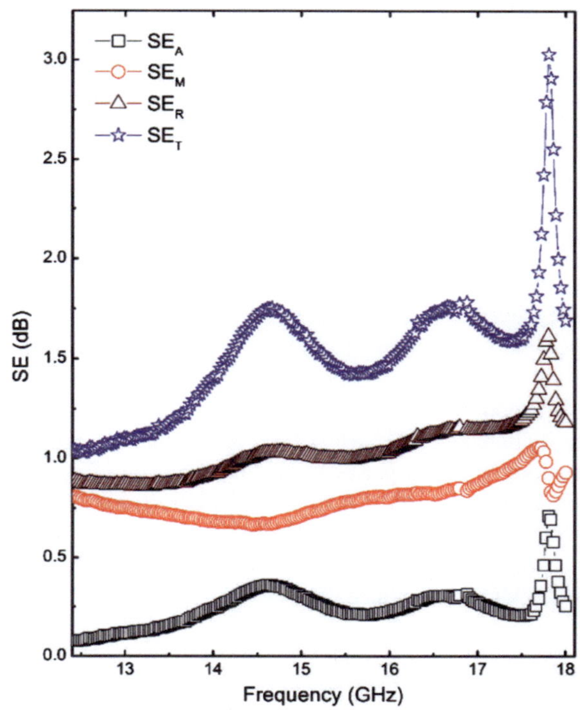

Fig. 3 Frequency dependent EMI shielding effectiveness of $La_{0.7}Sr_{0.3}MnO_3$ thin film

in shielding effectiveness. As the conductivity improves, the reflection loss also increases and its correlation with shielding effectiveness can be seen in Fig. 3. This reveals that higher value of conductivity is required to enhance the surface reflections in the film. One more important part is skin depth in which both conductivity as well as film thickness plays a significant role and its variation will directly impart an effect on absorption of EMI shielding. At high frequencies, the electromagnetic radiations penetrate near the surface section of an electrically conducting region known as the skin effect. Electric field which penetrates through a conducting region drops exponentially with increasing depth into the conductor. This depth where the field drops to 1/e times of its incident value is called the skin effect or skin depth (δ). For absorption of EM waves, skin depth has been calculated using angular frequency (ω), permeability (μ), and conductivity (σ_{ac}) with a relation $\delta = (2/\omega\mu\sigma_{ac})^{1/2}$ (Otto, 2009). Figure 4 (b) shows the variation of skin depth δ with frequency. Skin depth reduces by increasing frequency implies that mainly surface conduction is existing at higher frequency region (Ohlan et al., 2009). The conductivity and magnetic permeability dependent skin depth indicates that skin depth must be less for conducting and magnetically sound films. When an electromagnetic signal transmitted from the outer surface travels in the conductive shield, the signal experiences an exponential decrease of amplitude due to absorption thereby dissipated in the form of heat. Once a wave reached to another surface layer of shield, some part of signal gets reflected into the shielding material. If shielding thickness is greater than skin depth, then reflected part from the internal surface region will be absorbed by conductive layer of shield. In our case, film thickness is very small and reflection was found to be primary mechanism. Better shielding by absorption of signal and multiple internal reflections may be achieved by improving film thickness and magnetic permeability. Good reflection loss can be achieved by highly conductive atmosphere in the film.

4 Conclusion

We developed an approach for the deposition of LSMO thin film using sol-gel derived spin coating technique. This method is reliable and scalable at industrial level. The film was deposited on quartz substrate and was annealed at 600 °C. The XRD confirmed single phase polycrystalline nature of film without impurity. Permittivity behavior of LSMO film revealed that dielectric loss occurs in whole frequency region. Negative permeability indicated that magnetic energy is radiated out from film. An observed EMI shielding effectiveness (SE) obtained is 3.01 dB at 17.8 GHz which is less for the technological application. The SE may be improved by optimizing film thickness and conductivity. This study opened a new view of functionality in LSMO compounds.

Fig. 4 Frequency dependent: (a) ac-conductivity $\sigma)_{ac}$) and (b) skin depth (δ) of $La_{0.7}Sr_{0.3}MnO_3$ thin film

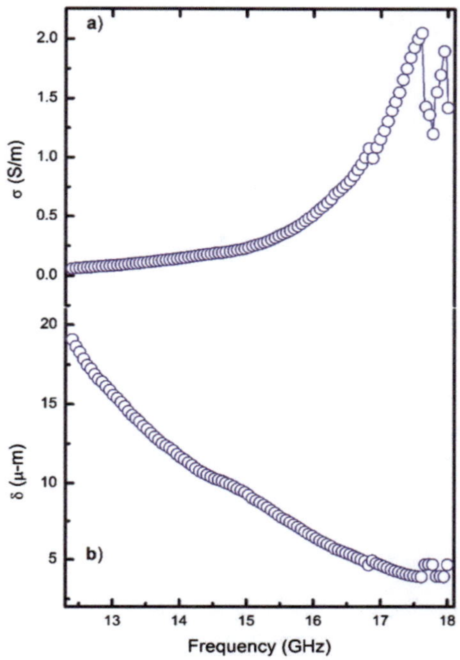

Acknowledgements Authors viz H.R. Reshi and V. Shelke are grateful to University Grants Commission (UGC) for financial support. The authors are also thankful to Dr. Mukul Gupta, UGC-DAE Consortium for Scientific Research, Indore for facilitating characterization techniques of this work.

References

Andres, A. De, Garcia-Hermandez, M., Martinez, J.L. and Prieto, C. (1999). Low-temperature magnetoresistance in polycrystalline manganites: connectivity versus grain size *Appl. Phys. Lett.*, **74:** 3884.

Cheng, Y.L., Dai, J.M., Zhu, X.B., Wu, D.J., Yang, Z.R. and Sun, Y.P. (2009). Enhanced Microwave Absorption Properties of Intrinsically Core/shell Structured $La_{0.6}Sr_{0.4}MnO_3$Nanoparticles, *J. Nanoscale Res. Lett.*, **4:** 1153.

Das, N.C., Chaki, T.K., Khastgir, D. and Chakraborty, A. (2000). Electromagnetic interference shielding effectiveness of carbon black and carbon fibre filled EVA and NR based composites, *Composites Part A*, **31:** 1069–1081.

Gao, N., Li, W.P., Wang, W.S., Liu, D.P., Cui, Y.M., Guo, L. and Wang, G.S. (2020). Balancing Dielectric Loss and Magnetic Loss in Fe-NiS_2/NiS/PVDF Composites toward Strong Microwave Reflection Loss, *ACS Appl. Mater. Interfaces*, **12:** 14416–14424.

Gao, S., Yang, S.H., Wang, H.Y., Wang, G.S. and Yin, P.G., (2020). Excellent electromagnetic wave absorbing properties of two dimensional carbon-based nanocomposite supported by transition metal carbides Fe_3C, *Carbon*, **162:** 438–444.

Haghiri-Gosnet, A.M. and Renard, J.P. (2003). CMR manganites: physics, thin films and devices, *J. Phys. D: Appl. Phys.*, **36:** R127.
Joo, J. and Epstein, A.J. (1994). Electromagnetic radiation shielding by intrinsically conducting polymers, *Appl. Phys. Lett.*, **65:** 2278.
Joo, J. and Lee, C.Y. (2000). High frequency electromagnetic interference shielding response of mixtures and multilayer films based on conducting polymers, *J. Appl. Phys.*, **88:** 513.
Kim, H.M., Kim, K., Lee, C.Y., Joo, J., S. Cho, J., Yoon, H.S., Pejakovic, D.A., Yoo, J.W. and Epstein, A.J. (2004). Electrical conductivity and electromagnetic interference shielding of multiwalled carbon nanotube composites containing Fe catalyst, *Appl. Phys. Lett.*, **84:** 589.
Kim, S.S., Ahn, J.M. and Kim, K.H. (2004). Magnetic and microwave absorbing properties of Co–Fe thin films plated on hollow ceramic microspheres of low density, *J. Mag. Mag. Mater.*, **271:** 39–45.
Kimura, S., Kato, T., Hyodo, T., Shimizu, Y., Egashira, M. (2007). Electromagnetic wave absorption properties of carbonyl iron-ferrite/PMMA composites fabricated by hybridization method, *J. Magn. Magn. Mater.*, **312:** 181–186.
Li, G., Hu, G.G., Zhou, H.D., Fan, X.J. and Li X.G. (2002). Attractive microwave-absorbing properties of $La_{1-x}Sr_xMnO_3$ manganite powders, *Mater. Chem. Phys.*, **75:** 101–104.
Liu, J., Che, R., Chen, H., Zhang, F., Xia, F., Wu, Q. and Wang, M. (2012). Microwave Absorption Enhancement of Multifunctional Composite Microspheres with Spinel Fe_3O_4 Cores and Anatase TiO_2 Shells, *Small,* **8:** 1214–1221.
Lv, H., Yang, Z., Wang, P.L., G. Ji, Song, Zheng, J.L.H., Zeng and Xu, Z. J. (2018). A Voltage-Boosting Strategy Enabling a Low-Frequency, Flexible Electromagnetic Wave Absorption Device, *Adv. Mater.*, **30:** 1706343.
Mazumdar, D., Shelke, V., Iliev, Jesse, M., S., Kumar, A., Kalinin, S.V., Baddrof, A.P. and Gupta, A. (2010). Nanoscale Switching Characteristics of Nearly Tetragonal $BiFeO_3$ Thin Films. *Nano Lett.* **10:** 2555–2561.
Nicholson, A.M. and Ross, G.F. (1970). Measurement of the Intrinsic Properties of Materials by Time-Domain Techniques, *IEEE Trans. Instrum. Meas.*, **19:** 377–382.
Ohlan, A., Singh, K., Chandra, A., Singh, V.N. and Dhawan, S.K. (2009). Conjugated polymer nanocomposites: Synthesis, dielectric, and microwave absorption studies, *J. Appl. Phys.*, **106:** 044305.
Otto, H.W. (2009). Electromagnetic Compatibility Engineering, Wiley, New Jersey.
Park, K.Y., Lee, S.E., Kim, C.G. and Han, J.H. (2006). Fabrication and electromagnetic characteristics of electromagnetic wave absorbing sandwich structures, *Compos. Sci. Technol.*, **66:** 576–584.
Park, S.H., Thielemann, P., Asbeck, P. and Bandaru, P.R. (2009). Enhanced dielectric constants and shielding effectiveness of, uniformly dispersed, functionalized carbon nanotube composites, *Appl. Phys. Lett.*, **94:** 243111.
Pawar, R.P., Jadhav, R.N. and Puri, V. (2012). Effect of Ca on microwave properties of strontium manganite synthesized by microwave processing, *Elect, Mater. Lett.*, **8:** 321–324.
Reshi, H.A., Pillai, S., Bhuwal, D. and Shelke, V. (2013). Nanostructure Induced Metal-Insulator Transition and Enhanced Low-Field Magnetoresistance in $La_{0.7}Sr_{0.3}MnO_3$ Systems. *J. Nanosci. Nanotechnol.*, **13:** 4608–4615.
Reshi, H.A., Pillai, S. and Shelke, V. (2014). Comparative study on multifunctional behaviour of rare earth manganites with micro and nano grain size, *J. Mater. Sci. Mater. Electron.* **25:** 3795–3800.
Reshi, H.A. Pillai, S., Yadav, R.S., Para, T.A., Deshpande, U.P., Shripathi, T. and Shelke, V. (2015). Kondo-like electronic transport and ferromagnetic cluster-glass behavior in $La_{0.7}Sr_{0.3}MnO_3$ nanostructures, *RSC Adv.* **5:** 85950–85956.
Reshi, H.A., Singh, A.P., Pillai, S., Para, T.A., Dhawan, S.K. and Shelke, V. (2016). X-band frequency response and electromagnetic interference shielding in multiferroic BiFeO3 nanomaterials**,** Appl. Phys. Lett. **109:** 142904.

Reshi, H.A., Singh, A.P., Pillai, S., Yadav, R.S., Dhawan, S.K. and Shelke, V. (2015). Nanostructured $La_{0.7}Sr_{0.3}MnO_3$ compounds for effective electromagnetic interference shielding in the X-band frequency range, *J. Mater. Chem.* C, **3:** 820–827.

Seo, M.A., J. Yim, H., Ahn, Y.H., Rotermund, F., Kim, D.S, Lee, S. and Lim, H. (2008). Terahertz electromagnetic interference shielding using single-walled carbon nanotube flexible films, *Appl. Phys. Lett.*, **93:** 213905.

Shelke, V., Mazumdar, D., Srinivasan, G., Kumar, A., Jesse, S., Kalinin, S., Baddorf, A. and Gupta, A. (2011). Reduced coercive field in $BiFeO_3$ thin films through domain engineering, *Adv. Mater.* **23:** 669–72.

Shelke, V., Mazumdar, D., Jesse, S., Kalinin, S., Baddorf, A. and Gupta, A. (2012). Ferroelectric domain scaling and switching in ultrathin $BiFeO_3$ films deposited on vicinal substrates, *New J. Phys.* **14:** 053040.

Singh, B.P., Prasanta, Choudhary, V. Saini, P., Pande, S., Singh, V.N. and Mathur, P.B. (2013). Enhanced microwave shielding and mechanical properties of high loading MWCNT–epoxy composites *J. Nanopart. Res.*, **15:** 1554.

Singh, A.P., Mishra, M., Sambyal, P., Gupta, B.K., Singh, B.P., Chandra, A. and Dhawan, S.K. (2014). Encapsulation of γ-Fe_2O_3 decorated reduced graphene oxide in polyaniline core–shell tubes as an exceptional tracker for electromagnetic environmental pollution, *J. Mater. Chem. A*, **2:** 3581–3593.

Vaz, C.A.F., Hoffman, J., Ahn, C.H. and Ramesh, R. (2010). Magnetoelectric coupling effects in multiferroic complex oxide composite structures, *Adv. Mater.*, **22:** 2900–18.

Vaz, C.A.F., Hoffman. J., Segal, Y., Reiner, J.W., Grober, R.D., Zhang, Z., Ahn, C.H. and Walker, F.J. (2010). Origin of the Magnetoelectric Coupling Effect in $Pb(Zr_{0.2}Ti_{0.8})O_3/La_{0.8}Sr_{0.2}MnO_3$ Multiferroic Heterostructures, *Phys. Rev. Lett.*, **104:** 127202 .

Wang. Z.J., Usuki. H., Kumagai. T. and Kokawa. H. (2006). Microstructure and electrical properties of $La_{0.7}Sr_{0.3}MnO_3$ thin films deposited by metallo-organic decomposition method, *J. Cryst. Growth*, **293:** 68–73.

Wang, T., Liu, Z., Lu, M., Wen, B., Ouyang, Q., Chen, Y., Zhu, C., Gao, P., C.M, Li, Cao and Qi, L. (2013). Graphene-Fe_3O_4 nanohybrids: Synthesis and excellent electromagnetic absorption properties, *J. Apply. Phys.*, **113:** 024314.

Yan, C.H., Huang, Y.H., Wang, A.M., Zhu, T, Liao, C.S., Cheng, F.X., Zhang, Z.L. and Xu, G.X. (2000). Microstructure and transport properties of sol–gel fabricated $La_{0.7}Sr_{0.3}MnO_3$ nanocrystalline films on Si(100) substrates, *Thin Solid Films*, **366:** 302–305.

Yang, R.B., Hsu, S.D. and Lin, C.K. (2009). Frequency-dependent complex permittivity and permeability of iron-based powders in 2–18 GHz, *J. Appl. Phys.*, **105:** 07A527.

Yang, F., Kemik, N., Biegalski, M.D., Christen, H.M., Arenholz, E. and Takamura, Y. (2010). Strain engineering to control the magnetic and magnetotransport properties of $La_{0.67}Sr_{0.33}MnO_3$ thin films, *Appl. Phys. Lett.*, **97:** 092503.

Yussuf, A.A., Sbarski, I., J. Hayes, P., Solomon, M. and Tran, N. (2005). Microwave welding of polymeric-microfluidic devices, *J. Micromech. Microeng.*, **15:** 1692.

Zarbali, M., Goktas, A., Mutlu, I.H., Kazan, S., A. Sale, G. and Mikailzade, F. (2012). Structure and Magnetic Properties of $La_{0.66}Sr_{0.33}MnO_3$ Thin Films Derived Using Sol-Gel Technique, *J. Supercond. Nov. Magn.*, **25:** 2767–2770.

Zhang, X.F., Dong, X.L., Huang, H., Liu, Y.Y., Wang, W.N., X. Zhu, Lv, G., B., Lei, J.P. and Lee, C.G. (2006). Microwave absorption properties of the carbon-coated nickel nanocapsules, *Appl. Phys. Lett.*, **89:** 053115.

Zhang, X.J., Zhu, J.Q., Yin, P.G., Guo, A.P., Huang, A.P., Guo, L. and Wang, G.S. (2018). Tunable High-Performance Microwave Absorption of $Co_{1-x}S$ Hollow Spheres Constructed by Nanosheets within Ultralow Filler Loading, *Adv. Funct. Mater.*, **28:** 1800761.

Zhang, Y.T., Zhang, M.Y. and Li, S. (2010). Effect of polarization-memory in $SrTiO_3$/ $La_{0.9}Sr_{0.1}MnO_3SrTiO_3/La_{0.9}Sr_{0.1}MnO_3$ multilayer on Si substrate, *Appl. Phys. Lett.*, **97:** 052903.

Zou, Y.H., Liu, H.B., Yang, L.Z. and Chen, Z. (2006). The influence of temperature on magnetic and microwave absorption properties of Fe/graphite oxide nanocomposites, *J. Magn. Magn. Mater.*, **302:** 343–347.

Zou, Z., A. Xuan, G., Yan, Z.G., Wu, Y.X. and Li, N. (2010). Preparation of Fe_3O_4 particles from copper/iron ore cinder and their microwave absorption properties, *Chem. Eng. Sci.*, **65:** 160–164.

MIX
Papier aus verantwortungsvollen Quellen
Paper from responsible sources
FSC® C105338

If you have any concerns about our products,
you can contact us on
ProductSafety@springernature.com

In case Publisher is established outside the EU,
the EU authorized representative is:
Springer Nature Customer Service Center GmbH
Europaplatz 3, 69115 Heidelberg, Germany

Printed by Libri Plureos GmbH
in Hamburg, Germany